T0211692

Lecture Notes in Computer Science 9304

Commenced Publication in 1973
Founding and Former Series Editors:
Gerhard Goos, Juris Hartmanis, and Jan van Leeuwen

More information about this series at http://www.springer.com/series/7407

Florin Manea · Dirk Nowotka (Eds.)

Combinatorics on Words

10th International Conference, WORDS 2015
Kiel, Germany, September 14–17, 2015
Proceedings

 Springer

Editors
Florin Manea
Universität Kiel
Kiel
Germany

Dirk Nowotka
Universität Kiel
Kiel
Germany

ISSN 0302-9743   ISSN 1611-3349   (electronic)
Lecture Notes in Computer Science
ISBN 978-3-319-23659-9   ISBN 978-3-319-23660-5   (eBook)
DOI 10.1007/978-3-319-23660-5

Library of Congress Control Number: 2015947798

LNCS Sublibrary: SL1 – Theoretical Computer Science and General Issues

Springer Cham Heidelberg New York Dordrecht London

Printed on acid-free paper

Springer International Publishing AG Switzerland is part of Springer Science+Business Media
(www.springer.com)

Preface

This volume contains the proceedings of the 10th International Conference on WORDS, which took place at Kiel University, Germany, September 14-17, 2015. WORDS is the main conference series devoted to the mathematical theory of words, and it takes place every two years. The first conference in the series was organized in 1997 in Rouen, France, with the events that followed taking place in Rouen, Palermo, Turku, Montreal, Marseille, Salerno, Prague, and Turku.

The main object in the scope of the conference, words, are finite or infinite sequences of symbols over a finite alphabet. They appear as natural and basic mathematical model in many areas, theoretical or applicative. Accordingly, the WORDS conference is open to both theoretical contributions related to combinatorial, algebraic, and algorithmic aspects of words, as well as to contributions presenting applications of the theory of words, for instance, in other fields of computer science, linguistics, biology and bioinformatics, or physics.

For the second time in the history of WORDS, after the 2013 edition, a refereed proceedings volume was published in Springer's Lecture Notes in Computer Science series. In addition, a local proceedings volume was published in the Kiel Computer Science Series of the Kiel University. Being a conference at the border between theoretical computer science and mathematics, WORDS tries to capture in its two proceedings volumes the characteristics of the conferences from both these worlds. While the Lecture Notes in Computer Science volume is dedicated to original contributions that fulfil the high quality standards required by a formal publication in computer science, the local proceedings volume allows, in the spirit of mathematics conferences, the publication of less formally prepared contributions, informing on current research and work in progress, surveying some areas connected to the core topics of WORDS, or presenting relevant previously published results. All the papers, the ones published in the Lecture Notes in Computer Science proceedings volume or the ones from the local proceedings volume, were refereed to high standards by the members of the Program Committee. Following the conference, a special issue of the Theoretical Computer Science journal will be edited.

We received 22 submissions, from 15 countries. From these, 14 papers were selected to be published in this refereed proceedings volume. In addition to the contributed talks, the conference program included six invited talks given by leading experts in the areas covered by the WORDS conference: Jörg Endrullis (Amsterdam), Markus Lohrey (Siegen), Jean Néraud (Rouen), Dominique Perrin (Paris), Michaël Rao (Lyon), Thomas Stoll (Nancy). Extended abstracts of these talks are included in this proceedings volume. WORDS 2015 was the tenth conference in the series, so we were extremely happy to welcome, as invited speaker at this anniversary event, Jean Néraud, one of the initiators of the series and the main organizer of the first two iterations of this conference. We thank all the invited speakers and all the authors of submitted papers for their contributions to the success of the conference.

We are grateful to the members of the Program Committee for their work that led to the selection of the contributed talks, and, implicitly, of the papers published in this volume. They were assisted in their task by a series of external referees, gratefully acknowledged below. The submission and reviewing process was performed via the Easychair system; we thank Andrej Voronkov for this system which facilitated the work of the Program Committee and the editors considerably. We express our gratitude to the representatives of Springer for their collaboration. Finally, we thank the Organizing Committee of WORDS 2015 for ensuring the smooth run of the conference.

September 2015

Dirk Nowotka
Florin Manea

Organization

Program Committee

Maxime Crochemore	King's College London, UK
James Currie	University of Winnipeg, Canada
Stepan Holub	Charles University in Prague, Czech Republic
Juhani Karhumäki	University of Turku, Finland
Manfred Kufleitner	University of Stuttgart, Germany
Gad Landau	University of Haifa, Israel
Dirk Nowotka	Kiel University, Germany (PC-Chair)
Wojciech Plandowski	University of Warsaw, Poland
Antonio Restivo	University of Palermo, Italy
Michel Rigo	University of Liège, Belgium
Mikhail Volkov	Ural State University, Russia
Luca Zamboni	University Lyon 1, France

Additional Reviewers

Allouche, Jean-Paul
Amit, Mika
Badkobeh, Golnaz
Bucci, Michelangelo
Charlier, Emilie
De Luca, Alessandro
Dekking, Michel
Epifanio, Chiara
Ferenczi, Sebastien
Fici, Gabriele

Glen, Amy
Hadravova, Jana
Leroy, Julien
Manea, Florin
Mantaci, Sabrina
Mercas, Robert
Plandowski, Wojciech
Prodinger, Helmut
Puzynina, Svetlana
Rozenberg, Liat

Saarela, Aleksi
Sebastien, Labbe
Sheinwald, Dafna
Smyth, William F.
Stipulanti, Manon
Szabados, Michal
Sýkora, Jiří
Widmer, Steven
Zaroda, Artur

Abstracts of Invited Talks

Degrees of Transducibility

Jörg Endrullis[1], Jan Willem Klop[1,2], Aleksi Saarela[3],
and Markus Whiteland[3]

[1]Department of Computer Science, VU University Amsterdam,
Amsterdam, The Netherlands
[2]Centrum voor Wiskunde en Informatica (CWI), Amsterdam, The Netherlands
[3]Department of Mathematics and Statistics & FUNDIM,
University of Turku, Turku, Finland

Abstract. Our objects of study are infinite sequences and how they can be transformed into each other. As transformational devices, we focus here on Turing Machines, sequential finite state transducers and Mealy Machines. For each of these choices, the resulting transducibility relation $\geq$ is a preorder on the set of infinite sequences. This preorder induces equivalence classes, called *degrees*, and a partial order on the degrees.

For Turing Machines, this structure of degrees is well-studied and known as *degrees of unsolvability*. However, in this hierarchy, all the computable streams are identified in the bottom degree. It is therefore interesting to study transducibility with respect to weaker computational models, giving rise to more fine-grained structures of degrees. In contrast with the degrees of unsolvability, very little is known about the structure of degrees obtained from finite state transducers or Mealy Machines.

Equality Testing of Compressed Strings

Markus Lohrey

Universität Siegen, Siegen, Germany
lohrey@eti.uni-siegen.de

Abstract. This paper gives a survey on efficient algorithms for checking equality of grammar-compressed strings, i.e., strings that are represented succinctly by so called straight-line programs.

This research is supported by the DFG-project LO 748/10-1.

On the Contribution of WORDS to the Field of Combinatorics on Words

Jean Néraud

Université de Rouen, Rouen, France
jean.neraud@univ-rouen.fr, jean.neraud@wanadoo.fr,
neraud.jean@free.fr

Abstract. We propose some notes about the history and the features of the conference WORDS, our goal being to testify how the conference may be embedded in the development of the field of Combinatorics on Words.

litis, Université de Rouen, UFR Sciences et Techniques, Campus du Madrillet, 76800 Saint Etienne du Rouvray, France.

Codes and Automata in Minimal Sets

Dominique Perrin

LIGM, Université Paris Est, Paris, France
dominique.perrin@esiee.fr

Abstract. We explore several notions concerning codes and automata in a restricted set of words S. We define a notion of S-degree of an automaton and prove an inequality relating the cardinality of a prefix code included in a minimal set S and its S-degree.

Decidability of Abelian-Power-Freeness and Generalizations

Michaël Rao

LIP, CNRS, ENS de Lyon, UCBL, Université de Lyon, Lyon, France
michael.rao@ens-lyon.fr

Abstract. Avoidability of structures and patterns has been extensively studied in theoretical computer science since the work of Thue in 1906-1912 on avoidability of repetitions in words.

The avoidability of abelian repetitions has been studied since a question from Erdős in 1957. He asked whether it is possible to avoid abelian squares in an infinite word over an alphabet of size 4. Keränen answered positively to Erdős' question in 1992 by giving a morphism whose fixed point is abelian-square-free. Moreover, Dekking showed that it is possible to avoid abelian cubes on a ternary alphabet and abelian fourth powers over a binary alphabet.

Erdős also asked if it is possible to avoid arbitrarily long usual squares on a binary alphabet. This question was answered positively by Entringer, Jackson and Schatz in 1974. Mäkelä asked two questions in 2002 about the avoidability of long abelian squares (resp., cubes) on a ternary (resp., binary) alphabet.

The notion of k-abelian repetition has been introduced recently by Karhumäki *et al.* as a generalization of both repetition and abelian repetition. One can avoid 3-abelian squares (resp., 2-abelian squares) on a ternary (resp., binary) alphabet. Following Erdős' and Mäkelä's questions, one can also ask whether it is possible to avoid long k-abelian powers on a binary alphabet.

We present an overview on the results on avoidability of k-abelian repetitions on small alphabets. We present techniques to decide whether a morphic word avoid abelian and k-abelian repetitions. These techniques allow us to prove some new results on avoidability. In particular, we will show that:

- Long abelian squares are avoidable on a ternary alphabet, answering positively to a weak version of the first question from Mäkelä.
- Additive squares are avoidable over $\mathbb{Z}^2$.

(Based on a joint work with Matthieu Rosenfeld.)

Thue–Morse Along Two Polynomial Subsequences

Thomas Stoll

Institut Elie Cartan de Lorraine, UMR 7502,
Université de Lorraine/CNRS, 54506 Vandœuvre-lès-Nancy Cedex,
Nancy, France
thomas.stoll@univ-lorraine.fr
http://iecl.univ-lorraine.fr/∼Thomas.Stoll/

Abstract. The aim of the present article is twofold. We first give a survey on recent developments on the distribution of symbols in polynomial subsequences of the Thue–Morse sequence $\mathbf{t} = (t(n))_{n \geq 0}$ by highlighting effective results. Secondly, we give explicit bounds on

$$\min\{n : (t(pn), t(qn)) = (\varepsilon_1, \varepsilon_2)\},$$

for odd integers p, q, and on
$$\min\{n : (t(n^{h_1}), t(n^{h_2})) = (\varepsilon_1, \varepsilon_2)\}$$

where $h_1, h_2 \geq 1$, and $(\varepsilon_1, \varepsilon_2)$ is one of $(0, 0)$, $(0, 1)$, $(1, 0)$, $(1, 1)$.

Work supported by the ANR-FWF bilateral project MuDeRa "Multiplicativity: Determinism and Randomness" (France-Austria) and the joint project "Systèmes de numération : Propriétés arithmétiques, dynamiques et probabilistes" of the Université de Lorraine and the Conseil Régional de Lorraine.

Contents

Degrees of Transducibility 1
Jörg Endrullis, Jan Willem Klop, Aleksi Saarela, and Markus Whiteland

Equality Testing of Compressed Strings 14
Markus Lohrey

On the Contribution of WORDS to the Field of Combinatorics on Words ... 27
Jean Néraud

Codes and Automata in Minimal Sets 35
Dominique Perrin

Thue–Morse Along Two Polynomial Subsequences 47
Thomas Stoll

Canonical Representatives of Morphic Permutations 59
Sergey V. Avgustinovich, Anna E. Frid, and Svetlana Puzynina

Linear-Time Computation of Prefix Table for Weighted Strings 73
Carl Barton and Solon P. Pissis

New Formulas for Dyck Paths in a Rectangle 85
José Eduardo Blažek

Ambiguity of Morphisms in a Free Group 97
Joel D. Day and Daniel Reidenbach

The Degree of Squares is an Atom 109
Jörg Endrullis, Clemens Grabmayer, Dimitri Hendriks,
and Hans Zantema

Words with the Maximum Number of Abelian Squares 122
Gabriele Fici and Filippo Mignosi

Arithmetics on Suffix Arrays of Fibonacci Words 135
Dominik Köppl and Tomohiro I

Prefix-Suffix Square Completion 147
Marius Dumitran and Florin Manea

Square-Density Increasing Mappings 160
Florin Manea and Shinnosuke Seki

Mechanical Proofs of Properties of the Tribonacci Word 170
   Hamoon Mousavi and Jeffrey Shallit

On Arithmetic Progressions in the Generalized Thue-Morse Word. 191
   Olga G. Parshina

A Square Root Map on Sturmian Words (Extended Abstract) 197
   Jarkko Peltomäki and Markus Whiteland

Specular Sets 210
   Valérie Berthé, Clelia De Felice, Vincent Delecroix,
   Francesco Dolce, Julien Leroy, Dominique Perrin,
   Christophe Reutenauer, and Giuseppina Rindone

On the Tree of Ternary Square-Free Words 223
   Elena A. Petrova and Arseny M. Shur

Author Index 237

Degrees of Transducibility

Jörg Endrullis[1]([⊠]), Jan Willem Klop[1,2],
Aleksi Saarela[3], and Markus Whiteland[3]

[1] Department of Computer Science,
VU University Amsterdam, Amsterdam, The Netherlands
joerg@few.vu.nl
[2] Centrum Voor Wiskunde En Informatica (CWI), Amsterdam, The Netherlands
jwk@cs.vu.nl
[3] Department of Mathematics and Statistics and FUNDIM,
University of Turku, Turku, Finland
{amsaar,markus.whiteland}@utu.fi

Abstract. Our objects of study are infinite sequences and how they can be transformed into each other. As transformational devices, we focus here on Turing Machines, sequential finite state transducers and Mealy Machines. For each of these choices, the resulting transducibility relation $\geq$ is a preorder on the set of infinite sequences. This preorder induces equivalence classes, called *degrees*, and a partial order on the degrees.

For Turing Machines, this structure of degrees is well-studied and known as *degrees of unsolvability*. However, in this hierarchy, all the computable streams are identified in the bottom degree. It is therefore interesting to study transducibility with respect to weaker computational models, giving rise to more fine-grained structures of degrees. In contrast with the degrees of unsolvability, very little is known about the structure of degrees obtained from finite state transducers or Mealy Machines.

1 Introduction

In recent times, computer science, logic and mathematics have extended the focus of interest from finite data types to include infinite data types, of which the paradigm notion is that of infinite sequences of symbols, or *streams*. As Democritus in his adagium Panta Rhei already observed, streams are ubiquitous. Indeed they appear in functional programming, formal language theory, in the mathematics of dynamical systems, fractals and number theory, in business (financial data streams) and in physics (signal processing).

The title of this paper is inspired by the well-known 'degrees of unsolvability', described in Shoenfield [12]. Here sets of natural numbers are compared by means of transducibility using Turing Machines (TMs). The ensuing hierarchy of degrees of unsolvability has been widely studied in the 60's and 70's of the last century and later. We use the notion of degrees of unsolvability as a guiding analogy. In our case, we will deal with streams, noting that a set of natural numbers (as the subject of degrees of unsolvability) is also a stream over the alphabet $\{0, 1\}$ via its characteristic function. However, Turing Machines are too strong

© Springer International Publishing Switzerland 2015
F. Manea and D. Nowotka (Eds.): WORDS 2015, LNCS 9304, pp. 1–13, 2015.
DOI: 10.1007/978-3-319-23660-5_1

for our purposes, since typically we are interested in computable streams and they would all be identified by transducibility via Turing Machines.

We are therefore interested in studying transducibility of streams with respect to less powerful devices. A reduction of the computational power results in a finer structure of degrees. For transforming streams, a few choices present themselves: (sequential) finite state transducers (FSTs) or Mealy Machines (MMs). There are other possibilities, for instance: morphisms, many-one-reducibility, 1-reducibility and tt-reducibility (truth-table reducibility). The last three are more interesting in the context of degrees of unsolvability (Turing degrees).

Let us now describe the contents of this paper. In Sect. 2, we start with the formal definition of the three main notions of degrees, as generated by transducibility using Turing Machines, by sequential finite state transducers, and by Mealy Machines. The latter two machine models will be defined formally, for Turing Machines we will suppose familiarity, making a definition superfluous. At the end of this preliminary section, we briefly mention the possible employment of infinitary rewriting as an alternative way of phrasing the various transductions. The next section (Sect. 3) can be considered to be the heart of this paper, with a comparison between Turing degrees and Transducer degrees (arising from Turing Machines and finite state transducers, respectively). Here we mention a dozen of the main properties of Turing degrees, all without the jump operator, and compare these with the situation for Transducer degrees. This yields a number of open questions. In Sect. 4 we zoom in on an interesting area in the partial order of Transducer degrees, namely the area of 'rarefied ones'[1] streams. It turns out that the degree structure of this restricted area is already surprisingly rich. Remarkably the structure of degrees of these streams requires neither insight about finite state transducers, nor about infinite sequences. We conclude with an extensive list of questions about Transducer degrees, see Sect. 5.

A general word of warning may be in order. While we think that the questions arising from finite state transducibility of streams are fascinating, they seem to be challenging, some might even be intractable with the current state of the art.

2 Preliminaries

We briefly introduce the dramatis personae: (sequential) finite state transducers, Mealy Machines and Turing Machines. For a thorough introduction of finite automata and transducers, we refer the reader to [1,11].

Let Σ be an alphabet. We use ε to denote the empty word. We use Σ^* to denote the set of finite words over Σ, and let $\Sigma^+ = \Sigma^* \setminus \{\varepsilon\}$. The set of infinite sequences over Σ is $\Sigma^\omega = \{\sigma \mid \sigma : \mathbb{N} \to \Sigma\}$ and we let $\Sigma^\infty = \Sigma^* \cup \Sigma^\omega$. In this paper, we consider only sequences over finite alphabets. Without loss of generality we assume that the alphabets are of the form $\Sigma_n = \{0, 1, \ldots, n-1\}$ for some $n \in \mathbb{N}$. Then there are countably many alphabets and countably many

---

[1] The name 'rarefied ones' for the stream $01001000100001\cdots$ occurs in [[7] p.208] in the context of dynamical systems.

finite state transducers over these alphabets. We write $\mathbf{S}$ for the set of all streams over these alphabets, that is $\mathbf{S} = \bigcup_{n \in \mathbb{N}} \Sigma_n^\omega$.

2.1 Finite State Transducers and Mealy Machines

Sequential finite state transducers, also known as *deterministic generalised sequential machines (DGSMs)*, are finite automata with input letters and output words along the edges.

Definition 2.1. A *(sequential) finite state transducer (FST)* $A = \langle \Sigma, \Gamma, Q, q_0, \delta, \lambda \rangle$ consists of

(i) a finite input alphabet Σ,
(ii) a finite output alphabet Γ,
(iii) a finite set of states Q,
(iv) an initial state $q_0 \in Q$,
(v) a transition function $\delta : Q \times \Sigma \to Q$, and
(vi) an output function $\lambda : Q \times \Sigma \to \Gamma^*$.

Whenever Σ and Γ are clear from the context we write $A = \langle Q, q_0, \delta, \lambda \rangle$.

A finite state transducer reads an input stream letter by letter and produces a prefix of the output stream in each step.

Definition 2.2. Let $A = \langle \Sigma, \Gamma, Q, q_0, \delta, \lambda \rangle$ be a finite state transducer. We homomorphically extend the transition function δ to $Q \times \Sigma^* \to Q$ by

$$\delta(q, \varepsilon) = q \qquad\qquad \delta(q, au) = \delta(\delta(q, a), u)$$

for $q \in Q$, $a \in \Sigma$, $u \in \Sigma^*$, and the output function λ to $Q \times \Sigma^\infty \to \Gamma^\infty$ by

$$\lambda(q, \varepsilon) = \varepsilon \qquad\qquad \lambda(q, au) = \lambda(q, a) \cdot \lambda(\delta(q, a), u)$$

for $q \in Q$, $a \in \Sigma$, $u \in \Sigma^\infty$.

A Mealy Machine is an FST that outputs precisely one letter in each step.

Definition 2.3. A *Mealy Machine (MM)* is an FST $A = \langle \Sigma, \Gamma, Q, q_0, \delta, \lambda \rangle$ such that $|\lambda(q, a)| = 1$ for every $q \in Q$ and $a \in \Sigma$.

For convenience, we sometimes consider the output function of a Mealy Machine as having type $\lambda : Q \times \Sigma \to \Gamma$.

2.2 Degrees of Transducibility

We define the partial orders of degrees of streams arising from Turing Machines, finite state transducers and Mealy Machines. First, we define transducibility relations $\geq_{\mathrm{TM}}$, $\geq_{\mathrm{FST}}$ and $\geq_{\mathrm{MM}}$ on the set of streams.

Definition 2.4. Let Σ, Γ be finite alphabets, and $\sigma \in \Sigma^\omega, \tau \in \Gamma^\omega$ streams. For an FST $A = \langle \Sigma, \Gamma, Q, q_0, \delta, \lambda \rangle$, we write $\sigma \geq_A \tau$ if $\tau = \lambda(q_0, \sigma)$.

(i) We write $\sigma \geq_{\mathrm{FST}} \tau$ if there exists an FST A such that $\sigma \geq_A \tau$.
(ii) We write $\sigma \geq_{\mathrm{MM}} \tau$ if there exists an MM A such that $\sigma \geq_A \tau$.
(iii) We write $\sigma \geq_{\mathrm{TM}} \tau$ if τ is computable by a TM with oracle σ.

Note that the relations $\geq_{\mathrm{TM}}, \geq_{\mathrm{FST}}$ and $\geq_{\mathrm{MM}}$ are preorders on **S**. Each of these preorders $\geq$ induces a partial order of 'degrees', the equivalence classes with respect to $\geq \cap \leq$. We denote equivalence using $\equiv$.

Definition 2.5. Let $\mathrm{T} \in \{\mathrm{FST}, \mathrm{MM}, \mathrm{TM}\}$. We define $\equiv_{\mathrm{T}}$ as the equivalence relation $\geq_{\mathrm{T}} \cap \leq_{\mathrm{T}}$. The *T-degree* $[\sigma]_{\mathrm{T}}$ of a stream $\sigma \in \mathbf{S}$ is the equivalence class of σ with respect to $\equiv_{\mathrm{T}}$, that is, $[\sigma]_{\mathrm{T}} = \{\tau \in \mathbf{S} \mid \sigma \equiv_{\mathrm{T}} \tau\}$. For a set of streams $X \subseteq \mathbf{S}$, we write $[X]_{\mathrm{T}}$ for the set of degrees $\{[\sigma]_{\mathrm{T}} \mid \sigma \in X\}$.

The *T-degrees of transducibility* is the partial order $\langle [\mathbf{S}]_{\mathrm{T}}, \geq_{\mathrm{T}} \rangle$ induced by the preorder $\geq_{\mathrm{T}}$ on **S**, that is, for $\sigma, \tau \in \mathbf{S}$ we have $[\sigma]_{\mathrm{T}} \geq_{\mathrm{T}} [\tau]_{\mathrm{T}} \iff \sigma \geq_{\mathrm{T}} \tau$. We introduce some notation:

– We use $\mathbf{0}_{\mathrm{T}}$ to denote the *bottom degree* of $\langle [\mathbf{S}]_{\mathrm{T}}, \geq_{\mathrm{T}} \rangle$, that is, the unique degree $\mathbf{a} \in [\mathbf{S}]_{\mathrm{T}}$ such that $\mathbf{a} \leq_{\mathrm{T}} \mathbf{b}$ for every $\mathbf{b} \in [\mathbf{S}]_{\mathrm{T}}$.
– A *minimal cover* of a degree $\mathbf{a}$ is a degree $\mathbf{b}$ such that $\mathbf{a} <_{\mathrm{T}} \mathbf{b}$ and there exists no degree strictly between $\mathbf{a}$ and $\mathbf{b}$.
– An *atom* is a minimal cover of the bottom degree $\mathbf{0}_{\mathrm{T}}$.

In the sequel, we will refer to

– TM-degrees $\langle [\mathbf{S}]_{\mathrm{TM}}, \geq_{\mathrm{TM}} \rangle$ as *Turing degrees*,
– FST-degrees $\langle [\mathbf{S}]_{\mathrm{FST}}, \geq_{\mathrm{FST}} \rangle$ as *Transducer degrees* and
– MM-degrees $\langle [\mathbf{S}]_{\mathrm{MM}}, \geq_{\mathrm{MM}} \rangle$ as *Mealy degrees*.

Machine Models via Infinitary Rewriting. As we have seen, degrees of transducibility depend on the machine used. It is worth remarking that describing such machine models, including the transduction of streams, can be conveniently phrased in the framework of rewriting, in particular infinitary rewriting [5,6], including infinitary λ-calculus.

Clearly, Turing Machines are tantamount to finite λ-terms, as to their expressive power to define computable functions. Interestingly, oracle Turing Machines can also be described in λ-calculus, this time in infinitary λ-calculus. For a set $X \subseteq \mathbb{N}$, we use $\underline{X}$ to denote the infinite λ-term obtained using iterated pairing that describes the characteristic function of X. For example, if $X = \{1, 2, 4, 7, \ldots\}$ then the infinite λ-term $\underline{X}$ is

$$\underline{X} = \langle \underline{0}, \langle \underline{1}, \langle \underline{1}, \cdots \rangle \rangle \rangle = \lambda z.\, z\underline{0}(\lambda z.\, z\underline{1}(\lambda z.\, z\underline{1}(\cdots)))$$

Here $\langle p, q \rangle = \lambda z.\, zpq$ is the usual pairing in λ-calculus. Turing reducibility is then a matter of infinitary rewriting $\twoheadrightarrow$: for $X, Y \subseteq \mathbb{N}$, X is Turing reducible to Y, $X \leq_{\mathrm{TM}} Y$, if there exists a finite λ-term M such that $M\underline{Y} \twoheadrightarrow \underline{X}$. Sequential finite state transducers and Mealy Machines can be described using restricted forms of λ-terms M or infinitary first-order rewriting.

3 Comparison

In this section, we compare the structure of degrees of transducibility arising from Turing Machines with that obtained from sequential finite state transducers. We will also mention a few facts about the degrees obtained from Mealy Machines. All three partial orders have very different structural properties, to wit:

- In contrast to the Mealy degrees, there exist atoms (minimal non-zero degrees) in the Turing degrees and Transducer degrees.
- The Turing degrees and Mealy degrees form semi-lattices in contrast to the Transducer degrees for which there exist pairs of degrees without supremum.

Turing Degrees. Our comparison will be guided by questions that have been studied for Turing degrees, and we start by recalling some of the classical results. The bottom degree $\mathbf{0}_{\text{TM}}$ of this hierarchy consists of all computable streams.

In the following theorem, we summarise a few known results about Turing degrees. For each result we indicate on the right using ✔, ✘ and ? whether the property holds, does not hold or is open for Transducer degrees, respectively. For further reading on Turing degrees we refer the reader to [8,9,12,13,15,16].

Theorem 3.1. *For Turing degrees we have:*

(i) (Kleene, Post) Every degree is countably infinite. ✔

(ii) (Kleene, Post) There are $2^{\aleph_0}$ distinct degrees. ✔

(iii) (Kleene, Post) For every degree $\mathbf{a}$, $\mathbf{a}\!\downarrow = \{\mathbf{b} \mid \mathbf{a} \geq \mathbf{b}\}$ is countable. ✔

(iv) (Kleene, Post) For every degree $\mathbf{a}$, the set $\mathbf{a}\!\uparrow = \{\mathbf{b} \mid \mathbf{b} \geq_{TM} \mathbf{a}\}$ has cardinality $2^{\aleph_0}$. ✔

(v) (Spector) There exists an atom. ✔

(vi) (Spector) Every degree has a minimal cover. ?

(vii) (Kleene, Post) Every finite set of degrees has a least upper bound. ✘

(viii) (Kleene, Post, Spector) No infinite ascending sequence of degrees has a least upper bound. ?

(ix) (Kleene, Post) There are pairs of degrees without greatest lower bound. ✔

(x) (Kleene, Post) For every degree $\neq \mathbf{0}$ there exists an incomparable degree. ✔

(xi) (Sacks) Every countable partially ordered set can be embedded. ?

(xii) (Sacks) The recursively enumerable degrees are dense: whenever $\mathbf{a} < \mathbf{c}$ for recursively enumerable degrees $\mathbf{a}, \mathbf{c}$, then there exists a recursively enumerable degree $\mathbf{b}$ such that $\mathbf{a} < \mathbf{b} < \mathbf{c}$. ✘

(xiii) (Simpson) The first-order theory of $[\mathbf{S}]_{TM}$ in the language $\langle \geq, = \rangle$ is recursively isomorphic to that of true second order arithmetic. ?

The items (viii) and (ix) of Theorem 3.1 are corollaries of the following famous result by Kleene, Post and Spector.

Theorem 3.2. (Kleene, Post, Spector [12]). *Let* $a_0 <_{TM} a_1 <_{TM} \cdots$ *be an infinite ascending sequence of Turing degrees. Then there exist Turing degrees* **b**, **c** *such that* **b** *and* **c** *are upper bounds for* $\{a_n\}_{n \in \mathbb{N}}$, *and there is no Turing degree that is both an upper bound for* $\{a_n\}_{n \in \mathbb{N}}$ *and a lower bound for* $\{b, c\}$.

The structure of Turing degrees is extremely complicated. Shore [13] discussed some conjectures about this structure due to Sacks, such as:

(C4) A partially ordered set P is embeddable in the Turing degrees if and only if P has at most continuum cardinality, and each downward cone is at most countable.

(C5) If S is a set of independent Turing degrees of cardinality less than continuum, then there exists a degree $d \notin S$ such that $S \cup \{d\}$ is an independent set of degrees.

The first conjecture is still open[2], and the second was shown to be independent of the axioms of ZFC set theory! We expect that the structure of Transducer degrees is more tractable, less complicated than the structure of Turing degrees.

Transducer Degrees. Sequential finite state transducers are less powerful than Turing machines, and consequently, the Transducer degrees are more fine-grained than the Turing degrees. The Transducer degrees provide an interesting complexity measure for streams. On the one hand, transducers are 'weak enough' to exhibit a rich structure within the computable streams (which trivialise in the bottom degree of the Turing degrees). On the other hand, finite state transduction generalises several usual transformations in dealing with streams, such as alphabet renaming, insertion and removal of elements, or applying a morphism that substitutes words for letters.

The structure of the Transducer degrees is largely unexplored territory with a large number of interesting open questions. An initial study of this partial order of degrees has been carried out in [3,4]. The bottom degree **0** of the hierarchy is formed by the ultimately periodic streams. There exist infinite ascending and infinite descending sequences, and thus the hierarchy is not well-founded. It is not difficult to see that there exists no maximal degree, and a set of degrees has an upper bound if and only if the set is countable. The cardinality results (i)–(iv) of Theorem 3.1 hold also for the Transducer degrees. In [4] it has been shown that the degree of the stream

$$\Pi = 1101001000100001000001\ldots$$

is an atom (a minimal non-zero degree) and hence Theorem 3.1(v) is valid for the Transducer degrees. We refer to Sect. 4 for more on atom degrees. However, it is *open* whether every degree has a minimal cover (compare with Theorem 3.1(vi)). Analogously to the degrees of unsolvability, we call a Transducer

---

[2] The conjecture was open at the time of Shore [13] and it has remained open to the best knowledge of the authors.

degree recursively enumerable if it contains a recursively enumerable stream. As a consequence of the degree of Π being an atom, it follows that the recursively enumerable Transducer degrees are not dense, and hence Theorem 3.1(xii) fails for the Transducer degrees. However, it is interesting and *open* whether there exist dense substructures (e.g. dense intervals).

Theorem 3.1(ix) holds for the Transducer degrees: there exist pairs of degrees without a greatest lower bound. In contrast to the Turing degrees, there also exist pairs of Transducer degrees without a least upper bound and thus Theorem 3.1(vii) fails. It is *open* whether there exist infinite ascending sequences of Transducer degrees with a least upper bound (Theorem 3.1(viii)). The validity of Theorem 3.1(x) for Transducer degrees (the existence of incomparable degrees) follows immediately from the fact that finite state transducers are weaker than Turing Machines. Theorem 3.1(xi) is *open* for Transducer degrees. It is even *open* whether every finite distributive lattice can be embedded. Finally, also the complexity of the first-order theory of Transducer degrees in the language $\langle \geq, = \rangle$ is *open* (compare with Theorem 3.1(xiii)).

Mealy Degrees. The hierarchy of degrees induced by transducibility via Mealy Machines, has been studied by Rayna in [10] and Belov in [2]. We briefly mention a few interesting facts about this hierarchy. The bottom degree **0** of consists of the ultimately periodic streams, just as for the Transducer degrees. Except for the common bottom degree, the Mealy degrees and Transducer degrees exhibit very different properties. In the Mealy degrees, every stream $\sigma \not\equiv \mathbf{0}$ admits an infinite descending chain, while there exist atom degrees in the Transducer degrees. In the Mealy degrees, the degree of a stream $\sigma \not\equiv \mathbf{0}$ is always strictly lower than the degree of every strict suffix of σ. In contrast, the Transducer degree of a stream is invariant under removal and insertion of finitely many elements. In the Mealy degrees, every finite set of degrees has a least upper bound.

4   Atoms and Polynomials

In this section, we want to highlight an intriguing connection between finite state transduction and number theory.

We will consider the following 'rarefied ones' streams: for $f : \mathbb{N} \to \mathbb{N}$ we use $\langle f \rangle \in \mathbf{2}^\omega$ to denote the sequence

$$\langle f \rangle = \prod_{i=0}^{\infty} 0^{f(i)} 1 = 0^{f(0)} 10^{f(1)} 10^{f(2)} \cdots ,$$

Note that for every stream $\sigma \in \{0, 1\}^\omega$ there exists $f : \mathbb{N} \to \mathbb{N}$ such that $\sigma = \langle f \rangle$, and for every stream $\sigma \in \mathbf{S}$ there exists $f : \mathbb{N} \to \mathbb{N}$ such that $\sigma \equiv_{\mathrm{FST}} \langle f \rangle$.

In general, it is difficult to characterise the set of transducts of a sequence $\langle f \rangle$. We will therefore consider the case where the function f is a polynomial. Surprisingly, even for this simple class of functions, there is a rich structure in the degrees, and we reach very soon a large terra incognita.

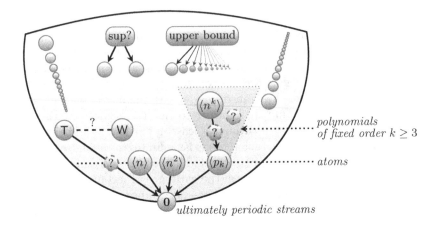

Fig. 1. *The partial order of Transducer degrees. Question marks indicate open problems. Here p_k is a polynomial of order k, see Section 4 for the form of this polynomial. The degree of $\langle p_k \rangle$ is an atom and all other polynomials of order k can be transduced to p_k. Note that $\langle n^k \rangle$ is not an atom for $k \geq 3$.*

Let us lead up to the situation as described in Fig. 1 by the following step-wise example; afterwards we present the technical key to establish these facts. In the sequel, when speaking about polynomials, we always mean polynomials with non-negative integer coefficients.

(i) *Linear functions.*

All linear functions $\langle an + b \rangle$ are equivalent to $\langle n \rangle$, and the degree of $\langle n \rangle$ is an atom. For example, the following transducer transforms $\langle 2n + 1 \rangle$ to $\langle n \rangle$:

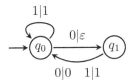

The way back from $\langle n \rangle$ to $\langle 2n + 1 \rangle$ is an easy exercise. The proof that the degree of $\langle n \rangle$ is an atom requires an understanding of the method explained below.

(ii) *Quadratic functions.*

Every quadratic function $\langle an^2 + bn + c \rangle$ transduces to $\langle n^2 \rangle$ and the degree of $\langle n^2 \rangle$ is an atom. This has been shown in [3] using the technical analysis described below. We expect that the same argument yields that $\langle n^2 \rangle$ also transduces to $\langle an^2 + bn + c \rangle$, and hence all quadratic polynomials have the same degree. We wonder whether there is a relation to the well-known geometrical fact that the graphs of quadratic polynomials, parabolas, coincide up to translation and scaling.

Let's work out a typical example which gives a feeling for the capabilities of transducers. We show that $\langle 2n^2 + n + 3 \rangle \geq_{\text{FST}} \langle n^2 \rangle$:

$$
\begin{aligned}
\langle 2n^2 + n + 3 \rangle &\equiv \langle 2n^2 + n \rangle && \text{subtracting a constant} \\
&\equiv \langle 2(n+1)^2 + (n+1) \rangle = \langle 2n^2 + 5n + 3 \rangle && \text{taking the tail} \\
&\geq \langle 1(2(2n)^2 + 5(2n) + 3) + && \text{merging even \& odd blocks} \\
&\quad\ 3(2(2n+1)^2 + 5(2n+1) + 3) \rangle && \text{multiplying odd blocks by 3} \\
&= \langle 32n^2 + 64n \rangle \equiv \langle 32(n+1)^2 \rangle && \text{adding a constant} \\
&\equiv \langle (n+1)^2 \rangle && \text{division by 32} \\
&\equiv \langle n^2 \rangle && \text{prefixing 1}
\end{aligned}
$$

(iii) *Cubic functions and higher order.*

For polynomials of order three and higher the picture becomes much more complicated. For $k \geq 3$, the degree of $\langle n^k \rangle$ is *not* an atom. But nevertheless there are polynomials $p_k(n)$ of order k that do have an atom degree, namely

$$
p_k(n) = (kn + 0)^k + (kn + 1)^k + \cdots + (kn + (k-1))^k .
$$

Moreover, all degrees of polynomials of order k are above or equal to this atom degree (and hence the atom degree is unique among them). For example, the unique atom for polynomials of order 3 is the degree of

$$
\begin{aligned}
p_3(n) &= (3n + 0)^3 + (3n + 1)^3 + (3n + 2)^3 \\
&= 81n^3 + 81n^2 + 45n + 9 .
\end{aligned}
$$

The results on polynomials of order 3 and higher are very recent and still unpublished. We include then as they indicate that there is a rich structure inside this 'polynomial subhierarchy'.

Let us briefly discuss the technical key observations underlying these results. A *block* is an occurrence of a word $100 \cdots 0$ in a stream. Finite state transducers can multiply (and divide) the length of a block by any non-negative rational number. A transducer can 'merge' consecutive blocks by erasing the 1 between the blocks. Moreover, transducers have a finite number of states, so they can multiply and merge in a periodic fashion. This is the essence of what we call 'weighted products', denoted by $\alpha \otimes f$. Here α is a tuple of weights and a weight is a tuple of rational numbers. For example, let us consider $f(n) = n$ and $\alpha = \langle \alpha_1, \alpha_2 \rangle$ with $\alpha_1 = \langle 1, 2, 3, 4 \rangle$, $\alpha_2 = \langle 0, 1, 1 \rangle$. Then:

f	0	1	2	3	4	5	6	7	8	9	$\cdots$
		$\times 1$	$\times 2$ $\times 3$		$\times 0$ $\times 1$		$\times 1$ $\times 2$ $\times 3$		$\times 0$ $\times 1$		
			$+4$		$+1$		$+4$		$+1$		
$\alpha \otimes f$		12			5		42		10		$\cdots$

Intuitively, the weight $\alpha_1 = \langle 1, 2, 3, 4 \rangle$ means that three consecutive blocks are merged, where the length of the blocks is multiplied by 1, 2 and 3, respectively, and finally 4 is added to the result. Likewise, the weight $\alpha_2 = \langle 0, 1, 1 \rangle$ means that two consecutive blocks are merged while being multiplied by 0 and 1, respectively, and 1 is added to the result.

Such transformations can always be realised by finite state transducers, that is, for every $f : \mathbb{N} \to \mathbb{N}$ and every tuple of weights α we have $\langle f \rangle \geq_{\mathrm{FST}} \langle \alpha \otimes f \rangle$. However, the crucial observation is the following: for a certain class of functions f this is 'all' that finite state transducers can do. This is the class of 'spiralling' functions; polynomials fall in this class.

Definition 4.1. A function $f : \mathbb{N} \to \mathbb{N}$ is called *spiralling* if

(i) $\lim_{n \to \infty} f(n) = \infty$, and
(ii) for every $m \geq 1$, the function $n \mapsto f(n) \mod m$ is ultimately periodic.

Functions with the property (ii) are called 'ultimately periodic reducible' in [14]. Note that polynomials with non-negative integer coefficients are spiralling.

For a tuple $\alpha = \langle \alpha_0, \ldots, \alpha_m \rangle$ we define its *rotation* by $\alpha' = \langle \alpha_1, \ldots, \alpha_m, \alpha_0 \rangle$.

Definition 4.2. A *weight* is a tuple $\langle a_0, \ldots, a_{k-1}, b \rangle \in \mathbb{Q}^{k+1}$ of rational numbers such that $a_0, \ldots, a_{k-1} \geq 0$. Given a weight $\alpha = \langle a_0, \ldots, a_{k-1}, b \rangle$ and a function $f : \mathbb{N} \to \mathbb{N}$ we define $\alpha \cdot f \in \mathbb{Q}$ by

$$\alpha \cdot f = a_0 f(0) + a_1 f(1) + \cdots + a_{k-1} f(k-1) + b.$$

The weight α is said to be *constant* whenever $a_j = 0$ for all $j \in \mathbb{N}_{<k}$.

Definition 4.3. For functions $f : \mathbb{N} \to \mathbb{N}$, and tuples $\alpha = \langle \alpha_0, \alpha_1, \ldots, \alpha_{m-1} \rangle$ of weights, the *weighted product* of α and f is a function $\alpha \otimes f : \mathbb{N} \to \mathbb{Q}$ that is defined by induction on n through the following scheme of equations:

$$(\alpha \otimes f)(0) = \alpha_0 \cdot f$$
$$(\alpha \otimes f)(n+1) = (\alpha' \otimes \mathcal{S}^{|\alpha_0|-1}(f))(n) \qquad (n \in \mathbb{N})$$

where $|\alpha_i|$ is the length of the tuple α_i, and $\mathcal{S}^k(f)$ is the k-th shift of f.

The following theorem from [3] characterises up to equivalence the transducts of spiralling sequences in terms of weighted products.

Theorem 4.4. ([3]). *Let $f : \mathbb{N} \to \mathbb{N}$ be spiralling, and $\sigma \in 2^\omega$. Then $\langle f \rangle \geq_{FST} \sigma$ if and only if $\sigma \equiv_{FST} \langle \alpha \otimes \mathcal{S}^{n_0}(f) \rangle$ for some $n_0 \in \mathbb{N}$, and a tuple of weights α.*

As an immediate consequence of this theorem we obtain that polynomials of degree k are closed under transduction in the following sense.

Proposition 4.5. ([3]). *Let $p(n)$ be a polynomial of degree k with non-negative integer coefficients, and let σ be a transduct of $\langle p(n) \rangle$ with $\sigma \notin \mathbf{0}$. Then $\sigma \geq_{FST} \langle q(n) \rangle$ for some polynomial $q(n)$ of degree k with non-negative integer coefficients.*

In [3], Proposition 4.5 and Theorem 4.4 are used to show that the degree of $\langle n^2 \rangle$ is an atom. The following theorem characterises transduction between spiralling sequences (without the 'up to equivalence' of Theorem 4.4).

Theorem 4.6. *Let $f, g : \mathbb{N} \to \mathbb{N}$ be spiralling functions. Then $\langle g \rangle \geq_{FST} \langle f \rangle$ if and only if there exist $n_0, m_0 \in \mathbb{N}$ and a tuple of weights $\boldsymbol{\alpha}$ such that*

$$\mathcal{S}^{n_0}(f) = \boldsymbol{\alpha} \otimes \mathcal{S}^{m_0}(g) .$$

We expect that several questions about the structure of Transducer degrees could be answered if we understood what preorder $\mathcal{S}^{n_0}(f) = \boldsymbol{\alpha} \otimes \mathcal{S}^{m_0}(g)$ induces on spiralling functions and, in particular, on polynomials.

Even among the polynomials there seems to be a rich structure. Theorem 4.6 can be used to obtain the following two results.

Theorem 4.7. *For $k \geq 3$, the degree of $\langle n^k \rangle$ is not an atom.*

Nevertheless, it turns out that for every $k \geq 1$ there exists a unique atom among the degrees of sequences $\langle p \rangle$ where p is a polynomial of order k.

Theorem 4.8. *Let $k \geq 1$. Let $a_0, \dots, a_{k-1} \geq 1$ and define*

$$p(n) = a_0(kn + 0)^k + a_1(kn + 1)^k + \dots + a_k(kn + (k-1))^k$$

Then for every polynomial q of order k with non-negative integer coefficients it holds that $\langle q \rangle \geq_{FST} \langle p \rangle$. Hence the degree of $\langle p \rangle$ is an atom (the unique atom among polynomials of order k).

5   A Plethora of Questions

We mention a few interesting open questions about the Transducer degrees:

(1) How many atom degrees exist? Are there continuum many?
(2) Does every degree have a minimal cover?
(3) Is every degree **a** the greatest lower bound of a pair of degrees ($\neq$ **a**)?
(4) Are there dense intervals? That is degrees **a** and **e** with **a** < **e** such that for all degrees **b, d** with **a** ≤ **b** < **d** ≤ **e** there exists **c** with **b** < **c** < **d**.
(5) Can every finite partial order be embedded in the hierarchy?
(6) Can every finite distributive lattice be embedded in the hierarchy?
(7) When does a pair of degrees have a supremum?
(8) When does a pair of degrees have an infimum?
(9) Are there infinite ascending sequences of degrees with least upper bound?
(10) Are there infinite descending sequences of degrees with greatest lower bound?
(11) What is the structure of degrees of polynomials of order k (for fixed $k \geq 1$) with non-negative integer coefficients. Is the number of degrees finite for every $k \geq 1$?
(12) Is there a degree that has precisely *two* degrees below itself? This is displayed in Fig. 2 on the right.

(13) Is there a degree that has precisely *three* degrees below itself: two incomparable degrees and the bottom degree? This is displayed in Fig. 2 on the left.

(14) How complex is the first-order theory in the language $\langle \geq, = \rangle$? (Compare with Theorem 3.1 item (xiii).)

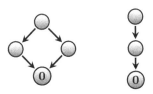

Fig. 2. Possible structures in the hierarchy: a diamond, and a line. The arrows $\rightarrow$ mean transducibility $\geq_{\mathrm{FST}}$.

We expect that some of these questions can be answered by better understanding what preorder Theorem 4.6 induces on spiralling functions and polynomials.

There are also intriguing decidability questions, for example:

(15) Is transducibility ($\geq_{\mathrm{FST}}$) decidable for automatic (or morphic) sequences?

(16) Is equivalence ($\equiv_{\mathrm{FST}}$) decidable for automatic (or morphic) sequences?

Moreover, there are challenging questions concerning concrete streams:

(17) Is the degree of Thue-Morse an atom?

(18) Consider the period doubling sequence $\sigma = 1011\ 1010\ 1011\ 1011\ 1011\cdots$ and drop every third element $\tau = 10\_1\ 1\_10\ \_01\_\ 10\_1\ 1\_11\cdots$. Do we have $\tau \geq_{\mathrm{FST}} \sigma$? If not, then Thue-Morse is not an atom.

(19) Are the degrees of Thue-Morse and Mephisto Waltz incomparable?

(20) Is it decidable whether an automatic sequence can be transduced to ($\geq_{\mathrm{FST}}$) or is equivalent to ($\equiv_{\mathrm{FST}}$) the Thue-Morse sequence?

References

1. Allouche, J.-P., Shallit, J.: Automatic Sequences: Theory, Applications Generalizations. Cambridge University Press, New York (2003)
2. Belov, A.: Some algebraic properties of machine poset of infinite words. ITA **42**(3), 451–466 (2008)
3. Endrullis, J., Grabmayer, C., Hendriks, D., Zantema, H.: The Degree of Squares is an Atom (2015)
4. Endrullis, J., Hendriks, D., Klop, J.W.: Degrees of streams. J. integers **11B**(A6), 1–40 (2011). Proceedings of the Leiden Numeration Conference 2010

5. Endrullis, J., Hendriks, D., Klop, J.W.: Highlights in infinitary rewriting and lambda calculus. Theor. Comput. Sci. **464**, 48–71 (2012)
6. Endrullis, J., Hendriks, D., Klop, J.W.: Streams are forever. Bull. EATCS **109**, 70–106 (2013)
7. Jacobs, K.: Invitation to mathematics. Princeton University Press (1992)
8. Kleene, S.C., Post, E.L.: The upper semi-lattice of degrees of recursive unsolvability. Ann. Math. **59**(3), 379–407 (1954)
9. Odifreddi, P.: Classical Recursion Theory. Studies in logic and the foundations of mathematics. North-Holland, Amsterdam (1999)
10. Rayna, G.: Degrees of finite-state transformability. Inf. Control **24**(2), 144–154 (1974)
11. Sakarovitch, J.: Elements of Automata Theory. Cambridge (2003)
12. Shoenfield, J.R.: Degrees of Unsolvability. North-Holland, Elsevier (1971)
13. Shore, R.A.: Conjectures and questions from Gerald Sacks's degrees of unsolvability. Arch. Math. Logic **36**(4–5), 233–253 (1997)
14. Siefkes, D.: Undecidable extensions of monadic second order successor arithmetic. Math. Logic Q. **17**(1), 385–394 (1971)
15. Soare, R.I.: Recursively enumerable sets and degrees. Bull. Am. Math. Soc. **84**(6), 1149–1181 (1978)
16. Spector, C.: On degrees of recursive unsolvability. Ann. Math. **64**, 581–592 (1956)

Equality Testing of Compressed Strings

Markus Lohrey[✉]

Universität Siegen, Siegen, Germany
lohrey@eti.uni-siegen.de

Abstract. This paper gives a survey on efficient algorithms for checking equality of grammar-compressed strings, i.e., strings that are represented succinctly by so called straight-line programs.

1   Introduction

The investigation of the computational complexity of algorithmic problems for succinct data started with the work of Galperin and Wigderson [10]. In that paper, a graph with 2^n vertices is represented by a Boolean circuit with $2n$ inputs, and there is an edge between $u \in \{0,1\}^n$ and $v \in \{0,1\}^n$ if and only if the circuit outputs 1 on input u, v. This kind of succinct representation was further investigated in [3,5,30,34]. It turned out that for circuit-encoded graphs, an upgrading theorem holds. Basically, it says that if a graph problem is hard for a certain complexity class, then the succinct version of the problem is hard for the exponentially larger class (precise assumptions on the underlying reductions have to be made).

In this paper, we are concerned with another succinct representation that allows to encode long strings: *straight-line programs*, briefly SLPs. An SLP is a context-free grammar that produces a single string. The length of this string can be exponential in the size of the SLP. Thus, SLPs allow exponential compression in the worst case. There exist several grammar-based compressors that compute from a given input string a small SLP for that string [6].

Another line of research studies algorithmic problems for SLP-compressed strings, see [25] for a survey. In this paper we deal with the equality problem for SLP-compressed strings: Given two SLPs $\mathcal{G}$ and $\mathcal{H}$, we want to check whether the strings produced by $\mathcal{G}$ and $\mathcal{H}$ are equal. We call this problem *compressed equality checking* in the following. Obviously, a simple decompress-and-check strategy that first produces the strings derived by $\mathcal{G}$ and $\mathcal{H}$ and then compares these strings symbol by symbol needs exponential time. Surprisingly, in 1994 three independent conference papers by Plandowski [31], Hirshfeld, Jerrum, and Moller [15] (see [16] for a long version), and Mehlhorn, Sundar, and Uhrig [27] (see [28] for a long version) were published, where polynomial time algorithms for compressed equality checking were presented. Since then, improvements concerning the running time have been achieved in [2,17,23]. The currently fastest

This research is supported by the DFG-project LO 748/10-1.

F. Manea and D. Nowotka (Eds.): WORDS 2015, LNCS 9304, pp. 14–26, 2015.
DOI: 10.1007/978-3-319-23660-5_2

and probably also simplest algorithm is due to Jeż [17] and has a quadratic running time (under some assumptions on the machine model). In Sect. 3 we outline Jeż's algorithm.

Let us remark that both, Plandowski [31] and Hirshfeld et al. [15,16], use Theorem 1 as a tool to solve another problem. Plandowski derives from Theorem 1 a polynomial time algorithm for testing whether two given morphisms (between free monoids) agree on a given context-free language. Hirshfeld et al. use Theorem 1 in order to check bisimilarity of two normed context-free processes (a certain class of infinite state systems) in polynomial time.

The algorithms from [2,16,17,23,28,31] are all sequential, and it is not clear whether any of them allows an efficient parallelization. In fact, it is open whether compressed equality checking belongs to NC or whether it is P-complete. On the other hand, recently a randomized parallel algorithm for compressed (dis)equality checking was presented in [22]. More precisely it was shown that compressed equality checking belongs to the class coRNC^2. The algorithm from [22] reduces compressed equality checking to a restricted form of polynomial identity testing over the polynomial ring $\mathbb{F}_2[x]$. For this restricted form, the identity testing algorithm of Agrawal and Biswas [1] combined with the parallel modular powering algorithm of Fich and Tompa [8] works in coRNC^2. In Sect. 4 we outline this algorithm.

2   Straight-Line Programs

For a string $s \in \Sigma^*$ we denote with $|s|$ the length of s. A factor of s is a string u such that there exist strings x, y with $s = xuy$.

A *straight-line program*, briefly *SLP*, is basically a context-free grammar that produces exactly one string. To ensure this, the grammar has to be acyclic and deterministic (every variable has a unique production where it occurs on the left-hand side. Formally, an SLP is a tuple $\mathcal{G} = (V, \Sigma, \text{rhs}, S)$, where V is a finite set of variables (or nonterminals), Σ is the terminal alphabet, $S \in V$ is the start variable, and rhs maps every variable to a right-hand side $\text{rhs}(A) \in (V \cup \Sigma)^*$. We require that there is a linear order $<$ on V such that $B < A$, whenever B occurs in $\text{rhs}(A)$. Every variable $A \in V$ derives to a unique string $\text{val}_{\mathcal{G}}(A)$ by iteratively replacing variables by the corresponding right-hand sides, starting with A. Finally, the string derived by $\mathcal{G}$ is $\text{val}(\mathcal{G}) = \text{val}_{\mathcal{G}}(S)$.

Let $\mathcal{G} = (V, \Sigma, \text{rhs}, S)$ be an SLP. The *size* of $\mathcal{G}$ is $|\mathcal{G}| = \sum_{A \in V} |\text{rhs}(A)|$, i.e., the total length of all right-hand sides. The SLP $\mathcal{G}$ is in *Chomsky normal form* if for every $A \in V$, $\text{rhs}(A)$ is either a symbol $a \in \Sigma$, or of the form BC, where $B, C \in V$. Every SLP can be transformed in linear time into an SLP in Chomsky normal form that derives the same string.

A simple induction shows that for every SLP $\mathcal{G}$ of size m one has $|\text{val}(\mathcal{G})| \leq \mathcal{O}(3^{m/3})$ [6, proof of Lemma 1]. On the other hand, it is straightforward to define an SLP $\mathcal{H}$ of size $2n$ such that $|\text{val}(\mathcal{H})| \geq 2^n$. This justifies to see an SLP $\mathcal{G}$ as a compressed representation of the string $\text{val}(\mathcal{G})$, and exponential compression rates can be achieved in this way.

An SLP can be also viewed as a multiplicative circuit over a free monoid Σ^*, where the variables are the gates that compute the concatenation of its inputs. This view can be generalized by replacing Σ^* by any finitely generated monoid, see [26]. In algebraic complexity theory, the term "straight-line program" is also used for algebraic circuits that compute (multivariate) polynomials. In such a circuit, every internal gate either computes the sum or the product of its inputs, and the input gates of the circuit are labelled with constants or variables. In Sect. 4 we will use this kind of straight-line programs, and we use the term "algebraic straight-line programs" to distinguish them from string-generating straight-line programs.

Example 1. Consider the SLP $\mathcal{G} = (V, \Sigma, \mathrm{rhs}, A_7)$ with $V = \{A_1, \ldots, A_7\}$, $\Sigma = \{a, b\}$, and the following right-hand side mapping: $\mathrm{rhs}(A_1) = b$, $\mathrm{rhs}(A_2) = a$, and $\mathrm{rhs}(A_i) = A_{i-1}A_{i-2}$ for $3 \leq i \leq 7$, Then $\mathrm{val}(\mathcal{G}) = abaababaabaab$, which is the 7th Fibonacci string. The SLP $\mathcal{G}$ is in Chomsky normal form and $|\mathcal{G}| = 12$.

One of the most basic tasks for SLP-compressed strings is *compressed equality checking*:

input: Two SLPs $\mathcal{G}$ and $\mathcal{H}$
question: Does $\mathrm{val}(\mathcal{G}) = \mathrm{val}(\mathcal{H})$ hold?

Clearly, a simple decompress-and-compare strategy is very inefficient. It takes exponential time to compute $\mathrm{val}(\mathcal{G})$ and $\mathrm{val}(\mathcal{H})$. Nevertheless a polynomial time algorithm exists. This was independently discovered by Hirshfeld, Jerrum, and Moller [15,16], Mehlhorn, Sundar, and Uhrig [27,28], and Plandowski [31]:

Theorem 1. *Compressed equality checking can be solved in polynomial time.*

In Sect. 3 we give an outline of the currently fastest (and probably also simplest) algorithm for compressed equality checking, which is due to Jeż [17]. In Sect. 4, we sketch a new approach from [22] that yields a randomized parallel algorithm for compressed equality checking.

3   Sequential Algorithms

The polynomial time compressed equality checking algorithms of Hirshfeld et al. [15,16] and Plandowski [31] use combinatorial properties of strings, in particular the periodicity lemma of Fine and Wilf [9]. This lemma states that if p and q are periods of a string w (i.e., $w[i] = w[i + p]$ and $w[j] = w[j + q]$ for all positions $1 \leq i \leq |w| - p$ and $1 \leq j \leq |w| - q$) and $p + q \leq |w|$ then also the greatest common divisor of p and q is a period of w. The algorithms from [16,31] achieve a running time of $\mathcal{O}(n^4)$, where $n = |\mathcal{G}| + |\mathcal{H}|$. An improvement to $\mathcal{O}(n^3)$ (for the more general problem of pattern matching), still using the periodicity lemma, was achieved by Lifshits [23].

In contrast to [16,23,31], the algorithm of Mehlhorn et al. [27,28] does not use the periodicity lemma of Fine and Wilf. Actually, in [28], Theorem 1 is not explicitly stated but follows immediately from the main result. Mehlhorn et al. provide an efficient data structure for a finite set of strings that supports the following operations:

- Set variable x to the symbol a.
- Set variable x to the concatenation of the values of variables y and z.
- Split the value of variable x into its length-k prefix and remaining part and store these strings in variables y and z.
- Check whether the values of variables x and y are identical.

The idea is to compute for each variable a signature which is a small number and that allows to do the equality test in constant time. The signature of a string is computed by iteratively breaking up the sequence into small blocks, which are encoded by integers using a pairing function. A single update operation $x := yz$ needs time $\mathcal{O}(\log n(\log m \log^* m + \log n))$ for the mth operation, where n is the length of the resulting string (hence, $\log(n) \leq m$). This leads to a cubic time algorithm for compressed equality checking. An improvement of the data structure from [28] can be found in [2].

The idea from [2,28] of recursively dividing a string into smaller pieces and replacing them by new symbols (integers in [2,28]) was taken up by Jeż, who came up with an extremely powerful technique for dealing with SLP-compressed strings (and the related problem of solving word equations [18]). It also yields the probably simplest proof of Theorem 1. In the rest of the section we briefly sketch this algorithm. We ignore some details. For instance, it is required in [17] that after each step, the terminal alphabet (which gets larger) is an initial segment of the natural numbers, which is ensured by using the radix sort algorithm, see [17] for more details.

Let $s \in \Sigma^+$ be a non-empty string over a finite alphabet Σ. We define the string block(s) as follows: Assume that $s = a_1^{n_1} a_2^{n_2} \cdots a_k^{n_k}$ with $a_1, \ldots, a_k \in \Sigma$, $a_i \neq a_{i+1}$ for all $1 \leq i < k$ and $n_i > 0$ for all $1 \leq i \leq k$. Then block(s) $= a_1^{(n_1)} a_2^{(n_2)} \cdots a_k^{(n_k)}$, where $a_1^{(n_1)}, a_2^{(n_2)}, \ldots, a_k^{(n_k)}$ are new symbols. For instance, for $s = aabbbaccb$ we have block(s) $= a^{(2)} b^{(3)} a^{(1)} c^{(2)} b^{(1)}$. For the symbol $a^{(1)}$ we will simply write a. Let us set block(ε) $= \varepsilon$.

For a partition $\Sigma = \Sigma_l \uplus \Sigma_r$ we denote with $s[\Sigma_l, \Sigma_r]$ the string that is obtained from s by replacing every occurrence of a factor ab in s with $a \in \Sigma_l$ and $b \in \Sigma_r$ by the new symbol $\langle ab \rangle$. For instance, for $s = abcbabcad$ and $\Sigma_l = \{a, c\}$ and $\Sigma_r = \{b, d\}$ we have $s[\Sigma_l, \Sigma_r] = \langle ab \rangle \langle cb \rangle \langle ab \rangle c \langle ad \rangle$. Since two different occurrences of factors from $\Sigma_l \Sigma_r$ must occupy disjoints sets of positions in s, the string $s[\Sigma_l, \Sigma_r]$ is well-defined.

Obviously, for all strings $s, t \in \Sigma^*$ we have

$$(s = t \iff \mathsf{block}(s) = \mathsf{block}(t)) \quad \text{and} \quad (s = t \iff s[\Sigma_l, \Sigma_r] = t[\Sigma_l, \Sigma_r]). \quad (1)$$

In the rest of this section, we assume that all SLPs $\mathcal{G} = (V, \Sigma, S, \mathrm{rhs})$ are in a kind of generalized Chomsky normal form: We require that for every variable $A \in V$, rhs(A) is either of the form $u \in \Sigma^+$, uBv with $u, v \in \Sigma^*$ and $B \in V$, or $uBvCw$ with $u, v, w \in \Sigma^*$ and $B, C \in V$. In other words, every right-hand side is non-empty and contains at most two occurrences of variables. In particular, we only consider SLPs that produce non-empty strings. This is not a crucial restriction for checking the equality val($\mathcal{G}$) = val($\mathcal{H}$), since we can first check easily in polynomial time, whether val($\mathcal{G}$) or val($\mathcal{H}$) produce the empty string.

For the following consideration, it is more convenient to have a single SLP $\mathcal{G}$ with two start variables S_1 and S_2; we write $\mathcal{G} = (V, \Sigma, \text{rhs}, S_1, S_2)$ for such an SLP. For Jeż's algorithm, it is important to assume that S_1 and S_2 do not occur in a right-hand side $\text{rhs}(A)$ $(A \in V)$, which can be easily enforced by renaming variables. Let us write $\text{val}_i(\mathcal{G}) = \text{val}_{\mathcal{G}}(S_i)$ for $i \in \{1,2\}$. Moreover, w.l.o.g. we always assume that $|\text{val}_1(\mathcal{G})| \leq |\text{val}_2(\mathcal{G})|$ (this property can be easily verified). The goal is to check $\text{val}_1(\mathcal{G}) = \text{val}_2(\mathcal{G})$ for a given SLP $\mathcal{G}$. Jeż's strategy [17] for checking this equality is to compute from $\mathcal{G}$ an SLP $\mathcal{H}$ such that $\text{val}_i(\mathcal{H}) = (\text{block}(\text{val}_i(\mathcal{G})))[\Sigma_l, \Sigma_r]$ for $i \in \{1,2\}$ and $\text{val}_1(\mathcal{H}) \leq c \cdot |\text{val}_1(\mathcal{G})|$ for some constant $c < 1$. This process is iterated. After at most $\log|\text{val}_1(\mathcal{G})| \in \mathcal{O}(|\mathcal{G}|)$ many iterations it must terminate with an SLP such that $\text{val}_1(\mathcal{G})$ has length one. Checking equality of the two strings produced by this SLP is easy. The main difficulty of this approach is to bound the size of the SLP during this process.

In the following, for an SLP $\mathcal{G}$ we denote with $|\mathcal{G}|_0$ (resp., $|\mathcal{G}|_1$) the total number of occurrences of terminal symbols (resp., nonterminal symbols) in all right-hand sides of $\mathcal{G}$. Thus, $|\mathcal{G}| = |\mathcal{G}|_0 + |\mathcal{G}|_1$. Moreover, let $\text{Var}(\mathcal{G})$ be the set of variables of $\mathcal{G}$. For block compression, we have:

Lemma 1. *There is an algorithm* CompressBlocks *that gets as input an SLP* $\mathcal{G} = (V, \Sigma, \text{rhs}, S_1, S_2)$ *and computes in time* $\mathcal{O}(|\mathcal{G}|)$ *an SLP* $\mathcal{H}$ *such that the following properties hold, where* $k = |V|$:

- $S_1, S_2 \in \text{Var}(\mathcal{H}) \subseteq V$,
- $\text{val}_i(\mathcal{H}) = \text{block}(\text{val}_i(\mathcal{G}))$ *for* $i \in \{1,2\}$,
- $|\mathcal{H}|_1 \leq |\mathcal{G}|_1 \leq 2k$ *and* $|\mathcal{H}|_0 \leq |\mathcal{G}|_0 + 4k$.

Note that this implies in particular that $\text{block}(\text{val}_i(\mathcal{G}))$ cannot contain too many different symbols. In fact, it is not hard to show that $\text{block}(\text{val}_i(\mathcal{G}))$ contains at most $|\mathcal{G}|$ many different symbols. For the proof of Lemma 1 one processes the SLP $\mathcal{G}$ bottom-up, i.e., if B occurs in the right-hand side of A, then A has to be processed before B. When A is processed, we remove from $\text{rhs}(A)$ the maximal prefix (resp., suffix) of the form a^n and insert in front of (resp., after) every occurrence of A in a right-hand side the block a^n. When all nonterminals (except of S_1 and S_2) are processed then every maximal block a^n in a right-hand side is replaced by the single letter $a^{(n)}$.

After block compression the two strings produced by the SLP $\mathcal{G}$ do not contain a factor of the form aa. For a string $s \in \Sigma^*$ that does not contain a factor of the form aa $(a \in \Sigma)$, a simple probabilistic argument shows that there exists a partition $\Sigma = \Sigma_l \uplus \Sigma_r$ such that $s[\Sigma_l, \Sigma_r]$ is by a constant factor $(\frac{3}{4}$ up to the additive constant $\frac{1}{4})$ shorter than s. Using a standard derandomization, this partition can be computed in linear time. Moreover, using a technique for counting digrams in SLP-compressed strings (see also [13]), one can compute the partition in linear time even if s is succinctly represented by an SLP $\mathcal{G}$. Once the partition $\Sigma = \Sigma_l \uplus \Sigma_r$ is computed, one can prove the following lemma in a similar way as Lemma 1.

Algorithm 1. CheckEquality

Data: SLP $\mathcal{G} = (V, \Sigma, \text{rhs}, S_1, S_2)$ such that $|\text{val}_1(\mathcal{G})| \leq |\text{val}_2(\mathcal{G})|$
while $|\text{val}_1(\mathcal{G})| > 1$ **do**
| $\mathcal{G} := \text{CompressPairs}(\text{CompressBlocks}(\mathcal{G}))$
end
check whether $\text{val}_1(\mathcal{G}) = \text{val}_2(\mathcal{G})$

Lemma 2. *There is an algorithm* CompressPairs *that gets as input an SLP* $\mathcal{G} = (V, \Sigma, \text{rhs}, S_1, S_2)$ *such that for* $i \in \{1, 2\}$, $\text{val}_i(\mathcal{G})$ *does not contain a factor* aa ($a \in \Sigma$) *and computes in time* $\mathcal{O}(|\mathcal{G}|)$ *a partition* $\Sigma = \Sigma_\ell \uplus \Sigma_r$ *and an SLP* $\mathcal{H}$ *such that the following properties hold, where* $k = |V|$:

- $S_1, S_2 \in \text{Var}(\mathcal{H}) \subseteq V$,
- $\text{val}_i(\mathcal{H}) = \text{val}_i(\mathcal{G})[\Sigma_l, \Sigma_r]$ *for* $i \in \{1, 2\}$,
- $|\mathcal{H}|_1 \leq |\mathcal{G}|_1 \leq 2k$ *and* $|\mathcal{H}|_0 \leq |\mathcal{G}|_0 + 4k$,
- $|\text{val}_1(\mathcal{H})| \leq \frac{1}{4} + \frac{3}{4}|\text{val}_1(\mathcal{G})|$.

Using Lemmas 1 and 2 we can prove Theorem 1: Assume that we have an SLP $\mathcal{G} = (V, \Sigma, \text{rhs}, S_1, S_2)$ over the terminal alphabet Σ, where $m := |\text{val}_1(\mathcal{G})| \leq |\text{val}_2(\mathcal{G})|$. Moreover, let $k = |V|$. Algorithm 1 checks whether $\text{val}_1(\mathcal{G}) = \text{val}_2(\mathcal{G})$. Correctness of the algorithm follows from observation (1). It remains to analyze the running time of the algorithm. By the last point from Lemma 2, the number of iterations of the while loop is bounded by $\mathcal{O}(\log(m)) \leq \mathcal{O}(|\mathcal{G}|)$. Let $\mathcal{G}_i$ be the SLP after i iterations of the while loop. The number of variables of $\mathcal{G}_i$ is at most k. Hence, by Lemmas 1 and 2 we have $|\mathcal{G}_i| \leq |\mathcal{G}| + 8ki \in \mathcal{O}(|\mathcal{G}|^2)$. Since the i-th iteration takes time $\mathcal{O}(|\mathcal{G}_i|)$, the total running time is $\mathcal{O}(|\mathcal{G}|^3)$.

There is a simple way to improve the running time to $\mathcal{O}(|\mathcal{G}|^2)$ (under some assumptions on the underlying machine model). In the above calculation we ignored the fact that block and pair compression also reduce the size of the SLP. To make use of this, we modify Algorithm 1 as follows. In every second iteration of the while loop, we choose the partition (Γ_l, Γ_r) for pair compression according to the following variant of Lemma 2:

Lemma 3. *There is an algorithm* CompressPairs' *with the same properties as algorithm* CompressPairs *from Lemma 2 except that the last property* $|\text{val}_1(\mathcal{H})| \leq \frac{1}{4} + \frac{3}{4}|\text{val}_1(\mathcal{G})|$ *in Lemma 2 is replaced by* $|\mathcal{H}|_0 \leq \frac{3}{4}k + 4k + \frac{3}{4}|\mathcal{G}|_0$.

The proof of this lemma is the same as for Lemma 2, except that the partition $\Sigma = \Sigma_\ell \cup \Sigma_r$ is chosen in such a way that the maximal factors from Σ^* in right-hand sides of $\mathcal{G}$ (there are at most $3k$ such factors since every right-hand side contains at most two variables) contain many factors of the form ab with $a \in \Sigma_l$, $b \in \Sigma_r$ (see Claim 1 in the proof of [19, Lemma 6] for a more precise statement).

Since Lemma 3 is used in every second iteration of the while loop, we get

$$|\mathcal{G}_{i+2}|_0 \leq \frac{3}{4}k + 4k + \frac{3}{4}(|\mathcal{G}_i|_0 + 12k) = \mathcal{O}(|\mathcal{G}|) + \frac{3}{4}|\mathcal{G}_i|_0$$

for every even i (we add the $12k$ since we apply CompressBlocks twice and CompressPairs once, before we apply CompressPairs$'$). A simple calculation shows that $|\mathcal{G}_i|_0 \in \mathcal{O}(|\mathcal{G}|)$ and hence $|\mathcal{G}_i| \in \mathcal{O}(|\mathcal{G}|)$ for all $i \geq 0$. The number of iterations of the while loop is still bounded by $\mathcal{O}(\log(m))$. This yields the running time $\mathcal{O}(\log(m) \cdot |\mathcal{G}|) \leq \mathcal{O}(|\mathcal{G}|^2)$.

This concludes our outline of Jeż's algorithm. We ignored some issues related to the machine model. More precisely, the time bound $\mathcal{O}(|\mathcal{G}|^2)$ only holds if the length m fit into a single machine word, see [17] for details. Let us finally remark that Jeż [17] obtains his result for the more general problem of fully compressed pattern matching, see Sect. 5.

Randomized algorithms for compressed equality checking are studied in [12, 32]. These algorithms are based on arithmetic modulo small prime numbers. The algorithm from [32] has a quadratic running time under the RAM model with *logarithmic cost measure*, which means that arithmetic operations on n-bit numbers need time $\mathcal{O}(n)$. If $\mathrm{val}(\mathcal{G}) = \mathrm{val}(\mathcal{H})$ then the algorithm will correctly output "yes"; if $\mathrm{val}(\mathcal{G}) \neq \mathrm{val}(\mathcal{H})$ then the algorithm may incorrectly output "yes" with a small error probability. In the next section, we will outline another randomized algorithm for compressed equality checking that allows an efficient parallelization.

4   A Parallel Algorithm

The polynomial time algorithms from [2,16,17,28,31] for compressed equality checking are all sequential, and it is not clear whether one of them allows a parallel implementation. It is in fact open, whether compressed equality checking belongs to the class NC of all problems that can be solved on a PRAM in polylogarithmic time using only polynomially many processors. In this section, we sketch a randomized parallel algorithm that was recently discovered in [22].

We use standard definitions concerning circuit complexity, see e.g. [35] for more details. In particular we will consider the class NC^i of all problems that can be solved by a circuit family $(\mathcal{C}_n)_{n \geq 1}$, where the size of $\mathcal{C}_n$ (the circuit for length-n inputs) is polynomially bounded in n, its depth is bounded by $\mathcal{O}(\log^i n)$, and $\mathcal{C}_n$ is built from input gates, NOT-gates and AND-gates and OR-gates of fan-in two. The class NC is the union of all classes NC^i. We assume circuit families to be logspace-uniform, which means that the mapping $a^n \mapsto \mathcal{C}_n$ can be computed in logspace.

To define a randomized version of NC^i, one uses circuit families with additional inputs. So, let the nth circuit $\mathcal{C}_n$ in the family have n normal input gates plus m random input gates, where m is polynomially bounded in n. For an input $x \in \{0,1\}^n$ one defines the acceptance probability as

$$\mathsf{Prob}[\mathcal{C}_n \text{ accepts } x] = \frac{|\{y \in \{0,1\}^m \mid \mathcal{C}_n(x,y) = 1\}|}{2^m}.$$

Here, $\mathcal{C}_n(x,y) = 1$ means that the circuit $\mathcal{C}_n$ evaluates to 1 if the ith normal input gate gets the ith bit of the input string x, and the ith random input gate gets the

i^{th} bit of the random string y. Then, the class RNC^i is the class of all problems A for which there exists a polynomial size circuit family $(\mathcal{C}_n)_{n \geq 0}$ of depth $\mathcal{O}(\log^i n)$ with random input gates that uses NOT-gates and AND-gates and OR-gates of fan-in two, such that for all inputs $x \in \{0,1\}^*$ of length n: (i) if $x \in A$, then $\mathrm{Prob}[\mathcal{C}_n \text{ accepts } x] \geq 1/2$, and (ii) if $x \notin A$, then $\mathrm{Prob}[\mathcal{C}_n \text{ accepts } x] = 0$. As usual, coRNC^i is the class of all complements of problems from RNC^i. Section B.9 in [14] contains several problems that are known to be in RNC, but which are not known to be in NC; the most prominent example is the existence of a perfect matching in a graph.

In this section, we will sketch a proof of the following result that was recently shown in [22]:

Theorem 2. *Compressed equality checking belongs to* coRNC^2.

Let us assume that we have a single SLP $\mathcal{G}$ in Chomsky normal form, and we want to check whether $\mathrm{val}_{\mathcal{G}}(X) = \mathrm{val}_{\mathcal{G}}(Y)$ for two variables X, Y. Without loss of generality we can assume that the terminal alphabet of $\mathcal{G}$ is $\{0,1\}$. In a first step, we compute the length $|\mathrm{val}_{\mathcal{G}}(A)|$ for every variable A. For this, one has to evaluate addition circuits over the natural numbers, which is possible in NC^2 (see [22] for details). If $|\mathrm{val}_{\mathcal{G}}(X)| \neq |\mathrm{val}_{\mathcal{G}}(X)|$, then we reject. So, let us assume that $|\mathrm{val}_{\mathcal{G}}(X)| = |\mathrm{val}_{\mathcal{G}}(X)|$. We omit the index $\mathcal{G}$ in the following.

As for many other randomized algorithms (also the RNC-algorithm for the existence of a perfect matching in a graph) we now shift the problem to an algebraic problem about polynomials. A string $w = a_0 a_1 \cdots a_n \in \{0,1\}^*$ with $a_i \in \{0,1\}$ can be encoded as the polynomial

$$p_w(x) = \sum_{i=0}^{n} a_i \cdot x^i \in \mathbb{F}_2[x]$$

over the field $\mathbb{F}_2$. Clearly, if $|u| = |v|$, then $u = v$ if and only if $p_u(x) + p_v(x) = 0$. Hence, it remains to check whether $p_{\mathrm{val}(X)}(x) + p_{\mathrm{val}(Y)}(x)$ is the zero polynomial.

The polynomial $p_{\mathrm{val}(X)}(x) + p_{\mathrm{val}(Y)}(x)$ has exponential degree, but it can be defined by a small algebraic circuit, or equivalently, a small algebraic straight-line program (ASLP). ASLPs are defined analogously to our SLPs for strings, but variables evaluate to polynomials from $\mathbb{F}_2[x]$ (or another polynomial ring, in general). In right-hand sides, the operations of (polynomial) addition and multiplication as well as the constants $0, 1, x$ can be used. We translate the SLP $\mathcal{G}$ into an ASLP $\mathcal{H}$ for the polynomial $p_{\mathrm{val}(X)}(x) + p_{\mathrm{val}(Y)}(x)$ as follows: If $\mathrm{rhs}(A) = a \in \{0,1\}$ in $\mathcal{G}$, then also $\mathrm{rhs}(A) = a$ in $\mathcal{H}$, and if $\mathrm{rhs}(A) = BC$ in $\mathcal{G}$, then $\mathrm{rhs}(A) = B + x^n \cdot C$ in $\mathcal{H}$, where $n = |\mathrm{val}_{\mathcal{G}}(B)|$ (this length can be precomputed in NC^2). Finally, we add a new start variable S to $\mathcal{H}$ and set $\mathrm{rhs}(S) = X + Y$. It is easy to check, that $\mathcal{H}$ indeed produces the polynomial $p_{\mathrm{val}(X)}(x) + p_{\mathrm{val}(Y)}(x)$.

Note that the right-hand side $B + x^n \cdot C$ contains a big power of x. On the other hand, n is only exponential in the input size and could be replaced by a chain of multiplications. Testing whether $\mathcal{H}$ produces the zero polynomial is an instance of *polynomial identity testing* (PIT) over the ring $\mathbb{F}_2[x]$. This is a famous problem in complexity theory, which is known to be in coRP (the class of complements of problems from randomized polynomial time), but for which no polynomial time algorithm is known. Moreover, proving PIT to be in P would prove circuit lower bounds that currently seem to be out of reach, see [20].

So far, we have put compressed equality checking into the class coRP only. To lower this bound to coRNC^2 we have to make use of the particular form of the ASLPs in our situation. The right-hand side $B + x^n \cdot C$ can be replaced by $B + D$, where D is a fresh variable with rhs$(D) = x^n \cdot C$. We now have obtained an ASLP $\mathcal{H}$, where every right-hand side has one of the following forms:

– a constant $a \in \{0, 1\}$,
– an addition $B + C$ of two variables B, C,
– a multiplication $x^n \cdot C$, where C is a variable and n is a number that is encoded in binary representation.

In [22] we called such an ASLP *powerful skew*: In a skew ASLP (or skew algebraic circuit), for every multiplication one of the two arguments has to be a constant or the variable x. The additional adjective "powerful" refers to the fact one of the arguments of every multiplication gate is a power of x, where the exponent is given in binary notation.

To test val$(\mathcal{H}) = 0$ for a powerful skew ASLP, we can use any of the randomized PIT-algorithms. In order to get a coRNC^2-algorithm, the identity testing algorithm of Agrawal and Biswas [1] is the right choice. This algorithm computes the polynomial val$(\mathcal{H})$ modulo a test polynomial $P(x) \in \mathbb{F}_2[x]$ of polynomial degree, which is randomly chosen from a suitable test space. Clearly, if val$(\mathcal{H}) = 0$, then also val$(\mathcal{H})$ mod $P(x) = 0$. On the other hand, by the specific choice of the test space, if val$(\mathcal{H})$ is not the zero polynomial, then also val$(\mathcal{H})$ mod $P(x)$ is not the zero polynomial with high probability. This part of the algorithm is the only place, were we use randomness.

It finally remains to compute val$(\mathcal{H})$ mod $P(x)$, which will be done in NC^2, using the modular powering algorithm of Fich and Tompa [8]. More precisely, Fich and Tompa proved in [8] that the following problem can be solved in NC^2 (we only present here the result for the polynomial ring $\mathbb{F}_2[x]$, but in [8] a more general version is shown):

input: polynomials $p(x), q(x) \in \mathbb{F}_2[x]$ and a binary encoded natural number n.
output: $p(x)^n$ mod $q(x)$

Using this result, we can replace in the ASLP $\mathcal{H}$ every power x^n by x^n mod $P(x)$ in NC^2. The resulting ASLP computes the same polynomial as $\mathcal{H}$ modulo $P(x)$. Moreover, the big powers x^n in right-hand sides of the form $x^n \cdot B$ are replaced by polynomials of polynomially bounded degree. This allows to compute the output polynomial explicitly in NC^2 using a standard reduction to matrix powering, see [22] for details. We still have to compute this output polynomial modulo $P(x)$, which can be done in NC^1 [7]. This concludes our proof sketch for Theorem 2.

The coRNC^2-algorithm for compressed equality checking easily generalizes to equality checking for SLP-compressed 2-dimensional pictures (and in fact, pictures of any dimension). Such a 2-dimensional picture is a rectangular array of symbols from a finite alphabet. To define 2-dimensional SLPs, one uses a horizontal and a vertical concatenation operation, which are both partially defined (for horizontal concatenation, the two pictures need to have the same height, and for vertical concatenation, the two pictures need to have the same width). This formalism was studied in [4], where it was shown that equality of SLP-compressed 2-dimensional pictures belongs to coRP using a reduction to PIT. Using the above technique, this bound was reduced to coRNC^2 in [22]. It is still open, whether equality of SLP-compressed 2-dimensional pictures can be checked in polynomial time.

5   Related Problems

A natural generalization of checking equality of two strings is pattern matching. In the classical *pattern matching problem* it is asked for given strings p (usually called the pattern) and t (usually called the text), whether p is a factor of t. There are many linear time algorithms for this problem on uncompressed strings. It is therefore natural to ask, whether a polynomial time algorithm for pattern matching on SLP-compressed strings exists. This problem is sometimes called *fully compressed pattern matching* and is defined as follows:

input: Two SLPs $\mathcal{P}$ and $\mathcal{T}$
question: Is val($\mathcal{P}$) a factor of val($\mathcal{T}$)?

The first polynomial time algorithm for fully compressed pattern matching was presented in [21] by Karpinski, Rytter, and Shinohara. Further improvements with respect to the running time were achieved in [11,17,23,29]. The algorithms from [11,21,23,29] use the periodicity lemma of Fine and Wilf, similarly to the solutions of Plandowski and Hirshfeld et al. for compressed equality checking. In contrast, Jeż's algorithm from [17] is based on his recompression technique and is a refinement of the algorithm sketched in the previous section. It is the currently fastest algorithm. Its running time is $\mathcal{O}((|\mathcal{T}| + |\mathcal{P}|) \cdot \log |\text{val}(\mathcal{P})|)$ under the assumption that $|\text{val}(\mathcal{P})|$ can be stored in a single machine word, otherwise an additional factor $\log(|\mathcal{T}| + |\mathcal{P}|)$ goes in.

Let us finally mention a result of Lifshits [24], which together with Theorem 1 gives an impression of the subtle borderline between tractability and intractability for problems on SLP-compressed strings. A function $f : \Sigma^* \to \mathbb{N}$ belongs to the counting class #P if there exists a nondeterministic polynomial time bounded Turing-machine M such that for every $x \in \Sigma^*$, $f(x)$ equals the number of accepting computation paths of M on input x. A function $f : \Sigma^* \to \mathbb{N}$ is #P-complete if it belongs to #P and for every #P-function $g : \Gamma^* \to \mathbb{N}$ there is a logspace computable mapping $h : \Gamma^* \to \Sigma^*$ such that $h \circ f = g$. Functions that are #P-complete are computationally very powerful. By a famous result of Toda [33], every language from the polynomial time hierarchy can be decided in deterministic polynomial time with the help of a #P-function, i.e., PH $\subseteq$ P$^{\#P}$. For two strings $u = a_1 \cdots a_n$ and $v = b_1 \cdots b_n$ of the same length n, the *Hamming-distance* $d_H(u, v)$ is the number of positions $i \in \{1, \ldots, n\}$ such that $a_i \neq b_i$.

Theorem 3 ([24]). *The mapping* $(\mathcal{G}, \mathcal{H}) \mapsto d_H(\mathrm{val}(\mathcal{G}), \mathrm{val}(\mathcal{H}))$, *where* $\mathcal{G}$ *and* $\mathcal{H}$ *are SLPs is* #P-*complete.*

6   Open Problems

The main open problem in the context of compressed equality checking is the precise complexity of this problem. Theorem 2 suggests that compressed equality checking is not P-complete (showing P ⊆ RNC would be a big surprise). Hence, we conjecture that compressed equality checking belongs to NC. For fully compressed pattern matching it is even open whether the problem belongs to coRNC (or RNC).

Another open problem is whether the quadratic running time of Jeż's algorithm for compressed equality checking can be further improved.

References

1. Agrawal, M., Biswas, S.: Primality and identity testing via chinese remaindering. J. Assoc. Comput. Mach. **50**(4), 429–443 (2003)
2. Alstrup, S., Brodal, G.S., Rauhe, T.: Pattern matching in dynamic texts. In: Proceedings of SODA 2000, pp. 819–828. ACM/SIAM (2000)
3. Balcázar, J.L.: The complexity of searching succinctly represented graphs. In: Fülöp, Z. (ed.) ICALP 1995. LNCS, vol. 944, pp. 208–219. Springer, Heidelberg (1995)
4. Berman, P., Karpinski, M., Larmore, L.L., Plandowski, W., Rytter, W.: On the complexity of pattern matching for highly compressed two-dimensional texts. J. Comput. Syst. Sci. **65**(2), 332–350 (2002)
5. Borchert, B., Lozano, A.: Succinct circuit representations and leaf language classes are basically the same concept. Inf. Process. Lett. **59**(4), 211–215 (1996)
6. Charikar, M., Lehman, E., Lehman, A., Liu, D., Panigrahy, R., Prabhakaran, M., Sahai, A., Shelat, A.: The smallest grammar problem. IEEE Trans. Inf. Theory **51**(7), 2554–2576 (2005)
7. Eberly, W.: Very fast parallel polynomial arithmetic. SIAM J. Comput. **18**(5), 955–976 (1989)
8. Fich, F.E., Tompa, M.: The parallel complexity of exponentiating polynomials over finite fields. In: Proceedings of STOC 1985, pp. 38–47. ACM (1985)
9. Fine, N.J., Wilf, H.S.: Uniqueness theorems for periodic functions. Proc. Am. Math. Soc. **16**, 109–114 (1965)
10. Galperin, H., Wigderson, A.: Succinct representations of graphs. Inf. Control **56**(3), 183–198 (1983)
11. Gasieniec, L., Karpinski, M., Plandowski, W., Rytter, W.: Efficient algorithms for Lempel-Ziv encoding (extended abstract). In: Karlsson, R., Lingas, A. (eds.) SWAT 1996. LNCS, vol. 1097, pp. 392–403. Springer, Heidelberg (1996)
12. Gasieniec, L., Karpinski, M., Plandowski, W., Rytter, W.: Randomized efficient algorithms for compressed strings: the finger-print approach (extended abstract). In: Hirschberg, D.S., Meyers, G. (eds.) CPM 1996. LNCS, vol. 1075, pp. 39–49. Springer, Heidelberg (1996)

13. Goto, K., Bannai, H., Inenaga, S., Takeda, M.: Fast q-gram mining on SLP compressed strings. In: Grossi, R., Sebastiani, F., Silvestri, F. (eds.) SPIRE 2011. LNCS, vol. 7024, pp. 278–289. Springer, Heidelberg (2011)

14. Greenlaw, R., Hoover, H.J., Ruzzo, W.L.: Limits to Parallel Computation: P-Completeness Theory. Oxford University Press, Oxford (1995)

15. Hirshfeld, Y., Jerrum, M., Moller, F.: A polynomial-time algorithm for deciding equivalence of normed context-free processes. In: Proceedings of FOCS 1994, pp. 623–631. IEEE Computer Society (1994)

16. Hirshfeld, Y., Jerrum, M., Moller, F.: A polynomial algorithm for deciding bisimilarity of normed context-free processes. Theor. Comput. Sci. **158**(1&2), 143–159 (1996)

17. Jeż, A.: Faster fully compressed pattern matching by recompression. In: Czumaj, A., Mehlhorn, K., Pitts, A., Wattenhofer, R. (eds.) ICALP 2012, Part I. LNCS, vol. 7391, pp. 533–544. Springer, Heidelberg (2012)

18. Jeż, A.: Recompression: a simple and powerful technique for word equations. In: Proceedings of STACS 2013. LIPIcs, vol. 20, pp. 233–244. Schloss Dagstuhl - Leibniz-Zentrum für Informatik (2013)

19. Jeż, A., Lohrey, M.: Approximation of smallest linear tree grammars. Technical report, arXiv.org (2014). http://arxiv.org/abs/1309.4958

20. Kabanets, V., Impagliazzo, R.: Derandomizing polynomial identity tests means proving circuit lower bounds. Comput. Complex. **13**(1–2), 1–46 (2004)

21. Karpinski, M., Rytter, W.: Pattern-matching for strings with short descriptions. In: Galil, Z., Ukkonen, E. (eds.) CPM 1995. LNCS, vol. 937, pp. 205–214. Springer, Heidelberg (1995)

22. König, D., Lohrey, M.: Parallel identity testing for skew circuits with big powers and applications. In: Italiano, G.F., Pighizzini, G., Sannella, D.T. (eds.) MFCS 2015. LNCS, vol. 9235, pp. 445–458. Springer, Heidelberg (2015)

23. Lifshits, Y.: Processing compressed texts: a tractability border. In: Ma, B., Zhang, K. (eds.) CPM 2007. LNCS, vol. 4580, pp. 228–240. Springer, Heidelberg (2007)

24. Lifshits, Y., Lohrey, M.: Querying and embedding compressed texts. In: Královič, R., Urzyczyn, P. (eds.) MFCS 2006. LNCS, vol. 4162, pp. 681–692. Springer, Heidelberg (2006)

25. Lohrey, M.: Algorithmics on SLP-compressed strings: a survey. Groups Complex. Cryptology **4**(2), 241–299 (2012)

26. Lohrey, M.: The Compressed Word Problem for Groups. SpringerBriefs in Mathematics. Springer, Heidelberg (2014)

27. Mehlhorn, K., Sundar, R., Uhrig, C.: Maintaining dynamic sequences under equality-tests in polylogarithmic time. In: Proceedings of SODA 1994, pp. 213–222. ACM/SIAM (1994)

28. Mehlhorn, K., Sundar, R., Uhrig, C.: Maintaining dynamic sequences under equality tests in polylogarithmic time. Algorithmica **17**(2), 183–198 (1997)

29. Miyazaki, M., Shinohara, A., Takeda, M.: An improved pattern matching algorithm for strings in terms of straight-line programs. In: Hein, J., Apostolico, A. (eds.) CPM 1997. LNCS, vol. 1264, pp. 1–11. Springer, Heidelberg (1997)

30. Papadimitriou, C.H., Yannakakis, M.: A note on succinct representations of graphs. Inform. Control **71**(3), 181–185 (1986)

31. Plandowski, W.: Testing equivalence of morphisms on context-free languages. In: van Leeuwen, J. (ed.) ESA 1994. LNCS, vol. 855, pp. 460–470. Springer, Heidelberg (1994)

32. Schmidt-Schauß, M., Schnitger, G.: Fast equality test for straight-line compressed strings. Inf. Process. Lett. **112**(8–9), 341–345 (2012)

33. Toda, S.: PP is as hard as the polynomial-time hierarchy. SIAM J. Comput. **20**(5), 865–877 (1991)
34. Veith, H.: Succinct representation, leaf languages, and projection reductions. Inf. Comput. **142**(2), 207–236 (1998)
35. Vollmer, H.: Introduction to Circuit Complexity. Springer, Heidelberg (1999)

On the Contribution of WORDS to the Field of Combinatorics on Words

Jean Néraud[✉]

Université de Rouen, Rouen, France
jean.neraud@univ-rouen.fr, jean.neraud@wanadoo.fr, neraud.jean@free.fr

Abstract. We propose some notes about the history and the features of the conference WORDS, our goal being to testify how the conference may be embedded in the development of the field of Combinatorics on Words.

The representation of numbers by sequences of symbols is inherent in mathematics. A noticeable step has certainly been reached by the introduction of the decimal representation. This notion appears at the tenth century in the documents written by the arabe mathematician Al-Uqlidisi, which was interested in the indian system of numeration. In the occident, the fractional representation of numbers delayed the introduction of the decimal representation until the seventeenth century, when the belgian mathematician Simon Stevin recommends it as a performing tool of calculation. In his document "La Disme", he predicts that these methods of calculation will be extended to unrestricted representations and even be applied to the so-called incommensurable numbers.

A systematic study of words as formal mathematical objects appeared at the beginning of the twentieth century, when three now famous papers were published by the norvegian mathematician Axel Thue [16–18]. Presently, thanks to Jean Berstel [1] and James F. Power [15], we get translations of these papers in a more recent terminology and in relation with more contemporary directions of research.

Actually, from the end of the nineteenth century, words have taken an important role in different domains of mathematics such as Groups, Semigroups, Formal Languages, Number Theory, Ergodic Theory. In another hand, constituting a unified treatment of words was more and more in demand: such a request was especially stimulated by Max Paul Schützenberger, who presented a series of challenging questions concerning the topic in his lectures of 1966.

In the growing importance of the field of Combinatorics on Words, a new fundamental step was reached in the eighties when were published the series of Lothaire's books [8,9] and the famous "Theory of Codes" from Jean Berstel and Dominique Perrin [2] (cf also [3]). The importance of the topic has been also supported by the publication of the third book from Lothaire [10], which

litis, Université de Rouen, UFR Sciences et Techniques, Campus du Madrillet, 76800 Saint Etienne du Rouvray, France.

© Springer International Publishing Switzerland 2015
F. Manea and D. Nowotka (Eds.): WORDS 2015, LNCS 9304, pp. 27–34, 2015.
DOI: 10.1007/978-3-319-23660-5_3

testifies that words have fundamental applications in many domains of computer science. Another important step was reached in 1991, when the terminology "Combinatorics on Words" has been introduced in the famous "Mathematics Subject Classification" as a subfield of "Discrete Mathematics in Relation to Computer Science" (68R15).

In view of sharing scientific results, international workshops play a complementary part beside the publication of books and full papers. They quickly provide a picture of the state-of-the-art and are special meeting places for the communauty. Actually, until the end of the nineties, due to the numerous varied topics in theoretical computer science, in most of the international conferences only a few sessions could be granted to Combinatorics on Words. As regard to the growing number of papers concerned by the topic, a specific international conference essentially devoted to words was strongly required.

The organization of "WORDS" was the response to such a request: with Aldo de Luca and Antonio Restivo, we drew the main features of the project, and opted in particular for a bi-annual series of meetings. The first conference WORDS was planned in Rouen, France, during September 1997. Thanks are due to the many researchers that supported that event and the subsequent conferences (they will recognize themselves); a special thought is due to Jean Berstel, for his invaluable investment in the project. A series of ten international workshops were organized:

WORDS 1997, Rouen, France, Jean Néraud chair,
WORDS 1999, Rouen, France, Jean Néraud chair,
WORDS 2001, Palermo, Italia, Felipo Mignosi chair,
WORDS 2003, Turku, Finland, Juhani Karhumäki chair,
WORDS 2005, Montréal, Canada, Srecko Brlek and Christophe Reutenauer co-chairs,
WORDS 2007, Marseille, France, Srecko Brlek and Julien Cassaigne co-chairs,
WORDS 2009, Salerno, Italia, Arturo Carpi and Clelia De Felice co-chairs,
WORDS 2011, Prague, Czech Republic, Štěpán Holub and Edita Pelandova co-chairs,
WORDS 2013, Turku, Finland, Juhani Karhumäki and Luca Zamboni co-chairs,
WORDS 2015, Kiel, Germany, Dirk Nowotka, chair.

For each meeting, papers selected to be presented are collected in written acts; moreover among them, some of the most stricking are published in a special issue of an international scientific revue [4–7, 11–14].

The tenth conference WORDS provides the opportunity to take a panoramic look at all the fascinating papers which were presented. Of course, drawing up an exhaustive list would be an impossible task: in what follows, only a limited number of results will be mentioned: according to the frequency of their presentations, we have opted for a classification into three thematic areas.

The topic of Unavoidable Patterns

The famous infinite word of The-Morse have the fundamental property that given a letter a, no factor of type $avava$ may appear in the word. It is also well-known that this word allows to construct an infinite cube-free word on a three letter alphabet. These properties have naturally opened a more general problem: given finite alphabet does an infinite word exists such that none of its factors may be a *repetition* of type $(uv)^k u$, with $u \neq \varepsilon$ and $k \geq 2$.

- *Avoidance of patterns* is a central question in the topic. In the meeting of 1997, Roman Kolpakov, Gregory Kucherov and Yuri Tarannikov present some properties of repetition-free binary words of minimal density. During that of 2003, James Currie draws an extensive survey concentrating on open problems. In the same workshop, Ina N. Rampersad, Jeffrey Shallit and Ming-Wei Wang are interested in infinite words avoiding long squares. In the conference of 2007, Arturo Carpi and Valerio D'Alonzo present a word avoiding near repeats. In their contribution to WORDS 2011, Elena A. Petrova and Arseny M. Shur will focuse on binary cube-free words that cannot be infinitely extended preserving cube-freeness: they prove that such words exist but can have arbitrarily long finite cube-free extensions both to one side and two sides. In WORDS 2013, Tero Harju presents an infinite square-free word on a three-letter alphabet that can be shuffled with itself to produce another infinite square-free word. In the same conference, Tomi Kärki gives an overview of the results concerning repetition-freeness in connection with the so-called *similarity relations*, which are relations on words of equal length induced by a symmetric and reflexive relation on letters.

- A natural question consists in examining the number of patterns that may appear in a finite word. In their contribution to WORDS 1997, the exact number of squares in the Fibonacci words is examined by Aviezri S. Fraenkel and Jamie Simpson: they prove that a word of length n contains at most $2n$ squares. The bound will be refined as $2n - O(\log n)$ by Lucian Illie in his talk of WORDS 2005.

- In their talk of WORDS 2009, Maxime Crochemore, Lucian Ilie and Liviu Tinta define *runs* as patterns of type $uu\alpha$ that are maximal in the sense where they cannot be extended from left or right to obtain the same type of pattern. Such objects play an important role in a lot of string matching algorithms: the authors show that, given a word of length n, the number of its runs is up-bounded by $1.029n$.

- The *repetition threshold* for k letters, which we denote by $RT(k)$, is the shortest word x such that there exists an infinite x^+-free word over a k-letter alphabet. Actually, repetitions of the Thue-Morse sequence have exponent at most 2 and we have $RT(2) = 2$. In the seventies, Françoise Dejean conjectured that for every $k > 2$ the following holds:

$$RT(k) = \begin{cases} 7/4 & \text{if } k = 3 \\ 7/5 & \text{if } k = 4 \\ k/k - 1 & \text{otherwise.} \end{cases}$$

A stronger version of this conjecture has been stated by Pascal Ochem in his contribution to WORDS 2005:

- For every $k \geq 5$, an infinite $(k/k - 1)^+$-free word over k letters exists with letter frequency $1/k + 1$
- For every $k \geq 6$, an infinite $(k/k - 1)^+$-free word over k letters exists with letter frequency $1/k - 1$.

Dejean's conjecture had been partially solved by several authors. The final proof was completed in 2009 by James Currie and Narad Rampersad for $15 \leq n \leq 26$, and independently by Michaël Rao for $8 \leq k \leq 38$. In WORDS 2009, Michaël Rao will present his proof. Moreover the technic that he applies allow also to prove Ochem's stronger version of the conjecture for $9 \leq k \leq 38$.

- Abelian patterns are also concerned: an *abelian square* consists in a pattern of type xy, where the word y is obtained by applying a permutation on the letters of x. In 1992, Veikko Keränen had solved a famous open problem in constructing an *abelian square free* word over a four-letter alphabet. In the meeting of 2007, he will present new abelian square-free morphisms and a powerful substitution over 4 letters.

Two words u, v are k-*abelian equivalents* if every word of length at most k occurs as a factor in u as many times as in v. A word is *strongly k-abelian nth-power* if it is k-abelian equivalent to a nth-power. In WORDS 2013, Mari Huova and Aleksi Saarela prove that strongly k-abelian nth-powers are unavoidable on any alphabet.

- Pattern avoidance by palindromes was the subject of the talk from Inna A. Mikhailova and Mikhail Volkov, in WORDS 2007.

A word is a *pseudopalindrome* if it is the fixed point of some involutary antimorphism ϕ of the free monoid (i.e. $\phi^2 = id$, $\phi(uv) = \phi(v)\phi(u)$) and the *pseudopalindromic closure* of a word w is the shortest pseudopalindrome having w as a prefix. In their contribution to WORDS 2009, Damien Jamet, Genevieve Paquin, Gwenaël Richomme and Laurent Vuillon present some several combinatorial properties of the fixed points under iterated by pseudopalindromic closure.

Factorization of words. Equations

Other important informations may be obtained by decomposing words into a convenient sequence of consecutive factors: $w = w_1 \cdots w_n$.

- In their talk of WORDS 1997, Juhani Karhumäki, Wojciech Plandowski and Wojciech Rytter investigate the properties of the so-called $\mathcal{F}$-factorization, where the preceding sequence $(w_1, \cdots, w_n)$ was assigned to satisfy a given property $\mathcal{F}$. From an algorithmic point of view, they examine the behavior of three fundamental properties of such factorization, namely *completeness, uniqueness* and *synchronization*.

- *Periodicity* is also clearly concerned by the notion of factorization. If for an integer $n \geq 2$, all the preceding words $w_1, \cdots, w_{n-1}$ are equal, the word w_n being one of their prefix, we say that the length of w_1 is a *period* of w. The famous theorem of Fine and Wilf states that if some powers of two words x, y have a common prefix of length $|x| + |y| - \gcd(|x|, |y|)$, then x and y themselves are powers of the same word. In WORDS 1997, Maria Gabriella Castelli, Filippo Mignosi and Antonio Restivo present an extension of Fine and Wilf's theorem for three periods. In their contribution to WORDS 2003, Sorin Constantinescu and Lucian Ilie prove a new extension of that theorem for arbitrary number of periods, and in WORDS 2007, Vesa Halava, Tero Harju, Tomi kärki, and Luca Q. Zamboni will study the so-called relational Fine and Wilf words. At least, in his contribution to WORDS 2007, Kalle Saari, examine periods of the factors of the Fibonacci word.

- A word w is *quasiperiodic* if another word x exists such that any position in w falls in an occurrence of x as a factor of w (unformaly, w may be completely "covered" by a set of occurences of the factor x). In the conference of 2013, Florence Levé and Gwenaël Richomme extend the work that they presented in WORDS 2007: in particular they present algorithms for deciding whether a morphism is strongly quasiperiodic on finite and infinite words.

- In WORDS 1999 Juhani Karhumäki and Ján Maňuch prove that if a non-periodic bi-infinite word possesses three disjoint factorizations on the words of a prefix-free set X, then a set Y with cardinality at most $|X| - 2$ exists such that $X \subseteq Y^*$.

Such a type of *defect effect* is strongly connected to *independent systems of equations*, as illustrated by the paper presented by Tero Harju and Dirk Nowotka in WORDS 2001, where the case of equations in three variables is investigated. Some properties of infinite systems of equations are also presented by Štěpán Holub and Juha Kortelainen in WORDS 2005.

- The famous Post Correspondence Problem (*PCP* for short) is also connected to decomposition of words. Given two morphisms h, g, it consists in examining wether the existence of a non-empty solution for the equation $h(x) = g(x)$ is decidable. In the meeting of 2005 Vesa Halava, Tero Harju, Juhani Karhumäki and Michel Latteux provide an extension of the decidability of the marked *PCP* to instances with unique blocks. The properties of new variants such as the circular-*PCP*, and the n-permutation *PCP* are also examined by Vesa Halava in his presentation of WORDS 2013. In the same meeting, in the topic of the *Dual-PCP*, the so-called periodicity forcing words is the aim of the talk from Joel D. Day, Daniel Reidenbach and Johannes C. Schneider.

Complexity issues

In the literature, with a word several notions of complexity can be associated, the most famous being certainly the *factor complexity*: given a word w, this complexity measures the number $p_w(n)$ of different factors of length n occuring in w. The famous characterization of Morse-Hedlund for ultimately periodic words has led to introduce the infinite *Surmian words* whose complexity is $p_w(n) = n + 1$, the best known example of them being certainly the famous *Fibonacci word*.

- Other notions of complexity for infinite words are defined by Sébastien Ferenczi and Zoltán Kása in their paper of WORDS 1997: the behavior of *upper (lower) total finite-word complexity* and *upper (lower) maximal finite-word complexity* are compared to the classical factor complexity, moreover new characterizations of Sturmian sequences are obtained.

-The *recurrence function* has been introduced by Morse and Hedlund: given a factor u, it associates with every non-negative integer n the size $R_u(n)$ of the smallest window that contains every factor of length n of u. During its talk of WORDS 1997, Julien Cassaigne introduces the *recurrence quotient* as $\limsup_{n \to \infty} \frac{R(n)}{n}$, moreover he computes it for Sturmian sequences.

- The notion of *special factor* allows to obtain a performing characterization of Sturmian words. In the case of finite words, it leads to introduce two parameters, namely R, L which, given a finite word w, represent the least integer such no right (left) special factors of lenght $\geq R$ ($\geq L$) may occur in w. In his paper of WORDS 1997, Aldo de Luca studies the connections between these parameters and the classical factor complexity.

- The study of the ratio $p(n)/n$ brings also noticeable informations on infinite words. In WORDS 1999, Alex Heinis shows that if $p(n)/n$ has a limit, then it is either equal to 1, or highter than and equal to 2. By using the Rauzy graphs, in WORDS 2001, Ali Aberkane will present characterizations of the words such that the limit is 1.

- An words is *balanced* if for any pairs (u, v), of factors with same length, and for any letter a, we have $||u|_a - |v|_a| \leq 1$ (where $|u|_a$ stands for the number of occurrences of the letter a in u). In the paper he presents in WORDS 2001, Boris Adamczewski defines the *balance function* as $\max_{a \in A} \max_{u,v \in F(w)} \{||u|_a - |v|_a|\}$: as regard to the so-called *primitive substitutions*, he investigates the connections between the asymptotic behavior of the balance function and the incidence matrix of such a substitution. In the workshop of 2007, several contributions to the topic are also presented: Nicolas Bédaride, Eric Domenjoud, Damien Jamet and Jean-Luc Remy study the number of balanced words of given length and height on a binary alphabet. During WORDS 2013, two talks relating investigations of the property of balancedness of the Arnoux-Rauzy word are presented: one is from Julien Cassaigne and the other one from Vincent Delecroix, Tomáš Hejda and Wolfgang Steiner.

- The *arithmetical complexity* of an infinite sequence is the number of all words of a given length whose symbols occur in the sequence at positions which constitute an arithmetical progression. The study of arithmetical complexity will also appear in the contribution of Julien Cassaigne and Anna Frid of WORDS 2005: they give a uniform $O(n^3)$ upper bound for the arithmetical complexity of a Sturmian word and provide explicit expressions for the arithmetical complexity of Sturmian words of slope between 1/3 and 2/3 (this is, in particular, the case of the infinite Fibonacci word). In this case the difference between the genuine arithmetical complexity function and the precedingly mentioned upper bound is itself bounded and ultimately 2-periodic.

- The *palindromic complexity* of an infinite word is the function which counts the number $P(n)$ of different palindromes of each length occurring as factors in the word. In their talk of WORDS 2005, Peter Baláži, Zuzana Masáková and Edita Pelantová provide an estimate of the palindromic complexity $P(n)$ for uniformly recurrent words; denoting by $p(n)$ the classical factor complexity this estimation is based on the equation: $P(n) + P(n+1) = p(n+1) - p(n) + 2$.

- The *m-binomial complexity* of an infinite word w maps an integer n to the number of m-binomial equivalence classes of factors of length n that occur in w. This relation of binomial equivalence is defined as follows: two words u, v are m-equivalent if, for any word x of length at most m, x appears in u and v with the same number of occurrences. In their contribution to WORDS 2013 Michel Rigo and Pavel Salimov compute the m-binomial complexity of famous words: the Sturmian words and the Thue-Morse word.

- Given a word v with several occurrences in an infinite word, the set of *return words* of v has for elements all distinct words beginning with an occurrence of v and ending just before the next occurrence of v. It had been proved that a word is Sturmian if and only if each of its factors has two returns: in the meeting of 2011, Svetlana Puzynina and Luca Zamboni prove that a word is Sturmian if and only if each of its factors has two or three abelian returns.

- Complexity is also implicitly present in the talk given by Shuo Tan and Jeffrey Shallit in WORDS 2013. Given an alphabet A, a subset X of A^n is *representable* if it occurs as the set of all factors of length n of a finite word. Clearly, the set A^n itself is represented by any De Bruinj word of order n. One of the questions which are presented by the authors consists in examining the length of a word needed to represent a given representable set X: they provide a lower and a upper bounds of the form α^{2^n} with $\alpha = \sqrt{2}$ for the lower bound and $\alpha = \sqrt[10]{4}$ for the upper-one.

Many other interesting topics were presented during the conference: in the sequel we can only mention some of them.

As regard to sets of words, in WORDS 1999, Jean Berstel and Luc Boasson prove that, given a finite set of words S, at most one (normalized) multiset P may exists such that S is the shuffle of the words in P, the multiset P being effectively computable. Codes were also the subject of talks from Véronique Bruyère and Dominique Perrin (WORDS 1997), Jean Néraud and Carla Selmi (WORDS 2001), Fabio Burderi (WORDS 2011).

Extensions of the classical concept of words were also the subject of a lot of presentations: the notion of partial word has been introduced in WORDS 1997 by Jean Berstel and Luc Boasson; in WORDS 2009, in the framework of binary words, Francine Blanchet Sadri and Brian Shirey will examine the relationship between such a notion and periodicity. Multidimensional words were the feature of the talks from Valérie Berthé and Robert Tijdeman in WORDS 1999, and in the meeting of 2005, they were also the subject of two presentations: one was given by Pierre Arnoux, Valérie Berthé, Thomas Fernique and Damien Jamet and the other one by Jean-Pierre Borel.

Connections with Semigroups Theory were the subject of the lectures of Sergei I. Adian in WORDS 2003.

Words in connection with Number Theory and Numeration Systems were also the feature of very interesting papers from Tom Brown (WORDS 1999), Petr Ambrož and Christiane Frougny (WORDS 2005), Daniel Dombek (WORDS 2011), Shigeki Akiyama, Victor Marsault and Jacques Sakarovitch (WORDS 2013).

Anyway each of the numerous results which will be presented in the ten conferences plays a noticeable part in the state-of-the-art. We will have reached our goal if the preceding notes may testify to the involvement of WORDS in the development of the field of Combinatorics on Words.

References

1. Berstel, J.: Axel Thue's papers on repetitions in words: a translation. http://igm. univ-mlv.fr/~berstel/Articles/1994ThueTranslation.pdf
2. Berstel, J., Perrin, D.: Theory of Codes. Academic Press, New York (1985)
3. Berstel, J., Perrin, D., Reutenauer, C.: Codes an Automata, Encyclopedia of Mathematics and its Applications, vol. 129. Cambridge University Press, Cambridge (2010)
4. Carpi, A., De Felice, C. (Guest Editors): Combinatorics on Words (WORDS 2009), 7th International Conference on Words, Fisciano, Italy, Theoretical Computer Science, vol. 412, Issue 27, pp. 2909–3032 (2011)
5. Brlek, S., Reutenauer, C. (Guest Editors): Combinatorics on words. Theor. Comput. Sci. **380**(3), 219–410 (2007)
6. Harju, T., Karhumäki, J., Restivo, A. (Guest Editors): Combinatorics on words. Theor. Comput. Sci. **339**(1), 1–166 (2005)
7. Karhumäki, J., Lepistö, A., Zamboni, L.: Combinatorics on Words, Theoretical Computer Science (to appear)
8. Lothaire, M.: Combinatorics on Words, 2nd edn. Cambridge University Press, Cambridge (1997). First edition 1983
9. Lothaire, M.: Algebraic Combinatorics on Words, Encyclopedia of Mathametics and its Applications, vol. 90. Cambridge University Press, Cambridge (2002)
10. Lothaire, M.: Applied Combinatorics on Words, Encyclopedia of Mathematics and its Applications, vol. 105. Cambridge University Press, Cambridge (2005)
11. Masáková, Z., Holub, Š.: Special Issue WORDS 2011. Int. J. Found. Comput. Sci. **23**(8), 1579–1728 (2012)
12. Néraud, J. (Guest Editor): WORDS. Theor. Comput. Sci. **218**(1), 1–216 (1999)
13. Néraud, J. (Guest Editor): WORDS. Theor. Comput. Sci. **273**(1–2), 1–306 (2002)
14. Néraud, J. (Guest Editor): WORDS. Theor. Comput. Sci. **307**(1), 1–216 (2003)
15. Power, J.F.: Thues 1914 paper: a translation. http://arxiv.org/pdf/1308.5858.pdf
16. Thue, A.: Über unendliche zeichenreihen. Nor. Vid. Selsk; Skr. I. Mat. Kl. Christiana **7**, 1–22 (1906)
17. Thue, A.: Über die gegenseitige lager gleicher teile gewisser zeichenreihen. Skr. I. Mat. Kl. Christiana **1**, 1–67 (1912)
18. Thue, A.: Problemeüber Veränderungen von Zeichenreihen nach gegebenen Regeln. Christiana Videnskabs-Selskabs Skrifter, I. Math. naturv. Kl. 10 (1914). Reprinted in [NSST77, pp. 493–524]

Codes and Automata in Minimal Sets

Dominique Perrin[(✉)]

LIGM, Université Paris Est, Paris, France
dominique.perrin@esiee.fr

Abstract. We explore several notions concerning codes and automata in a restricted set of words S. We define a notion of S-degree of an automaton and prove an inequality relating the cardinality of a prefix code included in a minimal set S and its S-degree.

1 Introduction

We have introduced in [1] the notion of tree set as a common generalization of Sturmian sets and of interval exchange sets. In this paper, we investigate several new directions concerning codes and automata in minimal sets.

Codes and automata in restricted sets of words have already been investigated several times. In particular, Restivo has investigated codes in sets of finite type [2] and Reutenauer has studied the more general notion of codes of paths in a graph [3]. We have initiated in [4] with several other authors, a systematic study of bifix codes in Sturmian sets, a subject already considered before in [5]. The overall conclusion of this study is that very surprising phenomena appear in this context in relation with subgroups of finite index of the free group, allowing one to obtain positive bases of the subgroups contained in a given minimal set.

In this paper, we investigate several notions concerning codes and automata in relation with a factorial set S. This includes a definition of minimal S-rank of an automaton, which is equal to 1 if and only if the automaton is synchronized. We prove a result which allows to compute the minimal S-rank when S is minimal (Theorem 3.1). We also show that for a recurrent set S and a strongly connected automaton $\mathcal{A}$, the set of elements of the transition monoid M of minimal S-rank is included in a $\mathcal{D}$-class of M called its S-minimal $\mathcal{D}$-class (Proposition 3.2). This regular $\mathcal{D}$-class is unique when S is minimal and it is related with the results of [6] and [7] on the regular $\mathcal{J}$-classes of free profinite semigroups.

We define the S-degree of a prefix code X included in S as the minimal S-rank of the minimal automaton of X^*. We show that the cardinality of a prefix code is bounded below by a linear function of its S-degree (Theorem 4.4).

Let X be a prefix code and let M be the transition monoid of the minimal automaton of X^*. We associate to X a permutation group denoted $G_X(S)$ which is the structure group of the S-minimal $\mathcal{D}$-class of M. We show that for any uniformly recurrent tree set S and any finite S-maximal bifix code X, the group $G_X(S)$ is equivalent to the representation of the free group on the cosets of the subgroup generated by X (Theorem 4.5).

© Springer International Publishing Switzerland 2015
F. Manea and D. Nowotka (Eds.): WORDS 2015, LNCS 9304, pp. 35–46, 2015.
DOI: 10.1007/978-3-319-23660-5_4

2 Neutral and Tree Sets

Let A be a finite alphabet. We denote by A^* the set of all words on A. We denote by ε or 1 the empty word. A set of words on the alphabet A and containing A is said to be *factorial* if it contains the factors of its elements. An *internal factor* of a word x is a word v such that $x = uvw$ with u, w nonempty.

2.1 Neutral Sets

Let S be a factorial set on the alphabet A. For $w \in S$, we denote $L_S(w) = \{a \in A \mid aw \in S\}$, $R_S(w) = \{a \in A \mid wa \in S\}$, $E_S(w) = \{(a, b) \in A \times A \mid awb \in S\}$, and further $\ell_S(w) = \mathrm{Card}(L_S(w))$, $r_S(w) = \mathrm{Card}(R_S(w))$, $e_S(w) = \mathrm{Card}(E_S(w))$.

We omit the subscript S when it is clear from the context. A word w is *right-extendable* if $r(w) > 0$, *left-extendable* if $\ell(w) > 0$ and *biextendable* if $e(w) > 0$. A factorial set S is called *right-extendable* (resp. *left-extendable*, resp. *biextendable*) if every word in S is right-extendable (resp. left-extendable, resp. biextendable).

A word w is called *right-special* if $r(w) \geq 2$. It is called *left-special* if $\ell(w) \geq 2$. It is called *bispecial* if it is both left-special and right-special. For $w \in S$, we denote

$$m_S(w) = e_S(w) - \ell_S(w) - r_S(w) + 1.$$

A word w is called *neutral* if $m_S(w) = 0$. We say that a set S is *neutral* if it is factorial and every nonempty word $w \in S$ is neutral. The *characteristic* of S is the integer $\chi(S) = 1 - m_S(\varepsilon)$.

A neutral set of characteristic 1, simply called a neutral set, is such that all words (including the empty word) are neutral.

The following is a trivial example of a neutral set of characteristic 2.

Example 2.1. Let $A = \{a, b\}$ and let S be the set of factors of $(ab)^*$. Then S is neutral of characteristic 2.

As a more interesting example, any Sturmian set is a neutral set [1] (by a Sturmian set, we mean the set of factors of a strict episturmian word, see [8]).

The following example is the classical example of a Sturmian set.

Example 2.2. Let $A = \{a, b\}$ and let $f : A^* \to A^*$ be the *Fibonacci morphism* defined by $f(a) = ab$ and $f(b) = a$. The infinite word $x = \lim_{n \to \infty} f^n(a)$ is the *Fibonacci word*. One has $x = abaababa \cdots$. The *Fibonacci set* is the set of factors of the Fibonacci word. It is a Sturmian set, and thus a neutral set.

The *factor complexity* of a factorial set S of words on an alphabet A is the sequence $p_n = \mathrm{Card}(S \cap A^n)$. The complexity of a Sturmian set is $p_n = n(\mathrm{Card}(A) - 1) + 1$. The following result (see [9]) shows that a neutral set has linear complexity.

Proposition 2.1 *The factor complexity of a neutral set on k letters is given by $p_0 = 1$ and $p_n = n(k - \chi(S)) + \chi(S)$ for every $n \geq 1$.*

Example 2.3. The complexity of the set of Example 2.1 is $p_n = 2$ for any $n \geq 1$.

A set of words $S \neq \{\varepsilon\}$ is *recurrent* if it is factorial and for any $u, w \in S$, there is a $v \in S$ such that $uvw \in S$. An infinite factorial set is said to be *minimal* or *uniformly recurrent* if for any word $u \in S$ there is an integer $n \geq 1$ such that u is a factor of any word of S of length n. A uniformly recurrent set is recurrent.

2.2   Tree Sets

Let S be a biextendable set of words. For $w \in S$, we consider the set $E(w)$ as an undirected graph on the set of vertices which is the disjoint union of $L(w)$ and $R(w)$ with edges the pairs $(a, b) \in E(w)$. This graph is called the *extension graph* of w. We sometimes denote $1 \otimes L(w)$ and $R(w) \otimes 1$ the copies of $L(w)$ and $R(w)$ used to define the set of vertices of $E(w)$. We note that since $E(w)$ has $\ell(w) + r(w)$ vertices and $e(w)$ edges, the number $1 - m_S(w)$ is the Euler characteristic of the graph $E(w)$.

A biextendable set S is called a *tree set* of characteristic c if for any nonempty $w \in S$, the graph $E(w)$ is a tree and if $E(\varepsilon)$ is a union of c trees. Note that a tree set of characteristic c is a neutral set of characteristic c.

Example 2.4. The set S of Example 2.1 is a tree set of characteristic 2.

A tree set of characteristic 1, simply called a tree set as in [1], is such that $E(w)$ is a tree for any $w \in S$.

As an example, a Sturmian set is a tree set [1].

Example 2.5. Let $A = \{a, b\}$ and let $f : A^* \to A^*$ be the morphism defined by $f(a) = ab$ and $f(b) = ba$. The infinite word $x = \lim_{n \to \infty} f^n(a)$ is the *Thue-Morse word*. The *Thue-Morse set* is the set of factors of the Thue-Morse word. It is uniformly recurrent but it is not a tree set since $E(\varepsilon) = A \times A$.

Let S be a set of words. For $w \in S$, let $\Gamma_S(w) = \{x \in S \mid wx \in S \cap A^+ w\}$. If S is recurrent, the set $\Gamma_S(w)$ is nonempty. Let

$$\text{Ret}_S(w) = \Gamma_S(w) \setminus \Gamma_S(w) A^+$$

be the set of *return words* to w.

Note that a recurrent set S is uniformly recurrent if and only if the set $\text{Ret}_S(w)$ is finite for any $w \in S$. Indeed, if N is the maximal length of the words in $\text{Ret}_S(w)$ for a word w of length n, any word in S of length $N + n$ contains an occurrence of w. The converse is obvious.

We will use the following result [1, Theorem 4.5]. We denote by F_A the free group on A.

Theorem 2.2 (Return Theorem). *Let S be a uniformly recurrent tree set. For any $w \in S$, the set $\text{Ret}_S(w)$ is a basis of the free group F_A.*

Note that this result implies in particular that for any $w \in S$, the set $\text{Ret}_S(w)$ has $\text{Card}(A)$ elements.

Example 2.6. Let S be the Tribonacci set. It is the set of factors of the infinite word $x = abacaba \cdots$ which is the fixed point of the morphism f defined by $f(a) = ab, f(b) = ac, f(c) = a$. It is a Sturmian set (see [8]). We have $\mathrm{Ret}_S(a) = \{a, ba, ca\}$.

3   Automata

All automata considered in this paper are deterministic and strongly connected and we simply call them automata. An automaton on a finite set Q of states is given by a partial map from $Q \times A$ into Q denoted $p \mapsto p \cdot a$, and extended to words with the same notation. For a word w, we denote by $\varphi_{\mathcal{A}}$ the map $p \in Q \mapsto p \cdot w \in Q$.

The *transition monoid* of the automaton $\mathcal{A}$ is the monoid M of partial maps from Q to itself of the form $\varphi_{\mathcal{A}}(w)$ for $w \in A^*$. The rank of an element m of M is the cardinality of its image, denoted $\mathrm{Im}(m)$.

Let $\mathcal{A}$ be an automaton and let S be a set of words. Denote by $\mathrm{rank}_{\mathcal{A}}(w)$ the rank of the map $\varphi_{\mathcal{A}}(w)$, also called the *rank* of w with respect to the automaton $\mathcal{A}$. The *S-minimal rank* of $\mathcal{A}$ is the minimal value of $\mathrm{rank}_{\mathcal{A}}(w)$ for $w \in S$. It is denoted $\mathrm{rank}_{\mathcal{A}}(S)$. A word of rank 1 is called *synchronizing*.

The following result gives a method to compute $\mathrm{rank}_{\mathcal{A}}(S)$ and thus gives a method to decide if $\mathcal{A}$ admits synchronizing words.

Theorem 3.1. *Let S be a recurrent set and let $\mathcal{A}$ be an automaton. Let w be in S and let $I = \mathrm{Im}(w)$. Then w has rank equal to $\mathrm{rank}_{\mathcal{A}}(S)$ if and only if $\mathrm{rank}_{\mathcal{A}}(wz) = \mathrm{rank}_{\mathcal{A}}(w)$ for any $z \in \mathrm{Ret}_S(w)$.*

Proof. Assume first that $\mathrm{rank}_{\mathcal{A}}(w) = \mathrm{rank}_{\mathcal{A}}(S)$. If z is in $\mathrm{Ret}_S(w)$, then wz is in S. Since $\mathrm{rank}_{\mathcal{A}}(wz) \le \mathrm{rank}_{\mathcal{A}}(w)$ and since $\mathrm{rank}_{\mathcal{A}}(w)$ is minimal, this forces $\mathrm{rank}_{\mathcal{A}}(wz) = \mathrm{rank}_{\mathcal{A}}(w)$.

Conversely, assume that w satisfies the condition. For any $r \in \mathrm{Ret}_S(w)$, we have $I \cdot r = \mathrm{Im}(wr) \subset \mathrm{Im}(w) = I$. Since $\mathrm{rank}_{\mathcal{A}}(wr) = \mathrm{rank}_{\mathcal{A}}(w)$, this forces $I \cdot r = I$. Since $\Gamma_S(w) \subset \mathrm{Ret}_S(w)^*$, this proves that

$$\Gamma_S(w) \subset \{z \in S \mid I \cdot z = I\}. \tag{3.1}$$

Let u be a word of S of minimal rank. Since S is recurrent, there exists words v, v' such that $wvuv'w \in S$. Then $vuv'w$ is in $\Gamma_S(w)$ and thus $I \cdot vuv'w = I$ by (3.1). This implies that $\mathrm{rank}_{\mathcal{A}}(u) \ge \mathrm{rank}_{\mathcal{A}}(vuv'w) = \mathrm{rank}_{\mathcal{A}}(w)$. Thus w has minimal rank in S.

Theorem 3.1 can be used to compute the S-minimal rank of an automaton in an effective way for a uniformly recurrent set S provided one can compute effectively the finite sets $\mathrm{Ret}_S(w)$ for $w \in S$.

Example 3.1. Let S be the Fibonacci set and let $\mathcal{A}$ be the automaton given by its transitions in Fig. 3.1 on the left. One has $\mathrm{Im}(a^2) = \{1, 2, 4\}$. The action on the 3-element sets of states of the automaton is shown on the right. By Theorem 3.1, we obtain $\mathrm{rank}_{\mathcal{A}}(S) = 3$.

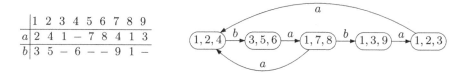

	1	2	3	4	5	6	7	8	9
a	2	4	1	–	7	8	4	1	3
b	3	5	–	6	–	–	9	1	–

Fig. 3.1. An automaton of S-degree 3.

We denote by $\mathcal{L}, \mathcal{R}, \mathcal{D}, \mathcal{H}$ the usual Green relations on a monoid M (see [10]). Recall that $\mathcal{R}$ is the equivalence on M defined by $m\mathcal{R}n$ if $mM = nM$. The $\mathcal{R}$-class of m is denoted $R(m)$. Symmetrically, one denotes by $\mathcal{L}$ the equivalence defined by $m\mathcal{L}n$ if $Mm = Mn$. It is well-known that the equivalences $\mathcal{R}$ and $\mathcal{L}$ commute. The equivalence $\mathcal{R}\mathcal{L} = \mathcal{L}\mathcal{R}$ is denoted $\mathcal{D}$. Finally, one denotes by $\mathcal{H}$ the equivalence $\mathcal{R} \cap \mathcal{L}$.

The following result is proved in [4] in a particular case (that is, for an automaton recognizing the submomoid generated by a bifix code).

Proposition 3.2. *Let S be a recurrent set and $\mathcal{A}$ be a strongly connected automaton. Set $\varphi = \varphi_{\mathcal{A}}$ and $M = \varphi(A^*)$. The set of elements of $\varphi(S)$ of rank $\mathrm{rank}_{\mathcal{A}}(S)$ is included in a regular $\mathcal{D}$-class of M.*

Proof. Set $d = \mathrm{rank}_{\mathcal{A}}(S)$. Let $u, v \in S$ be two words of rank d. Set $m = \varphi(u)$ and $n = \varphi(v)$. Let w be such that $uwv \in S$. We show first that $m\mathcal{R}\varphi(uwv)$ and $n\mathcal{L}\varphi(uwv)$.

For this, let t be such that $uwvtu \in S$. Set $z = wvtu$. Since $uz \in S$, the rank of uz is d. Since $\mathrm{Im}(uz) \subset \mathrm{Im}(z) \subset \mathrm{Im}(u)$, this implies that the images are equal. Consequently, the restriction of $\varphi(z)$ to $\mathrm{Im}(u)$ is a permutation. Since $\mathrm{Im}(u)$ is finite, there is an integer $\ell \geq 1$ such that $\varphi(z)^{\ell}$ is the identity on $\mathrm{Im}(u)$. Set $e = \varphi(z)^{\ell}$ and $s = tuz^{\ell-1}$. Then, since e is the identity on $\mathrm{Im}(u)$, one has $m = me$. Thus $m = \varphi(uwv)\varphi(s)$, and since $\varphi(uwv) = m\varphi(wv)$, it follows that m and $\varphi(uwv)$ are $\mathcal{R}$-equivalent.

Similarly n and $\varphi(uwv)$ are $\mathcal{L}$-equivalent. Indeed, let t' be such that $vt'uwv \in S$. Set $z' = t'uwv$. Then $\mathrm{Im}(vz') \subset \mathrm{Im}(z') \subset \mathrm{Im}(v)$. Since vz' is a factor of z^2 and z has rank d, it follows that $d = \mathrm{rank}(z^2) \leq \mathrm{rank}(vz') \leq \mathrm{rank}(v) = d$. Therefore, vz' has rank d and consequently the images $\mathrm{Im}(vz')$, $\mathrm{Im}(z')$ and $\mathrm{Im}(v)$ are equal. There is an integer $\ell' \geq 1$ such that $\varphi(z')^{\ell'}$ is the identity on $\mathrm{Im}(v)$. Set $e' = \varphi(z')^{\ell'}$. Then $n = ne' = n\varphi(z')^{\ell'-1}\varphi(tuwv) = nq\varphi(uwv)$, with $q = \varphi(z')^{\ell'-1}\varphi(t)$. Since $\varphi(uwv) = \varphi(uw)n$, one has $n\mathcal{L}\varphi(uwv)$. Thus m, n are $\mathcal{D}$-equivalent, and $\varphi(uwv) \in R(m) \cap L(n)$.

Set $p = \varphi(wv)$. Then $p = \varphi(w)n$ and, with the previous notation, $n = ne' = nq\varphi(u)p$, so $L(n) = L(p)$. Thus $mp = \varphi(uwv) \in R(m) \cap L(p)$, and by Clifford and Miller's Lemma, $R(p) \cap L(m)$ contains an idempotent. Thus the $\mathcal{D}$-class of m, p and n is regular.

The $\mathcal{D}$-class containing the elements of $\varphi(S)$ of rank $\mathrm{rank}_{\mathcal{A}}(S)$ is called the S-*minimal $\mathcal{D}$-class* of M. This $\mathcal{D}$-class appears in a different context in [11] (for a survey concerning the use of Green's relations in automata theory, see [12]).

Example 3.2. Let S be the Fibonacci set and let $\mathcal{A}$ be the automaton represented in Fig. 3.2 on the left. The S-minimal $\mathcal{D}$-class of the transition monoid of $\mathcal{A}$ is represented in Fig. 3.2 on the right.

	3	1	2
$1,2$	b	$^{*}ba$	$^{*}ba^2$
$1,2,3$	$^{*}ab$	*	*

Fig. 3.2. The automaton $\mathcal{A}$ and the S-minimal $\mathcal{D}$-class

Thus $\operatorname{rank}_{\mathcal{A}}(S) = 1$. We indicate with a $*$ the $\mathcal{H}$-classes containing an idempotent.

Let us recall some notions concerning groups in transformation monoids (see [4] for a more detailed presentation). Let M be a transformation monoid on a set Q. For $I \subset Q$, we denote

$$\operatorname{Stab}_M(I) = \{x \in M \mid Ix = I\}$$

or $\operatorname{Stab}(I)$ if the monoid M is understood. The *holonomy group* of M relative to I is the restriction of the elements of $\operatorname{Stab}_M(I)$ to the set I. It is denoted $\operatorname{Group}(I)$.

Let D be a regular $\mathcal{D}$-class in a transformation monoid M on a set Q. The holonomy groups of M relative to the sets Qm for $m \in D$ are all equivalent. The *structure group* of D is any of them.

Let $\mathcal{A}$ be an automaton with Q as set of states and let $I \subset Q$. Let w be a word such that $\varphi_{\mathcal{A}}(w) \in \operatorname{Stab}(I)$. The restriction of $\varphi_{\mathcal{A}}(w)$ to I is a permutation which belongs to $\operatorname{Group}(I)$. It is called the permutation *defined* by the word w on the set I.

Let $\mathcal{A}$ be a strongly connected automaton and let S be a recurrent set of words. The *S-group* of $\mathcal{A}$ is the structure group of its S-minimal $\mathcal{D}$-class. It is denoted $G_{\mathcal{A}}(S)$.

For the set $S = A^*$ and a strongly connected automaton, the group $G_{\mathcal{A}}(S)$ is a transitive permutation group of degree $d_X(S)$ (see [10, Theorem 9.3.10]). We conjecture that it holds for a uniformly recurrent tree set. It is not true for any uniformly recurrent set S, as shown in the following examples.

Example 3.3. Let S be the set of factors of $(ab)^*$ and let $\mathcal{A}$ be the automaton of Fig. 3.3. The minimal S-rank of $\mathcal{A}$ is 2 but the group $G_{\mathcal{A}}(S)$ is trivial.

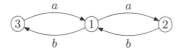

Fig. 3.3. An automaton of S-rank 2 with trivial S-group

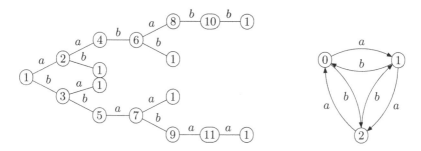

Fig. 3.4. An automaton of S-degree 3 with trivial S-group

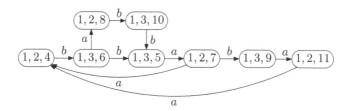

Fig. 3.5. The action on the minimal images

Example 3.4. Let S be the Thue-Morse set and let $\mathcal{A}$ be the automaton represented in Fig. 3.4 on the left. The word aa has rank 3 and image $I = \{1, 2, 4\}$.

The action on the images accessible from I is given in Fig. 3.5. All words with image $\{1, 2, 4\}$ end with aa. The paths returning for the first time to $\{1, 2, 4\}$ are labeled by the set $\mathrm{Ret}_S(aa) = \{b^2a^2, bab^2aba^2, bab^2a^2, b^2aba^2\}$. Thus $\mathrm{rank}_{\mathcal{A}}(S) = 3$ by Theorem 3.1. Moreover each of the words of $\mathrm{Ret}_S(a^2)$ defines the trivial permutation on the set $\{1, 2, 4\}$. Thus $G_{\mathcal{A}}(S)$ is trivial.

The fact that $d_{\mathcal{A}}(S) = 3$ and that $G_{\mathcal{A}}(S)$ is trivial can be seen directly as follows. Consider the group automaton $\mathcal{B}$ represented in Fig. 3.4 on the right and corresponding to the map sending each word to the difference modulo 3 of the number of occurrences of a and b. There is a reduction ρ from $\mathcal{A}$ onto $\mathcal{B}$ such that $1 \mapsto 0$, $2 \mapsto 1$, and $4 \mapsto 2$. This accounts for the fact that $d_{\mathcal{A}}(S) = 3$. Moreover, one may verify that any return word x to a^2 has equal number of a and b (if $x = uaa$ then $aauaa$ is in S, which implies that aua and thus uaa have the same number of a and b). This implies that the permutation $\varphi_B(x)$ is the identity, and therefore also the restriction of $\varphi_{\mathcal{A}}(x)$ to I. The same argument holds for Example 3.3 by considering the parity of the length.

4 Codes

A *code* is a set X such that for any $n, m \geq 0$ any $x_1, \ldots, x_n$ and $y_1, \ldots, y_m$ in X, one has $x_1 \cdots x_n = y_1 \cdots y_m$ only if $n = m$ and $x_1 = y_1, \ldots, x_n = y_n$. A *prefix code* is a set X of nonempty words which does not contain any proper prefix of its elements. A suffix code is defined symmetrically. A bifix code is a set which is both a prefix code and a suffix code.

Let S be a set of words. A prefix code $X \subset S$ is said to be *S-maximal* if it is not properly contained in any prefix code $Y \subset S$. The notion of an S-maximal suffix or bifix code are symmetrical.

It follows from results of [4] that for a recurrent set S, a finite bifix code $X \subset S$ is S-maximal as a bifix code if and only if it is S-maximal as a prefix code.

Given a set $X \subset S$, we denote $\lambda_S(X) = \sum_{x \in X} \lambda_S(x)$ where λ_S is the map defined by $\lambda_S(x) = e_S(x) - r_S(x)$. The following result is [9, Proposition 4].

Proposition 4.1. *Let S be a neutral set of characteristic c on the alphabet A, and let X be a finite S-maximal prefix code. Then $\lambda_S(X) = \mathrm{Card}(A) - c$.*

Symmetrically, one denotes $\rho_S(x) = e_S(x) - \ell_S(x)$. The dual of Proposition 4.1 holds for suffix codes instead of prefix codes with ρ_S instead of λ_S.

Note that when S is Sturmian, one has $\lambda_S(x) = \mathrm{Card}(A) - 1$ if x is left-special and $\lambda_S(x) = 0$ otherwise. Thus Proposition 4.1 expresses the fact that any finite S-maximal prefix code contains exactly one left-special word [4, Proposition 5.1.5].

Example 4.1. Let S be the Fibonacci set and let $X = \{aa, ab, b\}$. The set X is an S-maximal prefix code. It contains exactly one left-special word, namely ab. Accordingly, one has $\lambda_S(X) = 1$.

Let S be a factorial set and let $X \subset S$ be a finite prefix code. The *S-degree* of X is the S-minimal rank of the minimal automaton of X^*. It is denoted $d_X(S)$.

When X is a finite bifix code, the S-degree can be defined in a different way. A *parse* of a word w is a triple (s, x, p) such that $w = sxp$ with $s \in A^* \setminus A^*X$, $x \in X^*$ and $p \in A^* \setminus XA^*$. For a recurrent set S and an S-maximal bifix code X, $d_X(S)$ is the maximal number of parses of a word of S. A word $w \in S$ has $d_X(S)$ parses if and only if it is not an internal factor of a word of X (see [4]).

The following result is [13, Theorem 4.4].

Theorem 4.2 (Finite Index Basis Theorem). *Let S be a uniformly recurrent tree set and let $X \subset S$ be a finite bifix code. Then X is an S-maximal bifix code of S-degree d if and only if it is a basis of a subgroup of index d of F_A.*

Note that the result implies that any S-maximal bifix code of S-degree n has $d(\mathrm{Card}(A) - 1) + 1$ elements. Indeed, by Schreier's Formula, a subgroup of index d of a free group of rank r has rank $d(r - 1) + 1$.

Example 4.2. Let S be a Sturmian set. For any $n \geq 1$, the set $X = S \cap A^n$ is an S-maximal bifix code of S-degree n. According to theorem 4.2, it is a basis of the subgroup which is the kernel of the group morphism from F_A onto the additive group $\mathbb{Z}/n\mathbb{Z}$ sending each letter to 1.

The following statement generalizes [4, Theorem 4.3.7] where it is proved for a bifix code (and in this case with a stronger conclusion).

Theorem 4.3. *Let S be a recurrent set and let X be a finite S-maximal prefix code of S-degree n. The set of nonempty proper prefixes of X contains a disjoint union of $n - 1$ S-maximal suffix codes.*

Proof. Let P be the set of proper prefixes of X. Any word of S of rank n of length larger than the words of X has n suffixes which are in P.

We claim that this implies that any word in S is a suffix of a word with at least n suffixes in P. Indeed, let $x \in S$ be of minimal rank. For any $w \in S$, since S is recurrent, there is some u such that $xuw \in S$. Then xuw is of rank n and has n suffixes in P. This proves the claim.

Let Y_i for $1 \leq i \leq n$ be the set of $p \in P$ which have i suffixes in P. One has $Y_1 = \{\varepsilon\}$ and each Y_i for $2 \leq i \leq d$ is clearly a suffix code. It follows from the claim above that it is S-maximal. Since the Y_i are also disjoint, the result follows.

Corollary 1. *Let S be a recurrent neutral set of characteristic c, and let X be a finite S-maximal prefix code of S-degree n. The set P of proper prefixes of X satisfies $\rho_S(P) \geq n(\mathrm{Card}(A) - c)$.*

Proof. By Theorem 4.3, there exist $n-1$ pairwise disjoint S-maximal suffix codes Y_i ($2 \leq i \leq n$) such that P contains all Y_i. By the dual of Proposition 4.1, we have $\rho_S(Y_i) = \mathrm{Card}(A) - c$ for $2 \leq i \leq n$. Since $\rho_S(\varepsilon) = e_S(\varepsilon) - \ell_S(\varepsilon) = m_S(\varepsilon) + r_S(\varepsilon) - 1 = \mathrm{Card}(A) - c$, we obtain $\rho_S(P) \geq \rho_S(\varepsilon) + (n-1)(\mathrm{Card}(A) - c) = n(\mathrm{Card}(A) - c)$.

4.1 A Cardinality Theorem for Prefix Codes

Theorem 4.4. *Let S be a uniformly recurrent neutral set of characteristic c. Any finite S-maximal prefix code has at least $d_X(S)(\mathrm{Card}(A) - c) + 1$ elements.*

Proof. Let P be the set of proper prefixes of X. We may identify X with the set of leaves of a tree having P as set of internal nodes, each having $r_S(p)$ sons. By a well-known argument on trees, we have $\mathrm{Card}(X) = 1 + \sum_{p \in P}(r_S(p) - 1)$. Thus $\mathrm{Card}(X) = 1 + \rho_S(P)$. By Corollary 1, we have $\rho_S(P) \geq n(\mathrm{Card}(A) - c)$.

The next example shows that the prefix code can have strictly more than $d_X(S)(\mathrm{Card}(A) - c) + 1$ elements.

Example 4.3. Let S be the Fibonacci set. Let X be the S-maximal prefix code represented in Fig. 4.1. The states of the minimal automaton of X^* are represented on the figure. The automaton coincides with that of Example 3.1. Thus $d_X(S) = 3$ and $\mathrm{Card}(X) = 6$ while $d_X(S)(\mathrm{Card}(A) - 1) + 1 = 4$.

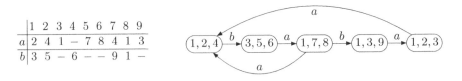

	1	2	3	4	5	6	7	8	9
a	2	4	1	–	7	8	4	1	3
b	3	5	–	6	–	–	9	1	–

Fig. 4.1. A prefix code of S-degree 3

If X is bifix, then it has $d_X(S)(\mathrm{Card}(A) - c) + 1$ elements by a result of [9]. The following example shows that an S-maximal prefix code can have $d_X(S)(\mathrm{Card}(A) - c) + 1$ elements without being bifix.

Fig. 4.2. The S-maximal prefix code X and the action on 2-subsets.

Example 4.4. Let S be the Fibonacci set and let

$$X = \{aaba, ab, ba\}.$$

The literal automaton of X^* is represented in Fig. 4.2 on the left. The prefix code X is S-maximal. The word ab has rank 2 in the literal automaton of X^*. Indeed, $\mathrm{Im}(ab) = \{1, 3\}$. Moreover $R_S(ab) = \{ab, aab\}$. The ranks of $abab$ and $abaab$ are also equal to 2, as shown in Fig. 4.2 on the right. Thus the S-degree of X is 2 by Proposition 3.1. The code X is not bifix since ba is a suffix of $aaba$.

4.2   The Group of a Bifix Code

The following result is proved in [4, Theorem 7.2.5] for a Sturmian set S. Recall that a *group code* of degree d is a bifix code Z such that $Z^* = \varphi^{-1}(K)$ for a surjective morphism φ from A^* onto a finite group G and a subgroup K of index d in G. Equivalently, a bifix code Z is a group code if it generates the submonoid $H \cap A^*$ where H is a subgroup of index d of the free group F_A.

The S-group of a prefix code, denoted $G_X(S)$, is the group $G_{\mathcal{A}}(S)$ where $\mathcal{A}$ is the minimal automaton of X^*.

Theorem 4.5. *Let Z be a group code of degree d and let S be a uniformly recurrent tree set S. The set $X = Z \cap S$ is an S-maximal bifix code of S-degree d and $G_X(S)$ is equivalent to the representation of F_A on the cosets of the subgroup generated by X.*

Proof. The first part is [14, Theorem 5.10], obtained as a corollary of the Finite Index Basis Theorem. To see the second part, let H be the subgroup generated by X of the free group F_A. Consider a word $w \in S$ which is not an internal factor of X. Let P be the set of proper prefixes of X which are suffixes of w. Then P has d elements since for each $p \in P$, there is a parse of w of the form (s, x, p). Moreover P is a set of representatives of the right cosets of H. Indeed, let $p, q \in P$ and assume that $p = uq$ with $u \in S$. If $p \in Hq$, then $u \in X^* \cap S$. Since p cannot have a prefix in X, we conclude that $p = q$. Since H has index d, this implies the conclusion.

Let $\mathcal{A} = (Q, i, i)$ be the minimal automaton of X^*. Set $I = Q \cdot w$. Let $\mathrm{Stab}(I)$ be the set of words $x \in A^*$ such that $I \cdot x = I$. Note that $\mathrm{Stab}(I)$ contains the set $\mathrm{Ret}_S(w)$ of right return words to w. For $x \in \mathrm{Stab}(I)$, let $\pi(x)$ be the permutation defined by x on I. By definition, the group $G_X(S)$ is generated by $\pi(\mathrm{Stab}(I))$. Since $\mathrm{Stab}(I)$ contains $\mathrm{Ret}_S(w)$ and since $\mathrm{Ret}_S(w)$ generates the free group F_A, the set $\mathrm{Stab}(I)$ generates F_A.

Let $x \in \mathrm{Stab}(I)$. For $p, q \in I$, let $u, v \in P$ be such that $i \cdot u = p$, $i \cdot v = q$. Let us verify that

$$p \cdot x = q \Leftrightarrow ux \in Hv. \tag{4.1}$$

Indeed, let $t \in S$ be such that $vt \in X$. Then, one has $p \cdot x = q$ if and only if $uxt \in X^*$ which is equivalent to $ux \in Hv$. Since $\mathrm{Stab}(I)$ generates F_A, Eq. (4.1) shows that the bijection $u \mapsto i \cdot u$ from P onto I defines an equivalence from $G_X(S)$ onto the representation of F_A on the cosets of H.

Example 4.5. Let S be the Fibonacci set and let $Z = A^2$ which is a group code of degree 2 corresponding to the morphism from A^* onto the additive $\mathbb{Z}/2\mathbb{Z}$ sending each letter to 1. Then $X = \{aa, ab, ba\}$. The minimal automaton of X^* is represented in Fig. 4.3 on the left. The word a has 2 parses and its image is the set $\{1, 2\}$. We have $\mathrm{Ret}_S(a) = \{a, ba\}$ and the action of $\mathrm{Ret}_S(a)$ on the minimal images is indicated in Fig. 4.3 on the right. The word a defines the permutation (12) and the word ba the identity.

Theorem 4.5 is not true for an arbitrary minimal set instead of a minimal tree set (see Example 3.4). The second part is true for an arbitrary finite S-maximal bifix code by the Finite Index Basis Theorem. We have no example where the second part is not true when X is S-maximal prefix instead of S-maximal bifix.

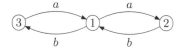

Fig. 4.3. The minimal automaton of X^* and the action on minimal images.

References

1. Berthé, V., De Felice, C., Dolce, F., Leroy, J., Perrin, D., Reutenauer, C., Rindone, G.: Acyclic, connected and tree sets. Monats. Math. **176**, 521–550 (2015)
2. Restivo, A.: Codes and local constraints. Theoret. Comput. Sci. **72**(1), 55–64 (1990)
3. Reutenauer, C.: Ensembles libres de chemins dans un graphe. Bull. Soc. Math. France **114**(2), 135–152 (1986)
4. Berstel, J., De Felice, C., Perrin, D., Reutenauer, C., Rindone, G.: Bifix codes and Sturmian words. J. Algebra **369**, 146–202 (2012)
5. Carpi, A., de Luca, A.: Codes of central Sturmian words. Theoret. Comput. Sci. **340**(2), 220–239 (2005)
6. Almeida, J., Costa, A.: On the transition semigroups of centrally labeled Rauzy graphs. Internat. J. Algebra Comput. **22**(2), 1250018 (2012). 25
7. Almeida, J., Costa, A.: Presentations of Schützenberger groups of minimal subshifts. Israel J. Math. **196**(1), 1–31 (2013)
8. Droubay, X., Justin, J., Pirillo, G.: Episturmian words and some constructions of de Luca and Rauzy. Theoret. Comput. Sci. **255**(1–2), 539–553 (2001)
9. Dolce, F., Perrin, D.: Enumeration formulæ in neutral sets. In: Potapov, I. (ed.) DLT 2015. LNCS, vol. 9168, pp. 215–227. Springer, Heidelberg (2015)
10. Berstel, J., Perrin, D., Reutenauer, C.: Codes and Automata. Cambridge University Press, Cambridge (2009)
11. Perrin, D., Schupp, P.: Automata on the integers, recurrence, distinguishability, and the equivalence of monadic theories. LICS **1986**, 301–304 (1986)
12. Colcombet, T.: Green's relations and their use in automata theory. In: Dediu, A.-H., Inenaga, S., Martín-Vide, C. (eds.) LATA 2011. LNCS, vol. 6638, pp. 1–21. Springer, Heidelberg (2011)
13. Berthé, V., De Felice, C., Dolce, F., Leroy, J., Perrin, D., Reutenauer, C., Rindone, G.: The finite index basis property. J. Pure Appl. Algebra **219**, 2521–2537 (2015)
14. Berthé, V., De Felice, C., Dolce, F., Leroy, J., Perrin, D., Reutenauer, C., Rindone, G.: Maximal bifix decoding. Discrete Math. **338**, 725–742 (2015)

Thue–Morse Along Two Polynomial Subsequences

Thomas Stoll[✉]

Institut Elie Cartan de Lorraine, UMR 7502,
Université de Lorraine / CNRS, 54506 Vandœuvre-lès-Nancy Cedex, Nancy, France
thomas.stoll@univ-lorraine.fr
http://iecl.univ-lorraine.fr/Thomas.Stoll/

Abstract. The aim of the present article is twofold. We first give a survey on recent developments on the distribution of symbols in polynomial subsequences of the Thue–Morse sequence $\mathbf{t} = (t(n))_{n \geq 0}$ by highlighting effective results. Secondly, we give explicit bounds on

$$\min\{n : (t(pn), t(qn)) = (\varepsilon_1, \varepsilon_2)\},$$

for odd integers p, q, and on

$$\min\{n : (t(n^{h_1}), t(n^{h_2})) = (\varepsilon_1, \varepsilon_2)\}$$

where $h_1, h_2 \geq 1$, and $(\varepsilon_1, \varepsilon_2)$ is one of $(0,0), (0,1), (1,0), (1,1)$.

Keywords: Thue–Morse sequence · Sum of digits · Polynomials

1 Introduction

The Thue–Morse sequence

$$\mathbf{t} = (t(n))_{n \geq 0} = 0, 1, 1, 0, 1, 0, 0, 1, 1, 0, 0, 1, 0, 1, 1, 0, \ldots$$

can be defined via

$$t(n) = s_2(n) \bmod 2, \tag{1}$$

where $s_2(n)$ denotes the number of one bits in the binary expansion of n, or equivalently, the sum of digits of n in base 2. This sequence can be found in various fields of mathematics and computer science, such as combinatorics on words, number theory, harmonic analysis and differential geometry. We refer the reader to the survey articles of Allouche and Shallit [2], and of Mauduit [14] for a concise introduction to this sequence. As is well-known, Thue–Morse is

Work supported by the ANR-FWF bilateral project MuDeRa "Multiplicativity: Determinism and Randomness" (France-Austria) and the joint project "Systèmes de numération : Propriétés arithmétiques, dynamiques et probabilistes" of the Université de Lorraine and the Conseil Régional de Lorraine.

F. Manea and D. Nowotka (Eds.): WORDS 2015, LNCS 9304, pp. 47–58, 2015.
DOI: 10.1007/978-3-319-23660-5_5

2-automatic and can be generated by the morphism $0 \mapsto 01$, $1 \mapsto 10$. It is also the prime example of an overlapfree sequence.

The overall distribution of the symbols 0 and 1 in Thue–Morse is trivial since the sequence consists exclusively of consecutive blocks of the forms 01 and 10, thus there are "as many 0's as 1's" in the sequence. The investigation of Thue–Morse along subsequences can be said to have started with an influential paper by Gelfond in 1967/68 [9]. He proved, via exponential sums techniques, that $\mathbf{t}$ is uniformly distributed along arithmetic progressions, namely,

$$\#\{n < N : \quad t(an + b) = \varepsilon\} \sim \frac{N}{2}, \qquad (\varepsilon = 0 \text{ or } 1)$$

with an explicit error term. Gelfond's result shows that there are "as many 0's as 1's" in the sequence also regarding arithmetic progressions. His result, however, gives no information on how long one actually has to wait to "see" the first, say, "1" along a specific arithmetic progression. Newman [19] showed that there is a weak preponderance of the 0's over the 1's in the sequence of the multiples of three. More precisely, he showed that

$$\#\{n < N : \quad t(3n) = 0\} = \frac{N}{2} + C(N),$$

with $c_1' N^{\log_4 3} < C(N) < c_2' N^{\log_4 3}$ for all $N \geq 1$ and certain positive constants c_1', c_2'. For the multiples of three one has to wait for 7 terms to "see the first 1", i.e.,

$$\min\{n : \quad t(3n) = 1\} = 7.$$

Morgenbesser, Shallit and Stoll [18] proved that for $p \geq 1$,

$$\min\{n : \quad t(pn) = 1\} \leq p + 4,$$

and this becomes sharp for $p = 2^{2r} - 1$ for $r \geq 1$ (Note that $3 = 2^{2 \cdot 1} - 1$ is exactly of that form). A huge literature is nowadays available for classes of arithmetic progressions where such Newman-type phenomena exist and many generalizations have been considered so far (see [3,4,8,10,12,22] and the references given therein). Still, a full classification is not yet at our disposal.

Most of the results that hold true for Thue–Morse in the number-theoretic setting of (1) have been proven for the sum of digits function in base q, where q is an integer greater than or equal to 2, and where the reduction in (1) is done modulo an arbitrary integer $m \geq 2$. We refrain here from the general statements and refer the interested readers to the original research papers.

Following the historical line, Gelfond [9] posed two challenging questions concerning the distribution of the sum of digits function along primes and along polynomial values instead of looking at linear subsequences. A third question was concerned with the simultaneous distribution when the sum of digits is taken to different bases; this question has been settled by Kim [13]. In recent years, this area of research gained much momentum due to an article by Mauduit and Rivat [15] who answered Gelfond's question for primes with an explicit error term. In a second paper [16], they also answered Gelfond's question for the sequence of squares. Their result implies that

$$\#\{n < N : \quad t(n^2) = \varepsilon\} \sim \frac{N}{2}.$$

In a very recent paper, Drmota, Mauduit and Rivat [6] showed that $\mathbf{t}$ along squares gives indeed a normal sequence in base 2 meaning that each binary block appears with the expected frequency. This quantifies a result of Moshe [17] who answered a question posed by Allouche and Shallit [1, Problem10.12.7] about the complexity of Thue–Morse along polynomial extractions.

We are still very far from understanding

$$\#\{n < N : \quad t(P(n)) = \varepsilon\},$$

where $P(x) \in \mathbb{Z}[x]$ is a polynomial of degree ≥ 3. Drmota, Mauduit and Rivat [7] obtained an asymptotic formula for $\#\{n < N : s_q(P(n)) = \varepsilon \pmod{m}\}$ whenever q is sufficiently large in terms of the degree of P. The case of Thue–Morse is yet out of reach of current technology. The currently best result is due to the author [23], who showed that there exists a constant $c = c(P)$ depending only on the polynomial P such that

$$\#\{n < N : \quad t(P(n)) = \varepsilon\} \geq cN^{4/(3 \deg P + 1)}. \tag{2}$$

This improves on a result of Dartyge and Tenenbaum [5] who had $N^{2/(\deg P)!}$ for the lower bound. The method of proof for (2) is constructive and gives an explicit bound on the minimal non-trivial n such that $t(n^h) = \varepsilon$ for fixed $h \geq 1$. Since $t(n^h) = 1$ for all $n = 2^r$, and $t(0^h) = 0$, we restrict our attention to

$$\mathcal{A} = \{n : \quad n \neq 2^r, \ r \geq 0\} = \{3, 5, 6, 7, 9, 10, 11, 12, 13, 14, 15, 17, \ldots\}.$$

From the proof of (2) follows that

$$\min(n \in \mathcal{A} : \quad t(n^h) = \varepsilon) \leq 64^{h+1} \left(8h \cdot 12^h\right)^{3h+1}.$$

Hence, there exists an absolute constant $c_1 > 0$ such that

$$\min(n \in \mathcal{A} : \quad t(n^h) = \varepsilon) \leq \exp(c_1 h^2). \tag{3}$$

With some extra work, a similar result can be obtained for a general polynomial $P(n)$ instead of n^h, where there corresponding constant will depend on the coefficients of P.

The joint distribution of the binary digits of integer multiples has been studied by J. Schmid [20] and in the more general setting of polynomials by Steiner [21]. The asymptotic formulas do not imply effective bounds on the first n that realizes such a system and it is the aim of this paper to prove effective bounds in the case of two equations for integer multiples and for monomials.

Our first result is as follows.

Theorem 1.   *Let $p > q \geq 1$ be odd integers. Then there exists an absolute constant $c_2 > 0$ such that*

$$\min(n : \; t(pn) = \varepsilon_1, \; t(qn) = \varepsilon_2) \leq \exp(c_2 \log p) \qquad (\varepsilon_1, \varepsilon_2 \in \{0, 1\}).$$

Remark 1. Note that for $p = 2^r + 1$ with $r \geq 1$ we have $t(pn) = 0$ for all $n < p-1$, so that there is no absolute bound for the minimal n.

There are examples that show that sometimes one has to "wait" quite some time to see all of the four possibilities for $(\varepsilon_1, \varepsilon_2)$ when the extraction is done along two monomial sequences. For instance, we have

$$\min\{n \in \mathcal{A} : \quad t(n^{130}) = \varepsilon_1)\} \leq 7 \quad \text{and} \quad \min\{n \in \mathcal{A} : \quad t(n^{53}) = \varepsilon_2)\} \leq 5$$

for $\varepsilon_1, \varepsilon_2 = 0, 1$ but

$$\min\left\{n : \quad (t(n^{130}), t(n^{53})) = (0, 0)\right\} = 113.$$

The construction that we will use to prove Theorem 1 will not be useful to study the minimal n along polynomial subsequences since in this case we would need to keep track of the binary digits sum of various binomial coefficients. Instead, we will use ideas from work of Hare, Laishram and the author [11] to show the following result.

Theorem 2.   *Let $h_1 > h_2 \geq 1$ be integers. Then there exists an absolute constant $c_3 > 0$ such that*

$$\min(n \in \mathcal{A} : \; t(n^{h_1}) = \varepsilon_1, \; t(n^{h_2}) = \varepsilon_2) \leq \exp(c_3 h_1^3) \qquad (\varepsilon_1, \varepsilon_2 \in \{0, 1\}).$$

Remark 2. The method also allows to treat general *monic* polynomials $P_1(x)$, $P_2(x) \in \mathbb{Z}[x]$ of different degree h_1, h_2 in place of x^{h_1}, x^{h_2}. Even more generally, we can deal with *non-monic* polynomials $P_1(x), P_2(x) \in \mathbb{Z}[x]$ provided h_1 is odd. As we will see in the proof (compare with the remark after (14)), the latter condition relies on the fact that for odd h the congruence $x^h \equiv a \pmod{16}$ admits a solution mod 16 for each odd a, while this is not true in general if h is even.

We write $\log_2$ for the logarithm to base 2. Moreover, for $n = \sum_{j=0}^{\ell} n_j 2^j$ with $n_j \in \{0, 1\}$ and $n_\ell \neq 0$ we write $(n_\ell, n_{\ell-1}, \cdots, n_1, n_0)_2$ for its digital representation in base 2 and set $\ell = \ell(n)$ for its length. To simplify our notation, we allow to fill up by a finite number of 0's to the left, i.e., $(n_\ell n_{\ell-1} \cdots n_1 n_0)_2 = (0 n_\ell n_{\ell-1} \cdots n_1 n_0)_2 = (0 \cdots 0 n_\ell n_{\ell-1} \cdots n_1 n_0)_2$.

The paper is structured as follows. In Sect. 2 we prove Theorem 1 and in Sect. 3 we show Theorem 2.

2   Thue–Morse at Distinct Multiples

The proof of Theorem 1 is based on the following lemma.

Lemma 1. *Let p, q be odd positive integers with $p > q \geq 1$ and let $(\varepsilon_1, \varepsilon_2)$ be one of $(0,0), (0,1), (1,0), (1,1)$. Then we have*

$$\min\{n : (t(pn), t(qn)) = (\varepsilon_1, \varepsilon_2)\} \leq C_{\varepsilon_1, \varepsilon_2}(p, q),$$

where

$$C_{0,0}(p,q) = C_{1,1}(p,q) = 4p, \quad C_{0,1}(p,q) = \frac{2^{23} p^{11}}{(p-q)^6} \quad C_{1,0}(p,q) = \frac{2^6 p^3}{(p-q)^2}.$$

Proof. Recall that for $1 \leq b < 2^k$ and $a, k \geq 1$, we have

$$s_2(a2^k + b) = s_2(a) + s_2(b), \qquad s_2(a2^k - b) = s_2(a-1) + k - s_2(b-1). \quad (4)$$

In the sequel we will make frequent use of these splitting formulas. We first deal with the two cases when $(\varepsilon_1, \varepsilon_2)$ is one of $(0,0), (1,1)$. If $2^k > p > q$ then $s_2(p(2^k - 1)) = s_2(q(2^k - 1)) = k$. Moreover, since $k \equiv 0$ or $1 \bmod 2$ and

$$2^k - 1 \leq 2^{\log p / \log 2 + 2} - 1 = 4p - 1 < 4p,$$

we get that $C_{0,0}(p,q), C_{1,1}(p,q) \leq 4p$. Finding explicit bounds for $C_{0,1}(p,q)$ and $C_{1,0}(p,q)$ is more involved. To begin with, we first claim that there exists $n_1 \geq 1$ with the following two properties:

(a) $\ell(pn_1) > \ell(qn_1)$,
(b) $pn_1 \equiv 1 \bmod 4$.

As for (a), we need to find two integers a, n_1 such that $2^a \leq pn_1$ and $2^a > qn_1$. This is equivalent to

$$\frac{2^a}{p} \leq n_1 < \frac{2^a}{q}. \quad (5)$$

For odd k, n either $kn \equiv 1 \pmod 4$ or $k(n + 2) \equiv 1 \pmod 4$, so provided $2^a \left(\frac{1}{q} - \frac{1}{p}\right) \geq 4$, we can find an odd n_1 that satisfies both (a) and (b). By taking a to be the unique integer with

$$\frac{4pq}{p-q} \leq 2^a < 2 \cdot \frac{4pq}{p-q}, \quad (6)$$

we get an n_1 with

$$n_1 < \frac{8p}{p-q}.$$

Now, define $n_2 = 2^{\ell(pn_1)} + 1$. Since both p and n_1 are odd we have $n_2 \leq pn_1$. Then

$$s_2(pn_1 n_2) = s_2(pn_1 2^{\ell(pn_1)} + pn_1) = 2s_2(pn_1) - 1 \equiv 1 \pmod 2, \quad (7)$$

since there is exactly one carry propagation from the most significant digit of pn_1 to the digit at digit place $\ell(pn_1)$ of $pn_1 2^{\ell(pn_1)}$ which stops immediately after one step because of property (b). On the other hand, (a) implies that

$$s_2(qn_1 n_2) = s_2(qn_1 2^{\ell(pn_1)} + qn_1) = 2s_2(qn_1) \equiv 0 \pmod 2, \tag{8}$$

because the terms qn_1 and $qn_1 2^{\ell(pn_1)}$ do not interfere and there is therefore no carry propagation while adding these two terms. We therefore can set

$$n = n_1 n_2 \leq \left(\frac{8p}{p-q}\right)^2 p$$

to get that $C_{1,0}(p,q) \leq 2^6 \cdot \frac{p^3}{(p-q)^2}$. For $(\varepsilon_1, \varepsilon_2) = (0,1)$, take m to be the unique odd integer with

$$pn_1 n_2 < 2^m \leq 4pn_1 n_2$$

and put

$$n_3 = 2^{2m} + 2^m - 1 \leq 2(4pn_1 n_2)^2 \leq 2^5 (pn_1)^4.$$

Then by (7),

$$\begin{aligned}
s_2(pn_1 n_2 n_3) &= s_2\left((pn_1 n_2 2^m + pn_1 n_2)2^m - pn_1 n_2\right) \\
&= s_2\left(pn_1 n_2 2^m + pn_1 n_2 - 1\right) + m - s_2\left(pn_1 n_2 - 1\right) \\
&= s_2\left(pn_1 n_2 - 1\right) + s\left(pn_1 n_2\right) + m - s_2\left(pn_1 n_2 - 1\right) \\
&\equiv m + 1 \equiv 0 \pmod 2.
\end{aligned}$$

A similar calculation shows by (8) that

$$s\left(qn_1 n_2 n_3\right) \equiv m \equiv 1 \pmod 2.$$

We set $n = n_1 n_2 n_3$ and get

$$n \leq \left(\frac{8p}{p-q}\right)^2 p \cdot 2^5 p^4 \cdot \frac{(8p)^4}{(p-q)^4} = 2^{23} \cdot \frac{p^{11}}{(p-q)^6}.$$

This completes the proof.   □

Proof (Theorem 1). This follows directly from Lemma 1 and

$$2^{23} \cdot \frac{p^{11}}{(p-q)^6} \leq 2^{17} p^{11}.$$

    □

3 Thue–Morse at Two Polynomials

This section is devoted to the proof of a technical result which implies Theorem 2. Considering the extractions of Thue–Morse along n^{h_1} and n^{h_2}, it is simple to get two and not too difficult to get three out of the four possibilities for $(\varepsilon_1, \varepsilon_2)$. However, to ensure that we see all of the four possibilities, we need a rather subtle construction. The difficulty is similar to that one to get $C_{0,0}(p,q)$ in the proof of Theorem 1. The idea is to shift two specific blocks against each other while all other terms in the expansions are non-interfering. Via this procedure, we will be able to keep track of the number of carry propagations. In the proof of Theorem 1 we have used the blocks pn_1 and qn_1. In the following, we will make use of $u_1 = 9 = (1001)_2$ and $u_2 = 1 = (0001)_2$. Then

$$s_2(u_1 + u_2) = 2,$$
$$s_2(u_1 + 2u_2) = 3,$$
$$s_2(u_1 + 2^2 u_2) = 3,$$
$$s_2(u_1 + 2^3 u_2) = 2,$$

and mod 2 we get the sequence $(0, 1, 1, 0)$. These particular expansions and additions will be of great importance in our argument (compare with (16)–(19)).

Lemma 2. *Let h_1, h_2 be positive integers with $h_1 > h_2 \geq 1$ and let $(\varepsilon_1, \varepsilon_2)$ be one of $(0,0), (0,1), (1,0), (1,1)$. Then we have*

$$\min\left\{n \in \mathcal{A} : (t(n^{h_1}), t(n^{h_2})) = (\varepsilon_1, \varepsilon_2)\right\} \leq C,$$

where

$$C = 32^{3h_1+5}\left(\frac{1}{15}\left(\frac{150}{h_1}8^{h_1}\right)^2\right)^{5h_1^2+h_1-5}.$$

Proof. Let $h \geq 1$ and put

$$l = \left\lfloor \log_2\left(\frac{1}{15}\left(\frac{150}{h}8^h\right)^2\right)\right\rfloor + 1, \tag{9}$$
$$a = (2^{lh} - 1)2^4 + 1,$$
$$b = \left\lfloor \frac{150}{h}8^h a^{1-\frac{1}{h}}\right\rfloor,$$
$$M = \left\lfloor 15a^{1-\frac{1}{h}}\right\rfloor + 1,$$
$$k = lh^2 + 3h + 4 - l.$$

It is straightforward to check that for all $h \geq 1$, we have

$$l \geq 12, \quad a \geq 65521, \quad b \geq 1200, \quad M \geq 16, \quad k \geq 7. \tag{10}$$

Obviously, we have $b \geq M$. Moreover,

$$a = 16 \cdot 2^{lh} - 15 \geq 15 \cdot 2^{lh} \geq \left(\frac{150}{h}8^h\right)^h$$

and therefore $a \geq b$. Furthermore, for $h \geq 2$, we have $aM > b^2$ since by

$$a > \left(\frac{1}{15} \left(\frac{150}{h} 8^h \right)^2 \right)^h$$

we have

$$aM - b^2 > a \cdot 15a^{1-\frac{1}{h}} - \left(\frac{150}{h} 8^h a^{1-\frac{1}{h}} \right)^2 > 0.$$

Let

$$T(x) = ax^5 + bx^4 + Mx^3 + Mx^2 - x + M$$

and write $T(x)^h = \sum_{i=0}^{5h} \alpha_i x^i$. Obviously, $\alpha_0 > 0$ and $\alpha_1 < 0$. We claim that for $h \geq 1$ we have $\alpha_i > 0$ for $2 \leq i \leq 5h$. To see this we write

$$T(x)^h = \left(ax^5 + bx^4 + Mx^3 + Mx^2 + M \right)^h + r(x), \tag{11}$$

with

$$r(x) = \sum_{j=1}^{h} \binom{h}{j} (-x)^j \left(ax^5 + bx^4 + Mx^3 + Mx^2 + M \right)^{h-j} = \sum_{j=1}^{5(h-1)} d_j x^j.$$

Since $a \geq b \geq M$ the coefficient of x^i in the first term in (11) is $\geq M^h$. On the other hand,

$$|d_j| < h2^h (5a)^{h-1},$$

and

$$M^h \geq \left(15a^{1-1/h} \right)^h > \left(3 \cdot 5a^{1-1/h} \right)^h > \left(2h^{1/h} (5a)^{1-1/h} \right)^h = |d_j|$$

which proves the claim. Next, we need a bound on the size of α_i. The coefficients $\alpha_i, 0 \leq i \leq 5h-2$, are bounded by the corresponding coefficients in the expansion of $(ax^5 + bx^4 + M(x^3 + x^2 + x + 1))^h$. Since $aM > b^2$ and $M \leq 16a^{1-1/h}$, each of these coefficients is bounded by

$$\left(a^{h-1}M \right) \cdot 6^h < a^{h-1}M8^h \leq a^{h-\frac{1}{h}} \cdot 16 \cdot 8^h,$$

and therefore

$$|\alpha_i| \leq a^{h-\frac{1}{h}} \cdot 16 \cdot 8^h, \qquad i = 0, 1, 2, \ldots, 5h - 2. \tag{12}$$

Moreover, we have

$$\alpha_{5h-1} = hba^{h-1} = \left\lfloor \frac{150}{h} 8^h a^{1-\frac{1}{h}} \right\rfloor ha^{h-1} \tag{13}$$

and

$$149 \cdot 8^h a^{h-\frac{1}{h}} \leq \alpha_{5h-1} \leq 150 \cdot 8^h a^{h-\frac{1}{h}},$$

which is true for all $a \geq 1$ and $h \geq 1$. Note that both the bound in (12) and the coefficient α_{5h-1} are increasing functions in h. From now on, suppose that $h \geq 2$. We further claim that

$$a^{h-\frac{1}{h}} \cdot 16 \cdot 8^h < 2^k, \qquad 9 \cdot 2^k \leq \alpha_{5h-1} < 10 \cdot 2^k, \tag{14}$$

which will give us the wanted overlap for the digital blocks of α_{5h-1} and α_{5h}. By (14) the binary expansion of α_{5h-1} is $(1001\cdots)_2$ and interferes with the digital block coming from $\alpha_{5h} = a^h$ which is $(\cdots 0001)_2$ since $a \equiv 1 \pmod{16}$. To prove (14), we show a stronger inequality that in turn implies (14), namely,

$$144 \cdot 8^h a^{h-\frac{1}{h}} < 9 \cdot 2^k \leq 149 \cdot 8^h a^{h-\frac{1}{h}}. \tag{15}$$

Passing to logarithms, this is equivalent to

$$(k - 3h - 4) - \delta \leq \left(h - \frac{1}{h} \right) \log_2 a < k - 3h - 4$$

with

$$\delta = \log_2 149 - \log_2 9 - 4 = 0.04924 \cdots > \frac{1}{25}.$$

We rewrite

$$\left(h - \frac{1}{h} \right) \log_2 a = \left(h - \frac{1}{h} \right) \left(lh + \log_2 \left(1 - \left(\frac{1}{2^{lh-4}} - \frac{1}{2^{lh}} \right) \right) \right),$$

which on the one hand shows that

$$\left(h - \frac{1}{h} \right) \log_2 a < \left(h - \frac{1}{h} \right) lh = lh^2 - l = k - 3h - 4,$$

and on the other hand by $-x/(1-x) < \log(1-x) < 0$ for $0 < x < 1$ that

$$\left(h - \frac{1}{h} \right) \log_2 a > lh^2 - l - \frac{1}{\log 2} \cdot \frac{(2^4 - 1) \cdot \left(h - \frac{1}{h} \right)}{2^{lh} - 2^4 + 1}.$$

Finally, we easily check that for all $h \geq 2$ and $l \geq 5$ we have

$$\frac{1}{\log 2} \cdot \frac{15 \left(h - \frac{1}{h} \right)}{2^{lh} - 15} < \frac{1}{25},$$

which finishes the proof of (15) and thus of (14).

After this technical preliminaries we proceed to the evaluation of the sum of digits. First note that for all $h \geq 1$ by construction none of $n = T(2^k)$, $T(2^{k+1})$, $T(2^{k+2})$, $T(2^{k+3})$ is a power of two and therefore $n \in \mathcal{A}$ in these four cases. Let $h = h_1 \geq 2$ and define a, b, M, l, k according to (9). To begin with, by (12)–(14) and the splitting formulas (4), we calculate

$$s_2(T(2^k)^{h_1}) = s_2 \left(\sum_{i=0}^{5h_1} \alpha_i 2^{ik} \right) = A_1 - 1 + A_2 + k + A_3, \tag{16}$$

where

$$A_1 = s_2(\alpha_{5h_1}) + s_2(\alpha_{5h_1-1}),$$

$$A_2 = s_2\left(\sum_{i=3}^{5h_1-2} \alpha_i\right) + s_2(\alpha_0),$$

$$A_3 = s_2(\alpha_2 - 1) - s_2(\alpha_1 - 1).$$

Note that the summand k in (16) comes from formula (4) due to the negative coefficient α_1, and the -1 comes from the addition of $(\cdots 0001)_2$ and $(1001)_2$ (the ending and starting blocks corresponding to α_{5h_1} and α_{5h_1-1}) which gives rise to exactly one carry. A similar calculation shows that

$$s_2(T(2^{k+1})^{h_1}) = A_1 + A_2 + (k+1) + A_3, \tag{17}$$

$$s_2(T(2^{k+2})^{h_1}) = A_1 + A_2 + (k+2) + A_3, \tag{18}$$

$$s_2(T(2^{k+3})^{h_1}) = A_1 + A_2 - 1 + (k+3) + A_3. \tag{19}$$

In (17) we add $(1001)_2$ to $(\cdots 00010)_2$ which gives no carry. The same happens for the addition of $(1001)_2$ to $(\cdots 000100)_2$ in (18). Finally, in (19) we again have exactly one carry in the addition of $(1001)_2$ to $(\cdots 0001000)_2$. If we look mod 2 this shows that

$$\left(t(T(2^k)^{h_1}), t(T(2^{k+1})^{h_1}), t(T(2^{k+2})^{h_1}), t(T(2^{k+3})^{h_1})\right)$$

is either $(0,0,1,1)$ or $(1,1,0,0)$. For $h_2 < h_1$ with $h_2 \geq 1$ all coefficients are non-interfering. To see this, consider the coefficients of $T(x)^{h_2} = \sum_{i=0}^{5h_2} \alpha_i' x^i$. By (12), they are clearly bounded in modulus by $a^{h_1-\frac{1}{h_1}} \cdot 16 \cdot 8^{h_1} < 2^k$ for $i = 0, 1, 2, \ldots, 5h_2 - 2$. Also, by (10) and $h_1 \geq 2$,

$$\alpha_{5h_2-1}' \leq 150 \cdot 8^{h_2} a^{h_2 - \frac{1}{h_2}} < \frac{150}{8} \cdot 8^{h_1} a^{h_1-1} < a^{h_1-\frac{1}{h_1}} \cdot 16 \cdot 8^{h_1} < 2^k,$$

and thus we don't have carry propagations in the addition of terms in the expansion of $T(2^k)^{h_2}$. Similarly, we show that

$$\left(t(T(2^k)^{h_2}), t(T(2^{k+1})^{h_2}), t(T(2^{k+2})^{h_2}), t(T(2^{k+3})^{h_2})\right)$$

is either $(0,1,0,1)$ or $(1,0,1,0)$. This yields in any case that

$$\{(t(T(2^{k+i})^{h_1}), t(T(2^{k+i})^{h_2})) : \quad i = 0, 1, 2, 3\} = \{(0,0), (0,1), (1,0), (1,1)\},$$

which are the four desired values. Finally,

$$T(2^{k+3}) \leq 2^{5(k+3)} \cdot 2^{lh_1+4}$$

$$\leq 2^{l(5h_1^2+h_1-5)} \cdot 2^{5(3h_1+4)+4}$$

$$\leq 32^{3h_1+5} \left(\frac{1}{15}\left(\frac{150}{h_1} 8^{h_1}\right)^2\right)^{5h_1^2+h_1-5},$$

which completes the proof. $\qquad\square$

Proof (Theorem 2). This follows from Lemma 2 and

$$32^{3h_1+5} \left(\frac{1}{15} \left(\frac{150}{h_1} 8^{h_1} \right)^2 \right)^{5h_1^2+h_1-5} \leq \exp\left(c_3 h_1^3\right)$$

for some suitable positive constant c_3.   □

Acknowledgements. I am pleased to thank Jeff Shallit for bringing to my attention the question on bounding Thue–Morse on two multiples. I also thank him and E. Rowland for several discussions.

References

1. Allouche, J.-P., Shallit, J.: Automatic Sequences: Theory, Applications Generalizations. Cambridge University Press, Cambridge (2003)
2. Allouche, J.-P., Shallit, J.: The ubiquitous Prouhet-Thue-Morse sequence. In: Ding, C., Helleseth, T., Niederreiter, H. (eds.) Sequences and Their Applications. Discrete Mathematics and Theoretical Computer Science, pp. 1–16. Springer, London (1999)
3. Boreico, I., El-Baz, D., Stoll, T.: On a conjecture of Dekking: the sum of digits of even numbers. J. Théor. Nombres Bordeaux **26**, 17–24 (2014)
4. Coquet, J.: A summation formula related to the binary digits. Invent. Math. **73**, 107–115 (1983)
5. Dartyge, C., Tenenbaum, G.: Congruences de sommes de chiffres de valeurs polynomiales. Bull. London Math. Soc. **38**(1), 61–69 (2006)
6. Drmota M., Mauduit C., Rivat J.: The Thue-Morse sequence along squares is normal, manuscript. http://www.dmg.tuwien.ac.at/drmota/alongsquares.pdf
7. Drmota, M., Mauduit, C., Rivat, J.: The sum of digits function of polynomial sequences. J. London Math. Soc. **84**, 81–102 (2011)
8. Drmota, M., Skałba, M.: Rarified sums of the Thue-Morse sequence. Trans. Am. Math. Soc. **352**, 609–642 (2000)
9. Gelfond, A.O.: Sur les nombres qui ont des propriétés additives et multiplicatives données. Acta Arith. **13**, pp. 259–265 (1967/1968)
10. Goldstein, S., Kelly, K.A., Speer, E.R.: The fractal structure of rarefied sums of the Thue-Morse sequence. J. Number Theor. **42**, 1–19 (1992)
11. Hare, K.G., Laishram, S., Stoll, T.: Stolarsky's conjecture and the sum of digits of polynomial values. Proc. Am. Math. Soc. **139**, 39–49 (2011)
12. Hofer, R.: Coquet-type formulas for the rarefied weighted Thue-Morse sequence. Discrete Math. **311**, 1724–1734 (2011)
13. Kim, D.-H.: On the joint distribution of q-additive functions in residue classes. J. Number Theor. **74**, 307–336 (1999)
14. Mauduit, C.: Multiplicative properties of the Thue-Morse sequence. Period. Math. Hungar. **43**, 137–153 (2001)
15. Mauduit, C., Rivat, J.: Sur un problème de Gelfond: la somme des chiffres des nombres premiers. Ann. Math. **171**, 1591–1646 (2010)
16. Mauduit, C., Rivat, J.: La somme des chiffres des carrés. Acta Math. **203**, 107–148 (2009)

17. Moshe, Y.: On the subword complexity of Thue-Morse polynomial extractions. Theor. Comput. Sci. **389**, 318–329 (2007)
18. Morgenbesser, J., Shallit, J., Stoll, T.: Thue-Morse at multiples of an integer. J. Number Theor. **131**, 1498–1512 (2011)
19. Newman, D.J.: On the number of binary digits in a multiple of three. Proc. Am. Math. Soc. **21**, 719–721 (1969)
20. Schmid, J.: The joint distribution of the binary digits of integer multiples. Acta Arith. **43**, 391–415 (1984)
21. Steiner W.: On the joint distribution of q-additive functions on polynomial sequences. In: Proceedings of the Conference Dedicated to the 90th Anniversary of Boris Vladimirovich Gnedenko (Kyiv, 2002), Theory Stochastic Process, **8**, pp. 336–357 (2002)
22. Stoll T.: Multi-parametric extensions of Newman's phenomenon. Integers (electronic), **5**, A14, p. 14 (2005)
23. Stoll, T.: The sum of digits of polynomial values in arithmetic progressions. Funct. Approx. Comment. Math. **47**, 233–239 (2012). part 2

Canonical Representatives of Morphic Permutations

Sergey V. Avgustinovich[1], Anna E. Frid[2]([✉]), and Svetlana Puzynina[1,3]

[1] Sobolev Institute of Mathematics, Novosibirsk, Russia
avgust@math.nsc.ru
[2] Aix-Marseille Université, Marseille, France
anna.e.frid@gmail.com
[3] LIP, ENS de Lyon, Université de Lyon, Lyon, France
s.puzynina@gmail.com

Abstract. An *infinite permutation* can be defined as a linear ordering of the set of natural numbers. In particular, an infinite permutation can be constructed with an aperiodic infinite word over $\{0, \ldots, q-1\}$ as the lexicographic order of the shifts of the word. In this paper, we discuss the question if an infinite permutation defined this way admits a *canonical representative*, that is, can be defined by a sequence of numbers from $[0, 1]$, such that the frequency of its elements in any interval is equal to the length of that interval. We show that a canonical representative exists if and only if the word is uniquely ergodic, and that is why we use the term *ergodic* permutations. We also discuss ways to construct the canonical representative of a permutation defined by a morphic word and generalize the construction of Makarov, 2009, for the Thue-Morse permutation to a wider class of infinite words.

1 Introduction

We continue the study of combinatorial properties of infinite permutations analogous to those of words. In this approach, infinite permutations are interpreted as equivalence classes of real sequences with distinct elements, such that only the order of elements is taken into account. In other words, an infinite permutation is a linear order in $\mathbb{N}$. We consider it as an object close to an infinite word, but instead of symbols, we have transitive relations $<$ or $>$ between each pair of elements.

Infinite permutations in the considered sense were introduced in [10]; see also a very similar approach coming from dynamics [6] and summarised in [3]. Since then, they were studied in two main directions: First, a series of results compared properties of infinite permutations with those of infinite words ([4,10,11] and others). Secondly, different authors studied permutations directly constructed

---

S. Puzynina—Supported by the LABEX MILYON (ANR-10-LABX-0070) of Université de Lyon, within the program "Investissements d'Avenir" (ANR-11-IDEX-0007) operated by the French National Research Agency (ANR).

© Springer International Publishing Switzerland 2015
F. Manea and D. Nowotka (Eds.): WORDS 2015, LNCS 9304, pp. 59–72, 2015.
DOI: 10.1007/978-3-319-23660-5_6

with the use of general words [7,13], as well as precise examples: the Thue-Morse word [14,18], other morphic words [17,19] or Sturmian words [15].

In the previous paper [5], we introduced the notion of an *ergodic* permutation, which means that a permutation can be defined by a sequence of numbers from [0,1] such that the frequency of its elements in any interval is equal to the length of the interval. We proved also that the minimal complexity (i.e., the number of subpermutations of length n) of an ergodic permutation is n, and the permutations of minimal complexity are Sturmian permutations in the sense of [15] (and close to the sense of [4]). So, the situation for ergodic permutations is similar to that for words. Note that for the permutations in general, this is not the case: The complexity of an aperiodic permutation can grow slower than any unbounded growing function [10].

In this paper, we focus on permutations generated by words. First of all, we prove that such a permutation is ergodic if and only if its generating word is uniquely ergodic, which explains the choice of the term. Then we generalize the construction of Makarov [14] and give a general method to construct the canonical representative sequence of any permutation generated by a fixed point of a primitive monotone separable morphism. We also discuss why this method cannot be directly extended further, and give some examples.

2 Basic Definitions

We consider finite and infinite words over a finite alphabet $\Sigma_q = \{0, 1, q-1\}$. A *factor* of an infinite word is any sequence of its consecutive letters. The factor $u[i] \cdots u[j]$ of an infinite word $u = u[0]u[1] \cdots u[n] \cdots$, with $u[k] \in \Sigma$, is denoted by $u[i..j]$; *prefixes* of a finite or an infinite word are as usual defined as starting factors.

The length of a finite word s is denoted by $|s|$. An infinite word $u = vww \cdots = vw^\omega$ for some non-empty word w is called *ultimately* ($|w|$-)*periodic*; otherwise it is called *aperiodic*.

When considering words on Σ_q, we refer to the *order* on finite and infinite words meaning lexicographic (partial) order: $0 < 1 < \ldots < q-1$, and $u < v$ if $u[0..i] = v[0..i]$ and $u[i+1] < v[i+1]$ for some i. For words such that one of them is the prefix of the other the order is not defined.

Now we recall the notion of the uniform frequency of letters and factors in an infinite word. For finite words v and w, we let $|v|_w$ denote the number of occurrences of w in v. The infinite word u has *uniform frequencies* of factors if, for every factor w of u, the ratio $\frac{|u[i..i+n]|_w}{n+1}$ has a limit $\rho_w(u)$ when $n \to \infty$ uniformly in k. For more on uniform frequencies in words we refer to [8].

To define infinite permutations, we will use sequences of real numbers. Analogously to a factor of a word, for a sequence $(a[n])_{n=0}^\infty$ of real numbers, any of its finite subsequences $a[i], a[i+1], \ldots, a[j]$ is called a *factor* and is denoted by $a[i..j]$. We define an equivalence relation $\sim$ on real infinite sequences with pairwise distinct elements as follows: $(a[n])_{n=0}^\infty \sim (b[n])_{n=0}^\infty$ if and only if for all i, j the conditions $a[i] < a[j]$ and $b[i] < b[j]$ are equivalent. Since we consider only sequences of pairwise distinct real numbers, the same condition can be defined by substituting ($<$) by ($>$): $a[i] > a[j]$ if and only if $b[i] > b[j]$. An *infinite permutation* is then defined as an equivalence class of real infinite sequences with pairwise distinct elements. So, an infinite permutation is a linear ordering of the set $\mathbb{N}_0 = \{0, \ldots, n, \ldots\}$. We denote it by $\alpha = (\alpha[n])_{n=0}^{\infty}$, where $\alpha[i]$ are abstract elements equipped by an order: $\alpha[i] < \alpha[j]$ if and only if $a[i] < a[j]$ or, which is the same, $b[i] < b[j]$ for every *representative* sequence $(a[n])$ or $(b[n])$ of α. So, one of the simplest ways to define an infinite permutation is by a representative, which can be any sequence of pairwise distinct real numbers.

Example 2.1. Both sequences $(a[n]) = (1, -1/2, 1/4, \ldots)$ with $a[n] = (-1/2)^n$ and $(b[n])$ with $b[n] = 1000 + (-1/3)^n$ are representatives of the same permutation $\alpha = \alpha[0], \alpha[1], \ldots$ defined by

$$\alpha[2n] > \alpha[2n+2] > \alpha[2k+3] > \alpha[2k+1]$$

for all $n, k \geq 0$.

A *factor* $\alpha[i..j]$ of an infinite permutation α is a finite sequence $(\alpha[i], \alpha[i+1], \ldots, \alpha[j])$ of abstract elements equipped by the same order as in α. Note that a factor of an infinite permutation can be naturally interpreted as a finite permutation: for example, if in a representative $(a[n])$ we have a factor $(2.5, 2, 7, 1.6)$, that is, the 4th element is the smallest, followed by the 2nd, 1st and 3rd, then in the permutation, it will correspond to a factor $\begin{pmatrix} 1\,2\,3\,4 \\ 3\,2\,4\,1 \end{pmatrix}$, which we will denote simply as (3241). Note that in general, we index the elements of infinite objects (words, sequences or permutations) starting with 0 and the elements of finite objects starting with 1.

A factor of a sequence (permutation) should not be confused with its subsequence $a[n_0], a[n_1], \ldots$ (subpermutation $\alpha[n_0], \alpha[n_1], \ldots$) which is defined as indexed with a growing subsequence (n_i) of indices.

Note, however, that in general, an infinite permutation cannot be defined as a permutation of $\mathbb{N}_0$. For instance, the permutation from Example 2.1 has all its elements between the first two ones.

3   Ergodic Permutations

Let $(a[i])_{i=0}^{\infty}$ be a sequence of real numbers from the interval $[0,1]$, representing an infinite permutation, a and p also be real numbers from $[0,1]$. We say that the *probability* that an element of $(a[i])$ is less than a exists and is equal to p if the ratio

$$\frac{\#\{a[j+k] | 0 \leq k < n, a[j+k] < a\}}{n}$$

has a limit p when $n \to \infty$ uniformly in j.

In other words, if we substitute all the elements from $(a[i])$ which are smaller than a by 1, and those which are bigger by 0, the above condition means that the uniform frequency of the letter 1 exists and equals p. So, in fact the probability to be smaller than a is the uniform frequency of the elements which are less than a.

We note that this is not exactly probability on the classical sense, since we do not have a random sequence. But we are interested in permutations where this "probability" behaves in certain sense like the probability of a random sequence uniformly distributed on $[0,1]$:

Definition 3.1. A sequence $(a[i])_{i=0}^{\infty}$ of real numbers is *canonical* if

- all the numbers are pairwise distinct;
- for all i we have $0 \leq a[i] \leq 1$;
- and for all a, the probability for any element $a[i]$ to be less than a is well-defined and equal to a for all $a \in [0,1]$.

Remark 3.2. The set $\{a[i] \mid i \in \mathbb{N}\}$ for a canonical sequence $(a[i])$ is dense on $[0,1]$.

Remark 3.3. In a canonical sequence, the frequency of the elements which fall into any interval $(t_1, t_2) \subseteq [0,1]$ exists and is equal to $t_2 - t_1$.

Remark 3.4. Symmetrically to the condition "the probability to be less than a is a" we can consider the equivalent condition "the probability to be greater than a is $1 - a$".

Definition 3.5. An infinite permutation $\alpha = (\alpha[i])_{i=1}^{\infty}$ is called *ergodic* if it has a canonical representative.

Example 3.6. For any irrational σ and for any ρ, consider the sequence of fractional parts $\{\rho + n\sigma\}$. It is uniformly distributed in $[0,1)$, so, the respective permutation is ergodic. In fact, such a permutation is a *Sturmian* permutation in the sense of [14]; in [4], the considered class of permutations is wider than that. It is easy to see that Sturmian permutations are directly related to Sturmian words [12,16].

Proposition 3.7. *An ergodic permutation α has a unique canonical representative.*

PROOF. Given α, for each i we define

$$a[i] = \lim_{n \to \infty} \frac{\#\{\alpha[k] \mid 0 \leq k < n, \alpha[k] < \alpha[i]\}}{n}$$

and see that, first, this limit must exist since α is ergodic, and secondly, $a[i]$ is the only possible value of an element of a canonical representative of α.   □

Note, however, that even if for some infinite permutation all the limits above exist, it does not imply the existence of the canonical representative. Indeed, there is another condition to fulfill: for different i the limits must be different.

4 Ergodic Permutations Generated by Words

Consider an aperiodic infinite word $u = u[0] \cdots u[n] \cdots$ over Σ_q and, as usual, define its nth shift $T^n u$ as the word obtained from u by erasing the first n symbols: $T^n u = u[n]u[n+1] \cdots$. We can also interpret a word u as a real number $0.u$ in the q-ary representation.

If the word u is aperiodic, then in the sequence $(0.T^n u)_{n=0}^{\infty}$ all the numbers are different and thus this sequence is a representative of a permutation which we denote by α_u. Clearly, $\alpha_u[i] < \alpha_u[j]$ if and only if $T^i u$ is lexicographically smaller than $T^j u$. A permutation which can be constructed like this is called *valid*; the structure of valid permutations has been studied in [13] (for the binary case) and [7] (in general).

Most of results of this paper were inspired by the following construction.

Example 4.1. The famous Thue-Morse word $0110100110010110 \cdots$ is defined as the fixed point starting with 0 of the morphism $f_{tm} : 0 \mapsto 01, 1 \mapsto 10$ [1,2]. The respective Thue-Morse permutation defined by the representative $(0.01101001 \cdots, 0.11010011 \cdots, 0.10100110 \cdots, 0.01001100 \cdots, \ldots)$ can also be defined by the following sequence, denoted by a_{tm}:

$$\frac{1}{2}, 1, \frac{3}{4}, \frac{1}{4}, \frac{5}{8}, \frac{1}{8}, \frac{3}{8}, \frac{7}{8}, \cdots,$$

that is the fixed point of the morphism $\varphi_{tm} : [0,1] \mapsto [0,1]^2$:

$$\varphi_{tm}(x) = \begin{cases} \frac{x}{2} + \frac{1}{4}, \frac{x}{2} + \frac{3}{4}, & \text{if } 0 \leq x \leq \frac{1}{2}, \\ \frac{x}{2} + \frac{1}{4}, \frac{x}{2} - \frac{1}{4}, & \text{if } \frac{1}{2} < x \leq 1. \end{cases}$$

It will be proved below that the latter sequence is canonical and thus the Thue-Morse permutation is ergodic. This construction and the equivalence of the two definitions was proved by Makarov in 2009 [15]; then the properties of the Thue-Morse permutation were studied by Widmer [18].

When is a valid permutation ergodic? The answer is simple and explains the choice of the term "ergodic".

Lemma 4.2. *A valid permutation α_u for a recurrent non-periodic word u is ergodic if and only if all the uniform frequencies of factors in u exist and are not equal to 0.*

Before proving the lemma, we prove the following proposition about words:

Proposition 4.3. *Let u be a recurrent aperiodic word and w and v some of its factors. Then in the orbit of w there can be the lexicographically maximal word from its closure starting with w, or the lexicographically minimal word from its closure starting with v, but not both at a time.*

PROOF. Suppose the opposite: let $T^k(u)$ be the maximal element of the orbit closure of u starting with w, and $T^l(u)$ be the minimal element of the orbit closure of u starting with v. Consider the prefix r of u of length $\max(k + |u|, l + |v|)$. Since u is recurrent, this prefix appears in it an infinite number of times, and since u is not ultimately periodic, there exists an extension p of r to the right which is right special: pa and pb are factors of u for some symbols $a \neq b$. Suppose that the prefix of u of the respective length is pa, and pb is a prefix of $T^n(u)$.

If $a < b$, then $u < T^n(u)$ and thus $T^k(u) < T^{k+n}(u)$, where $T^{k+n}(u)$ starts with w. A contradiction with the maximality of $T^k(u)$. If by contrary $a > b$, then $u > T^n(u)$ and thus $T^l(u) > T^{l+n}(u)$, where $T^{l+n}(u)$ starts with v. A contradiction with the minimality of $T^l(u)$. The proposition is proved. $\quad\square$.

PROOF OF LEMMA 4.2.

Suppose first that the frequency $\mu(w)$ of each factor w in u exists and is non-zero. We should prove that the corresponding valid permutation is ergodic. For every k we define

$$a[k] = \lim_{n \to \infty} \sum_{\substack{|v|=n, \\ v \leq w[k] \cdots w[k+n-1]}} \mu(v).$$

Clearly, such a limit exists and is in $[0,1]$, and by the definition, the probability that another element of the sequence $(a[i])$ is less than $a[k]$ is equal to $a[k]$.

It remains to prove that $a[k] \neq a[l]$ for $k \neq l$, that is, that the sequence $(a[n])$ is indeed a representative of a permutation.

Suppose the opposite: $a[k] = a[l]$ for $k \neq l$. Let $m \geq 0$ be the first position such that $w[k + m] \neq w[l + m]$: say, $w[k + m] < w[l + m]$. The only possibility for $a[l]$ and $a[k]$ to be equal is that $T^k(w) = w[k]w[k + 1] \cdots$ is the maximal word in the orbit closure of w starting with $w[k] \cdots w[k + m]$, and $T^l(w) = w[l]w[l + 1] \cdots$ is the minimal word in the orbit closure of w starting with $w[l] \cdots w[l + m]$. Due to Proposition 4.3, this is a contradiction. So, the values $a[k]$ are indeed all different, and thus the permutation is well-defined. Together with the condition on the probabilities we proved above, we get that the corresponding valid permutation is ergodic.

The proof of the converse is split into two parts. First we prove that for a valid ergodic permutation the frequencies of factors in the corresponding word must exist, then we prove that they are non-zero.

So, first we suppose that the frequencies of (some) factors of w do not exist. We are going to prove that the permutation is not ergodic, that is, that the canonical representative sequence $(a[n])$ is not well-defined.

Let us take the shortest and lexicographically minimal factor w whose frequency does not exist and consider the subsequence $(a[n_i])$ of the sequence $(a[n])$ corresponding to suffixes starting with w. The upper limit of $(a[n_i])$ should be equal to the sum of frequencies of the words of length $|w|$ less than or equal to w, but since the frequency of w is the only one of them that does not exist, this limit also does not exist. So, the sequence $(a[n])$ is not well-defined and hence the corresponding valid permutation is not ergodic.

The remaining case is that of zero frequencies: Suppose that w is the shortest and lexicographically minimal factor whose frequency is zero, and consider again the subsequence $(a[n_i])$ of the sequence $(a[n])$ corresponding to suffixes starting with w. The subsequence $(a[n_i])$ is infinite since u is recurrent, but all its elements must be equal: Their value is the sum of frequencies of words of length $|w|$ lexicographically less than w. So, the sequence $(a[n])$ does not correctly define a permutation, and hence in the case of zero frequencies the corresponding valid permutation is not ergodic.   □

We have seen above in Example 3.6 how the canonical representatives of permutations corresponding to Sturmian words are built.

Example 4.4. Let us continue the Thue-Morse example started above and prove that the representative a_{tm} is canonical. We should prove that the probability for any element $a[j]$ to be less than a is well-defined and equal to a. Let us prove by induction on k that the probability for an element to be in any binary rational interval $(d/2^k, (d+1)/2^k]$, where $0 \le d < 2^k$, is exactly $1/2^k$. Indeed, by the construction, the intervals $(0, 1/2]$ and $(1/2, 1]$ correspond to the zeros and ones in the original Thue-Morse word whose frequencies are $1/2$. The morphic image of any of these intervals is, consecutively, two intervals: $(0, 1/2] \mapsto (1/4, 2/4], (3/4, 4/4]$, and $(1/2, 1] \mapsto (2/4, 3/4], (0, 1/4]$. So, in both cases, the intervals are of the form $(d/2^2, (d+1)/2^2], d = 0, \ldots, 3$. Each of them is twice rarer than its pre-image; the four intervals cover $(0, 1]$ and do not intersect, so, the probability for a point $a[i]$ to be in each of them is $1/4$. But exactly the same argument works for any of these four intervals: its image is two intervals which are twice smaller and twice rarer than the pre-image interval. No other points appear in that shorter interval since each mapping corresponding to a position in the morphism is linear, and their ranges do not intersect. So, the probability for a point to be in an interval $(d/2^3, (d+1)/2^3]$ is $1/8$, and so on. By induction, it is true for any binary rational interval and thus for all interval subsets of $(0, 1]$: the frequency of elements in this interval is equal to its length. This proves that a_{tm} is indeed the canonical representative of the Thue-Morse permutation.

Remark 4.5. This example shows that the natural way of constructing the canonical representative of a valid permutation has little in common with frequencies of factors in the underlying word. The frequencies of symbols look important, but, for example, the frequency of 00 in the Thue-Morse word is $1/6$, whereas all the elements of the canonical representative are binary rationals.

Remark 4.6. In Lemma 4.2, we assumed that the word is recurrent. Indeed, if a word is not recurrent, the permutation can be ergodic. As an example, consider the word

$$01221211221121221\cdots,$$

that is, 0 followed by the Thue-Morse word on the alphabet $\{1, 2\}$. The respective permutation is still ergodic with the canonical representative $0, a_{tm} = 0, 1/2, 1, 3/4, 1/4, \ldots$.

Note also that this property depends on the order of symbols. For example, the permutation associated with the word

$$20110100110010110 \cdots = 2u_{tm}$$

is not ergodic since $a_t m[0]$ can be equal only to 1. On the other hand, it is well known that the first shift of the Thue-Morse word is the lexicographically largest element in its shift orbit closure. So, $a_t m[1]$ must also be equal to 1.

4.1   Morphisms on Words and Intervals

In this subsection, we generalize the above construction for the Thue-Morse word to a class of fixed points of morphisms: for any word from that class, we construct a morphism similar to the Thue-Morse interval morphism φ_{tm} defined in Example 4.1.

Let $\varphi : \{0, \ldots, q-1\}^* \mapsto \{0, \ldots, q-1\}^*$ be a morphism and $u = \varphi(u)$ be its aperiodic infinite fixed point starting with a letter a if it exists. In what follows we give a construction of the canonical representative a_u of the permutation α_u provided that the morphism φ is *primitive, monotone* and *separable*. We will now define what these properties mean.

Recall that the matrix A of a morphism φ is a $q \times q$-matrix whose element a_{ij} is equal to the number of occurrences of i in $\varphi(j)$. A matrix A and a morphism φ are called *primitive* if in some power A^n of A all the entries are positive, i.e., for every $b \in \{0, \ldots, q-1\}$ all the symbols of $\{0, \ldots, q-1\}$ appear in $\varphi^n(b)$ for some n. A classical Perron-Frobenius theorem says that a primitive matrix has a dominant positive *Perron-Frobenius eigenvalue* θ such that $\theta > |\lambda|$ for any other eigenvalue λ of A. It is also well-known that a fixed point of a primitive morphism is uniquely ergodic, and that the vector $\mu = (\mu(0), \ldots, \mu(q-1))^t$ of frequencies of symbols is the normalized Perron-Frobenius eigenvector of A:

$$A\mu = \theta\mu.$$

We say that a morphism φ is *monotone on an infinite word* u if for any $n, m > 0$ we have $T^n(u) < T^m(u)$ if and only if $\varphi(T^n(u)) < \varphi(T^m(u))$; here $<$ denotes the lexicographic order. A morphism is called *monotone* if it is monotone on all infinite words, or, equivalently, if for any infinite words u and v we have $u < v$ if and only if $\varphi(u) < \varphi(v)$.

Example 4.7. The Thue-Morse morphism φ_{tm} is monotone since $01 = f_{tm}(0) < f_{tm}(1) = 10$.

Example 4.8. The Fibonacci morphism $\varphi_f : 0 \to 01, 1 \to 0$ is not monotone since $01 = \varphi_f(0) > \varphi_f(10) = 001$, whereas $0 < 10$. At the same time, $\varphi_f^2 : 0 \to 010, 1 \to 01$ is monotone since for all $x, y \in \{0, 1\}$ we have $\varphi_f^2(0x) = 0100x' < 0101y' = \varphi_f^2(1y)$, where $x', y' \in \{0, 1\}^*$. So, to use our construction to the Fibonacci word $u_f = 01001010 \cdots$ which is the fixed point of φ_f, we should consider u_f as the fixed point of φ_f^2.

Example 4.9. As an example of a morphism which does not become monotone even when we consider its powers, consider $g : 0 \to 02, 1 \to 01, 2 \to 21$. It can be easily seen that $g^n(0) > g^n(1)$ for all $n \geq 1$.

The last condition we require from our morphism is to be *separable*. To define this property, consider the fixed point u as the infinite catenation of morphic images of its letters and say that the *type* $\tau(n)$ of a position n is the pair (a, p) such that $u[n] = \varphi(a)[p]$ in this "correct" decomposition into images of letters. So, there are $\sum_{a=0}^{q-1} |\varphi(a)|$ different types of positions in u. Also note that we index the elements of u starting with 0 and the elements of finite words $\varphi(a)$ starting from 1, so that, for example, $\tau(0) = (u[0], 1)$.

We say that a fixed point u of a morphism φ is *separable* if for every n, m such that $\tau(n) \neq \tau(m)$ the relation between $T^n(u)$ and $T^m(u)$ is uniquely defined by the pair $\tau(n), \tau(m)$. For a separable morphism φ we write $\tau(n) \preceq \tau(m)$ if and only if $T^n(u) \leq T^m(u)$.

Example 4.10. The Thue-Morse word is separable since for $\tau(n) = (0, 1)$ and $\tau(m) = (1, 2)$ we always have $T^n(u_{tm}) > T^m(u_{tm})$, i.e., all zeros which are first symbols of $f_{tm}(0) = 01$ give greater words than zeros which are second symbols of $f_{tm}(1) = 10$. Symmetrically, all ones which are first symbols of $f_{tm}(1) = 10$ give smaller words than ones which are second symbols of $f_{tm}(0) = 01$, that is, for $\tau(n) = (1, 1)$ and $\tau(m) = (0, 2)$ we always have $T^n(u_{tm}) < T^m(u_{tm})$.

Example 4.11. The fixed point

$$u = 0010010110010010110010110011 \cdots$$

of the morphism $0 \to 001, 1 \to 011$ is inseparable. Indeed, compare the following shifts: $T^2(u) = 1001011001 \cdots$, $T^5(u) = 1011 \cdots$ and $T^{17}(u) = 1001011011 \cdots$. We see that $T^2(u) < T^{17}(u) < T^5(u)$. At the same time, $\tau(2) = \tau(5) = (0, 3)$, and $\tau(17) = (1, 3)$.

Note that the class of primitive monotone separable morphisms includes in particular all morphisms considered by Valyuzhenich [17] who gave a formula for the permutation complexity of respective fixed points.

Similarly to morphisms on words, we define *a morphism on sequences of numbers* from an interval $[a, b]$ as a mapping $\varphi : [a, b]^* \mapsto [a, b]^*$. A fixed point of the morphism φ is defined as an infinite sequence $a[0], a[1], \ldots$ of numbers from $[a, b]$, such that $\varphi(a[0], a[1], \ldots) = a[0], a[1], \ldots$. Clearly, if a morphism φ has a fixed point, then there exists a number $c \in [a, b]$ such that $\varphi(c) = c, c[1], \ldots, c[k]$ for some $k \geq 1$ and $c[i] \in [a, b]$ for $i = 1, \ldots k$. Clearly, a fixed point of a morphism on sequences of numbers defines an infinite permutation (more precisely, its representative) if and only if all the elements of the sequence are distinct. The example of morphism defining an infinite permutation is given by the Thue-Morse permutation described in Example 4.1.

The rest of the section is organized as follows: First we provide the construction of a morphic ergodic permutation, then we give some examples, and finally we prove the correctness of the construction.

The construction of ergodic permutation corresponding to a separable fixed point of a monotone primitive morphism.

Now let us consider a separable fixed point u of a monotone primitive morphism φ over the alphabet $\{0, \ldots, q-1\}$, and construct the canonical representative a_u of the premutation α_u generated by it. To do it, we first look if u contains lexicographically minimal or maximal elements of the orbit with a given prefix. Note that due to Proposition 4.3, it cannot contain both of them. So, if u does not contain lexicographically maximal elements, we consider all the intervals to be half-open $[\cdot)$; in the opposite case, we can consider them to be half-open $(\cdot]$, like in the Thue-Morse case. Without loss of generality, in what follows we write the intervals $[\cdot)$, but the case of $(\cdot]$ is symmetric.

So, let $\mu = (\mu_0, \ldots, \mu_{q-1})$ be the vector of frequencies of symbols in u. Take the intervals $I_0 = [0, \mu_0), I_1 = [\mu_0, \mu_0 + \mu_1), \ldots, I_{q-1} = [1 - \mu_{q-1}, 1)$. An element e of a_u is in I_b if for another element of a_u the probability to be less than e is greater than the sum of frequences of letters less than b, and the probability to be greater than e is greater than the sum of frequences of letters greater than b. In other words, e is in I_b if and only if the respective symbol of u is b.

Now let us take all the $k = \sum_{a=0}^{q-1} |\varphi(a)|$ types of positions in u and denote them according to the order $\preceq$:

$$\tau_1 \prec \tau_2 \prec \cdots \prec \tau_k,$$

with $\tau_i = (a_i, p_i)$.

For each τ_i the frequency $l_i = \mu_{a_i}/\theta$, where θ is the Perron-Frobenius eigenvalue of φ, is the frequency of symbols of type τ_i in u. Indeed, the φ-images of a_i are Θ times rarer in u than a_i, and τ_i corresponds just to a position in such an image. Denote

$$J_1 = [0, l_1), J_2 = [l_1, l_1 + l_2), \ldots, J_k = [1 - l_k, 1);$$

so that in general, $J_i = [\sum_{m=1}^{i-1} l_m, \sum_{m=1}^{i} l_m)$. We will also denote $J_i = Ja_i, p_i$.

The interval J_i is the range of elements of a_u corresponding to the symbols of type τ_i in u. Note that all symbols of the same type are equal, and on the other hand, each symbol is of some type. For example, we have a collection of possible positions of 0 in images of letters, that is, a collection of types corresponding to 0, and all these types are less than any other type corresponding to any other symbol. So, the union of elements J_i corresponding to 0 is exactly I_0, and the same argument can be repeated for any greater symbol. In particular, each J_i is a subinterval of some I_a.

Now we define the morphism $\psi : [0,1]^* \mapsto [0,1]^*$ as follows: For $x \in I_a$ we have

$$\psi(x) = \psi_{a,1}(x), \ldots \psi_{a,|\varphi(a)|}.$$

Here $\psi_{a,p}$ is a linear mapping $\psi_{a,p} : I_a \mapsto J_{a,p}$: If $I_a = [x_1, x_2)$ and $J_{a,p} = [y_1, y_2)$, then

$$\psi_{a,p}(x) = \frac{y_2 - y_1}{x_2 - x_1}(x - x_1) + y_1. \tag{1}$$

Now we can define the starting point, that is, the value of a_1. Suppose that the first symbol of u is b; then $\varphi(b)$ starts with b, which means that $J_{b,1} \subset I_b$, and the mapping $\psi_{b,1}$ has a fixed point x: $\psi_{b,1}(x) = x$. We take a_1 to be this fixed point: $a_1 = x$. Note that if a_1 is the upper end of $J_{b,1}$, then we should take all the intervals to be $(\cdot]$; if it is the lower end, the intervals are $[\cdot)$; if it is in the middle of the interval, the ends are never attained. The situation when a_1 is an end of $J_{b,1}$ corresponds to the situation when there are the least or the greatest infinite words starting from some prefix in the orbit of u; as we have seen in Proposition 4.3, only one of these situations can appear at a time. In particular, in this situation, u is the least (or greatest) element of its orbit starting with b.

This construction may look bulky, but in fact, it is just a natural generalization of that for the Thue-Morse word. Indeed, in the Thue-Morse word, $\mu_0 = \mu_1 = 1/2$, $\theta = 2$, and the order of types is given in Example 4.10. So, $I_0 = [0, 1/2]$, $I_1 = [1/2, 1]$, $J_{0,1} = [1/4, 1/2]$, $J_{0,2} = [3/4, 1]$, $J_{1,1} = [1/2, 3/4]$, $J_{1,2} = [0, 1/4]$. Here the intervals are written as closed since at this stage we do not yet know whether we must take them $[\cdot)$ or $(\cdot]$. However, it becomes clear as soon as we consider the mapping $\psi_{0,1}$ which is the linear order-preserving mapping $I_0 \mapsto J_{0,1}$. Its fixed point is $1/2$, that is, the upper end of both intervals. Thus, the intervals must be chosen as $(\cdot]$. The mappings $\psi_{a,p}$ are explicitly written down in Example 4.1.

To give another example, consider the square of the Fibonacci morphism mentioned in Example 4.8.

Example 4.12. Consider the Fibonacci word as the fixed point of the square of the Fibonacci morphism: $\varphi_f^2 : 0 \to 010, 1 \to 01$. This morphism is clearly primitive; also, it is monotone as we have seen in Example 4.8, and separable: we can check that $(0, 3) \preceq (0, 1) \preceq (1, 1) \preceq (0, 2) \preceq (1, 2)$. In particular, this means that zeros which are first symbols of φ_f^2 are in the middle among other zeros. So, in what follows we can consider open intervals since their ends are never attained.

The Perron-Frobenius eigenvalue is $\theta = (3 + \sqrt{5})/2$, the frequencies of symbols are $\mu_0 = (\sqrt{5} - 1)/2$ and $\mu_1 = (3 - \sqrt{5})/2$. So, we have

$$I_0 = \left(0, \frac{\sqrt{5} - 1}{2}\right), I_1 = \left(\frac{\sqrt{5} - 1}{2}, 1\right),$$

and divide their lengths by θ to get the lengths of intervals corresponding to symbols from their images:

$$|J_{0,1}| = |J_{0,2}| = |J_{0,3}| = \frac{\mu_0}{\theta} = \sqrt{5} - 2, \ |J_{1,1}| = |J_{1,2}| = \frac{\mu_1}{\theta} = \frac{7 - 3\sqrt{5}}{2}.$$

The order of intervals is shown at Fig. 1.

Now the morphism ψ can be completely defined:

$$\psi(x) = \begin{cases} \psi_{0,1}(x), \psi_{0,2}(x), \psi_{0,3}(x) \text{ for } x \in I_0, \\ \psi_{1,1}(x), \psi_{1,2}(x) \text{ for } x \in I_1. \end{cases}$$

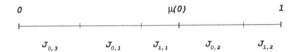

Fig. 1. Intervals for the Fibonacci permutation morphism

Here the mappings $\psi_{a,p} : I_a \mapsto J_{a,p}$ are defined according to (1). In particular, $\psi_{0,1} : (0, (\sqrt{5} - 1)/2) \mapsto (\sqrt{5} - 2, 2(\sqrt{5} - 2))$ has the fixed point $x = \psi_{0,1}(x) = (3 - \sqrt{5})/2$. This is the starting point a_1 of the fixed point a of ψ.

We remark that we could prove directly that the sequence a constructed above is exactly the canonical representative of the permutation associated with the Fibonacci word, using the fact that Fibonacci word belongs to the family of Sturmian words. However, we do not provide the proof for this example, since we now give a more general proof of the correctness of the general construction: the fixed point of the morphism ψ described above is indeed the canonical representative of our permutation.

Proof of correctness of the construction of the morphism ψ.

First we show that the fixed point of ψ is a representative of our permutation. Indeed, if $T^n(u) < T^m(u)$, and n and m are of different types, then, since the morphism is separable and by the construction, $a[n]$ and $a[m]$ are in different intervals $J_{a,p}$, and $a[n] < a[m]$. Now suppose that n and m are of the same type (a, p), that is, the nth (mth) symbol of u is the symbol number p of the image $\varphi(a)$, where a is the symbol number n' (m') of u, i.e., $u[n'] = a$, $u[n] = \varphi(a)[p]$, and applying the morphism φ to u sends $u[n']$ to $u[n-p+1..n-p+|\varphi(a)|]$. Then, since the morphism is monotone, $T^n(u) < T^m(u)$ if and only of $T^{n'}(u) < T^{m'}(u)$. Exactly the same condition is true for the relation $a_{[n]} < a_m$ if and only if $a_{n'} < a_{m'}$, since the mapping $\psi_{a,p}$ preserves the order. Now we can apply the same arguments to m' and n' instead of m and n, and so on. So, by the induction on the maximal power of φ involved, we also get that $T^n(u) < T^m(u)$ if and only if $a_{[n]} < a_{[m]}$. So, the sequence a is indeed a representative of the permutation generated by u.

It remains to prove that this representative is canonical. As above for the Thue-Morse word, it is done inductively on the intervals

$$\psi_{b_k,p_k}(\psi_{b_{k-1},p_{k-1}}(\ldots \psi_{b_1,p_1}(I_{b_1})\ldots)).$$

We prove that the probability for an element of a to be in this interval is equal to its length. For the intervals I_b, it is true by the construction as well as for their images. To make an induction step, we observe that the image of an interval under each $\psi_{b,p}$ is θ times smaller than the initial interval and corresponds to the situation which is θ times rarer. So, we have a partition of $(0, 1)$ to arbitrary small intervals for which the length is equal to the frequency of occurrences. This is sufficient to make sure that in fact, this is true for all intervals. $\quad\square$

Remark 4.13. In Example 4.12, we constructed a morphism for the Fibonacci permutation. However, it is not unique, and even not unique among piecewise linear morphisms. For example, the canonical representative b of each Sturmian permutation $\beta(\sigma, \rho)$ defined by $\beta_n = \{\sigma n + \rho\}$ for $n \geq 0$ is the fixed point of the following morphism $[0, 1]^* \mapsto [0, 1]^*$: $x \to \{2x - \rho\}, \{2x - \rho + \sigma\}$. Indeed, this is exactly a morphism which sends $\{\sigma n + \rho\}$ to $\{\sigma(2n) + \rho\}, \{\sigma(2n + 1) + \rho\}$. It is clearly piecewise linear as well as the function $\{\cdot\}$. Also, the same idea can be generalized to a k-uniform morphism for any $k \geq 2$.

Remark 4.14. We remark that the considerations used in the proof of the correctness of the construction are closely related to so-called Dumont-Thomas numeration systems [9].

References

1. Allouche, J.-P., Shallit, J.: Automatic Sequences – Theory, Applications, Generalizations. Cambridge University Press, Cambridge (2003)
2. Allouche, J.-P., Shallit, J.: The ubiquitous Prouhet-Thue-Morse sequence. In: Ding, C., Helleseth, T., Niederreiter, H. (eds.) Sequences and Their Applications. Discrete Mathematics and Theoretical Computer Science, pp. 1–16. Springer, London (1999)
3. Amigó, J.: Permutation Complexity in Dynamical Systems - Ordinal Patterns, Permutation Entropy and All That. Springer Series in Synergetics. Springer, Heidelberg (2010)
4. Avgustinovich, S.V., Frid, A., Kamae, T., Salimov, P.: Infinite permutations of lowest maximal pattern complexity. Theort. Comput. Sci. **412**, 2911–2921 (2011)
5. Avgustinovich, S.V., Frid, A.E., Puzynina, S.: Ergodic infinite permutations of minimal complexity. In: Potapov, I. (ed.) DLT 2015. LNCS, vol. 9168, pp. 71–84. Springer, Heidelberg (2015)
6. Bandt, C., Keller, G., Pompe, B.: Entropy of interval maps via permutations. Nonlinearity **15**, 1595–1602 (2002)
7. Elizalde, S.: The number of permutations realized by a shift. SIAM J. Discrete Math. **23**, 765–786 (2009)
8. Ferenczi, S., Monteil, T.: Infinite words with uniform frequencies, and invariant measures. Combinatorics, automata and number theory. Encyclopedia Math. Appl. **135**, 373–409 (2010). Cambridge University Press
9. Dumont, J.-M., Thomas, A.: Systèmes de numération et fonctions fractales relatifs aux substitutions. Theoret. Comput. Sci. **65**(2), 153–169 (1989)
10. Fon-Der-Flaass, D.G., Frid, A.E.: On periodicity and low complexity of infinite permutations. Eur. J. Combin. **28**, 2106–2114 (2007)
11. Frid, A.: Fine and Wilf's theorem for permutations. Sib. Elektron. Mat. Izv. **9**, 377–381 (2012)
12. Lothaire, M.: Algebraic Combinatorics on Words. Cambridge University Press, Cambridge (2002)
13. Makarov, M.: On permutations generated by infinite binary words. Sib. Elektron. Mat. Izv. **3**, 304–311 (2006)
14. Makarov, M.: On an infinite permutation similar to the Thue-Morse word. Discrete Math. **309**, 6641–6643 (2009)

15. Makarov, M.: On the permutations generated by Sturmian words. Sib. Math. J. **50**, 674–680 (2009)
16. Morse, M., Hedlund, G.: Symbolic dynamics II: Sturmian sequences. Amer. J. Math. **62**, 1–42 (1940)
17. Valyuzhenich, A.: On permutation complexity of fixed points of uniform binary morphisms. Discr. Math. Theoret. Comput. Sci. **16**, 95–128 (2014)
18. Widmer, S.: Permutation complexity of the Thue-Morse word. Adv. Appl. Math. **47**, 309–329 (2011)
19. Widmer, S.: Permutation complexity related to the letter doubling map. In: WORDS 2011 (2011)

Linear-Time Computation of Prefix Table for Weighted Strings

Carl Barton[1] and Solon P. Pissis[2]($\boxtimes$)

[1] The Blizard Institute, Barts and The London School of Medicine and Dentistry,
Queen Mary University of London, London, UK
c.barton@qmul.ac.uk

[2] Department of Informatics, King's College London, London, UK
solon.pissis@kcl.ac.uk

Abstract. The *prefix table* of a string is one of the most fundamental data structures of algorithms on strings: it determines the longest factor at each position of the string that matches a prefix of the string. It can be computed in time linear with respect to the size of the string, and hence it can be used efficiently for locating patterns or for regularity searching in strings. A *weighted string* is a string in which a set of letters may occur at each position with respective occurrence probabilities. Weighted strings, also known as *position weight matrices*, naturally arise in many biological contexts; for example, they provide a method to realise approximation among occurrences of the same DNA segment. In this article, given a weighted string x of length n and a constant *cumulative weight threshold* $1/z$, defined as the minimal probability of occurrence of factors in x, we present an $\mathcal{O}(n)$-time algorithm for computing the prefix table of x.

1 Introduction

An *alphabet* Σ is a finite non-empty set of size σ, whose elements are called *letters*. A *string* on an alphabet Σ is a finite, possibly empty, sequence of elements of Σ. The zero-letter sequence is called the *empty string*, and is denoted by ε. The *length* of a string x is defined as the length of the sequence associated with the string x, and is denoted by $|x|$. We denote by $x[i]$, for all $0 \leq i < |x|$, the letter at index i of x. Each index i, for all $0 \leq i < |x|$, is a position in x when $x \neq \varepsilon$. It follows that the i-th letter of x is the letter at position $i - 1$ in x.

The *concatenation* of two strings x and y is the string of the letters of x followed by the letters of y; it is denoted by xy. A string x is a *factor* of a string y if there exist two strings u and v, such that $y = uxv$. Consider the strings x, y, u, and v, such that $y = uxv$, if $u = \varepsilon$ then x is a *prefix* of y, if $v = \varepsilon$ then x is a *suffix* of y. Let x be a non-empty string and y be a string, we say that there exists an *occurrence* of x in y, or more simply, that x *occurs in* y, when x is a factor of y. Every occurrence of x can be characterised by a position in y; thus we say that x occurs at the *starting position* i in y when $y[i..i + |x| - 1] = x$.

Single nucleotide polymorphisms, as well as errors introduced by wet-lab sequencing platforms during the process of DNA sequencing, can occur in some

© Springer International Publishing Switzerland 2015
F. Manea and D. Nowotka (Eds.): WORDS 2015, LNCS 9304, pp. 73–84, 2015.
DOI: 10.1007/978-3-319-23660-5_7

positions of a DNA sequence. In some cases, these uncertainties can be accurately modelled as a *don't care* letter. However, in other cases they can be more subtly expressed, and, at each position of the sequence, a probability of occurrence can be assigned to each letter of the nucleotide alphabet; this process gives rise to a *weighted string* or a *position weight matrix*. For instance, consider a IUPAC-encoded [1] DNA sequence, where the ambiguity letter M occurs at some position of the sequence, representing either base A or base C. This gives rise to a weighted DNA sequence, where at the corresponding position of the sequence, we can assign to each of A and C an occurrence probability of 0.5.

A weighted string x of length n on an alphabet Σ is a finite sequence of n sets. Every $x[i]$, for all $0 \leq i < n$, is a set of ordered pairs $(s_j, \pi_i(s_j))$, where $s_j \in \Sigma$ and $\pi_i(s_j)$ is the probability of having letter s_j at position i. Formally, $x[i] = \{(s_j, \pi_i(s_j)) | s_j \neq s_\ell$ for $j \neq \ell$, and $\sum_j \pi_i(s_j) = 1\}$. A letter s_j *occurs* at position i of a weighted string x if and only if the *occurrence probability* of letter s_j at position i, $\pi_i(s_j)$, is greater than 0. A string u of length m is a *factor* of a weighted string if and only if it occurs at starting position i with *cumulative occurrence probability* $\prod_{j=0}^{m-1} \pi_{i+j}(u[j]) > 0$. Given a *cumulative weight threshold* $1/z \in (0,1]$, we say that factor u is *valid*, or equivalently that factor u has a valid occurrence, if it occurs at starting position i and $\prod_{j=0}^{m-1} \pi_{i+j}(u[j]) \geq 1/z$. For clarity of presentation, in the rest of this article, a set of ordered pairs in a weighted string is denoted by $[(s_0, \pi_i(s_0)), \ldots, (s_{\sigma-1}, \pi_i(s_{\sigma-1}))]$.

A great deal of research has been conducted on weighted strings for pattern matching [2,4], for computing various types of regularities [5,7,9], for indexing [2, 8], and for alignments [3]. The efficiency of most of the proposed algorithms relies on the assumption of a given *constant* cumulative weight threshold defining the minimal probability of occurrence of factors in the weighted string.

Similar to the standard setting [6,10], we can define the prefix table for a weighted string. Given a weighted string x of length n and a constant cumulative weight threshold $1/z$, we define the *prefix table* WP of x as follows:

$$WP[i] = \begin{cases} |u| & \text{if } i = 0 \text{ and } u \text{ is the longest valid prefix of } x. \\ |v| & \text{if } 0 < i < n \text{ and } v \text{ is the longest valid prefix of} \\ & x \text{ with a valid occurrence at } i. \end{cases}$$

For large alphabets it makes sense to perform a simple filtering on x to filter out letters with occurrence probability less than $1/z$. This is required as if the alphabet is not of fixed size, we may have many letters with low occurrence probability that are not of interest. We simply read the entire string and keep only those letters with probability greater than or equal to $1/z$; these are at most z for each position, so still constant. We are thus left with a string of size $\mathcal{O}(zn)$, and the entire stage takes time $\mathcal{O}(\sigma n)$. For clarity of presentation, in the rest of this article, we assume that the string resulting from this stage is the input weighted string x, and consider the following problem.

WEIGHTEDPREFIXTABLE
Input: a weighted string x of length n and a constant integer $z > 0$
Output: prefix table WP of x

Example 1. Let $x = $ aab$[(a, 0.5), (b, 0.5)][(a, 0.5), (b, 0.5)]$bab and $z = 4$. Then we have

$$i : 0\ 1\ 2\ 3\ 4\ 5\ 6\ 7$$
$$\mathsf{WP}[i] : 8\ 1\ 0\ 5\ 1\ 0\ 1\ 0.$$

$\mathsf{WP}[3] = 5$ since aabab has a valid occurrence at position 0 with probability $1/4 \geq 1/4$ and a valid occurrence at position 3 with probability $1/4 \geq 1/4$.

The main contribution of this article is the following.

Theorem 1. *Problem* WEIGHTEDPREFIXTABLE *can be solved in time* $\mathcal{O}(n)$.

An $\mathcal{O}(n)$-time bound, for $z = \mathcal{O}(1)$, for pattern matching on weighted strings was shown in [2]. This is useful in the context of molecular biology for finding IUPAC-encoded nucleotide or peptide sequences such as *cis*-elements in nucleotide sequences or small domains and motifs in protein sequences [11]. A direct application of Theorem 1 is a simple $\mathcal{O}(n)$-time algorithm for the same problem. However, similar to the standard setting, we anticipate that this structure will be of use for other problems such as regularity searching on weighted strings [5].

2   Properties and Auxiliary Data Structures

Fact 1. *Any factor of a valid factor of x is also valid.*

We perform a colouring stage on x, similar to the one before the construction of the weighted suffix tree [8], which assigns a colour to every position in x according to the following scheme:

- mark position i *black* (b), if *none* of the occurring letters at position i has probability of occurrence greater than $1 - 1/z$.
- mark position i *grey* (g), if *one* of the occurring letters at position i has probability of occurrence greater than $1 - 1/z$.
- mark position i *white* (w), if *one* of the occurring letters at position i has probability of occurrence 1.

This stage can be trivially performed in time $\mathcal{O}(zn)$.

Lemma 1 ([8]). *Any valid factor of x contains at most $\lceil \log z / \log(\frac{z}{z-1}) \rceil$ black positions.*

From Lemma 1 we know that any valid factor contains a *constant* number of black positions; for the rest of the article, we denote this constant by $\ell = \lceil \log z / \log(\frac{z}{z-1}) \rceil$. We can also see that any valid factor of a weighted string is *uniquely* determined by the letters it has at black positions: *any* white or grey position is common to all valid factors that contain the same positions. Hence an occurrence of a valid factor of x can be memorised as a tuple $< i, j, s >$, where i is the starting position in x, $j \geq i$ is the ending position in x, and s is a set of ordered pairs (p, c) denoting that letter $c \in \Sigma$ occurs at black position p, $i \leq p \leq j$. This representation requires space $\mathcal{O}(\ell)$ per occurrence by Lemma 1;

Table 1. Weighted string x and $z = 2$, colouring, array FP with cumulative occurrence probabilities between black positions, and array BP with black positions indices; the letter t at position 10 has already been filtered out since $0.4 < 1/z$

i	0	1	2	3	4	5	6	7	8	9	10
$x[i]$	a	c	t	t	(a, 0.5)	t	c	(a, 0.5)	t	t	(a, 0.6)
					(c, 0.5)			(t, 0.5)			(t, 0.4)
Colour	w	w	w	w	b	w	w	b	w	w	g
FP	1	1	1	1	0	1	1	0	1	1	0.6
BP	4	4	4	4	7	7	7	11	11	11	11

it can also be used to efficiently compute the occurrence probability of any factor of x in time proportional to the number of black positions it contains.

We start by computing an array BP of integers, such that BP$[i]$ stores the smallest position greater than i that is marked black; otherwise (if no such position exists) we set BP$[i] = n$. We next view x as a sequence of the form $u_0 g_0 u_1 g_1 \ldots u_{k-1} u_{k-1} u_k$, such that u_0, u_k are (possibly empty) strings on Σ, $u_1, \ldots, u_{k-1}$ are non-empty strings on Σ, and $g_0, \ldots, g_{k-1}$ are maximal sequences consisting only of black positions. We compute an array FP of factor probabilities, such that FP$[i + q]$ stores the occurrence probability of $u_j[0 .. q]$, for all $0 \le q < |u_j|$, starting at position i of x; otherwise (if no such factor starts at position i) we set FP$[i] = 0$. For an example, see Table 1. Both arrays BP and FP can be trivially computed in time $\mathcal{O}(n)$.

Given a range of positions in x, the black positions indices within this range, and the letters at black positions, that is, given a tuple $< i, j, s >$, we can easily compute the occurrence probability of the respective factor: we take the probability of any factor between black positions, the probability of the letters at black positions, the probability of any leading or trailing segments, and, finally, we multiply them all together. Notice that we can deal with any leading segments by computing the differences between prefixes via division. This takes time proportional to the number of black positions within the factor, and is thus $\mathcal{O}(\ell)$ for any valid factor.

Example 2. Consider the weighted string x in Table 1 and $z = 2$. We wish to determine the probability of the factor starting at position 2 and ending at position 7, with the two black positions as letters a and a; that is factor $< 2, 7, \{(4, a), (7, a)\} >$. We determine the probability of factor tt starting at position 2 by taking FP$[3]/$FP$[1] = 1$. We take the probability of factor tc starting at position 5, which is given by FP$[6] = 1$. The probability of the two black positions are 0.5 and 0.5. By multiplying all of these together we get the occurrence probability of factor $< 2, 7, \{(4, a), (7, a)\} >$ which is 0.25.

A *maximal factor* of a weighted string is a valid factor that cannot be extended to the left or to the right and preserve its validity (for a formal definition, see [2]).

Lemma 2 ([2]). *At most z maximal factors start at any position of x.*

Let $\mathsf{lcve}(u, v)$ denote the length of the longest common valid prefix (or longest common valid extension) of two weighted strings u and v.

Proposition 1. *Given a factor v of x with two valid occurrences $< i, j, s >$ and $< i', j', s' >$, then $\mathsf{lcve}(x[i \mathbin{..} n - 1], x[i' \mathbin{..} n - 1])$ can be computed in time $\mathcal{O}(\ell + \ell z m)$, where $m = \mathsf{lcve}(x[i \mathbin{..} n - 1], x[i' \mathbin{..} n - 1]) - |v|$.*

Proof. Let p_i and $p_{i'}$ be the probability of occurrence of the common valid factor v, starting at positions i and i' of x, respectively. We can compute p_i and $p_{i'}$ in time $\mathcal{O}(\ell)$ using $< i, j, s >$ and $< i', j', s' >$ with the technique outlined above (using array FP). We may then proceed by comparing letters from positions $j + 1$ and $j' + 1$ onwards. We have a number of cases to consider. Suppose that of the two positions we compare:

- *no* position is black; then we carry on comparing the letters and updating p_i and $p_{i'}$ accordingly.
- *one* position is black and the other position is either white or grey; we check if the single letter at the white or grey position occurs at the black position and update p_i and $p_{i'}$ accordingly.
- *both* positions are black; we consider all occurring letters that match and continue extending all corresponding valid factors. By Lemma 2, the number of these combinations of letters at black positions is at most z; by Lemma 1 the number of black positions in each combination is at most ℓ.

We terminate this procedure when we have no match or the probability threshold is violated. It is clear to see that the work at any position is no more than $\mathcal{O}(\ell z)$.   □

The main idea of our algorithm is based on the efficient computation of an auxiliary array P of n integers, which for some position i stores the length of the longest common prefix (or longest common extension) of strings $x[i \mathbin{..} n - 1]$ and $x[0 \mathbin{..} n - 1]$ as if in all black positions of x we had a *don't care* letter—a letter that does not belong in Σ and it matches itself and any other letter in Σ.

Computing array P in a naïve way could take as much as $\mathcal{O}(n^2)$ time; the transitive properties used in the standard setting [10] do not hold due to don't care letters. We will show here that it is possible to compute array P in time $\mathcal{O}(\ell n)$. The *critical* observation for this computation is that no entry in WP can be greater than the length of the longest valid prefix of x containing at most ℓ black positions: this is clear from Lemma 1. This means that we only need to compute the values of P for this longest valid prefix; this will allow us to efficiently compute the values of WP for all positions later on.

We now describe the method for the efficient computation of P. Let x' be the string obtained by replacing in x: (a) each black position with a *unique* letter $\$_h$ *not* in Σ; and (b) each grey position with the only letter in that position with probability of occurrence greater than $1 - 1/z$. Hence x' is of the form $x'_0 g_0 x'_1 g_1 \ldots x'_{k-1} g_{k-1} x'_k$, such that x'_0, x'_k are (possibly empty) strings on Σ, $x'_1, \ldots, x'_{k-1}$ are non-empty strings on Σ, and $g_0, \ldots, g_{k-1}$ are maximal sequences consisting only of letters $\$_h$. Let $\mathsf{lce}(u, v)$ denote the length of the longest common prefix of strings u and v. We now make use of the critical observation. We compute the standard prefix table of $x'_j\$x'$, such that $\$ \notin \Sigma$ and $\$ \neq \$_h$, *only if* x'_j is a factor of the longest valid prefix of x. Given these prefix tables we can easily compute, for each such x'_j, $0 \leq j \leq \ell$, the following array of size n:

$$\pi_{x'_j}[i] = \begin{cases} \mathsf{lce}(x'[i\mathinner{\ldotp\ldotp}n-1], x'_j) & \text{if } x'[i] \in \Sigma \\ 0 & \text{otherwise.} \end{cases}$$

We are now in a position to compute array P by making use of the above computed arrays. For any position i, $x'[i] \in \Sigma$, we start by checking the length of the longest common prefix between $x'[i\mathinner{\ldotp\ldotp}n-1]$ and x'_j. We note two cases about the mismatch occurring one position past the longest common prefix:

1. If the mismatch is caused by two letters from Σ, then there is a *legitimate* mismatch and we know the correct value of $\mathsf{P}[i]$;
2. If the mismatch is caused by (at least) one letter $\$_h$ not from Σ, then we need to check if it can be extended further.

By skipping the maximal sequence of letters $\$_h$ of x' (Case 2), we can reach the beginning of another one of the x'_j factors, $0 \leq j \leq \ell$, that we have computed a prefix table for, so we may continue extending the common factor until a legitimate mismatch (Case 1) or the end of the string is reached. For any position i, $x'[i] \notin \Sigma$, we similarly skip letters $\$_h$ until we reach the beginning of a x'_j factor.

If, for any position i, we have encountered more than ℓ letters $\$_h$ *either* in the prefix of x' or in $x'[i\mathinner{\ldotp\ldotp}n-1]$, we terminate the extension. If, for any position i, the length of the extension becomes greater than the length of the longest valid prefix of x (trivially computed), we terminate the extension and set $\mathsf{P}[i] = \infty$.

Example 3. Consider the weighted string x in Table 2 and $z = 64$. The longest valid prefix of x is equal to x. Let $x' =$g$\$_0g\$_1\$_2$gagcg$\$_3\$_4g\$_5$c, where $\$_0, \$_1, \$_2$, $\$_3, \$_4$, and $\$_5$ are unique letters not in Σ. x' is of the form $x'_0 g_0 x'_1 g_1 x'_2 g_2 x'_3 g_3 x'_4$: $x'_0 = $ g, $x'_1 = $ g, $x'_2 = $ gagcg, $x'_3 = $ g, and $x'_4 = $ c; and $g_0, \ldots, g_3$ are maximal sequences consisting only of letters not in Σ. Table 2 illustrates the computation

Table 2. Computation of array P

i	0	1	2	3	4	5	6	7	8	9	10	11	12	13	14
$x[i]$	g	(a, 0.5) (c, 0.5)	g	(g, 0.5) (t, 0.5)	(a, 0.5) (t, 0.5)	g	a	g	c	g	(g,0.5) (t,0.5)	(c,0.5) (t,0.5)	g	(a,0.5) (t,0.5)	c
$x'[i]$	g	$\$_0$	g	$\$_1$	$\$_2$	g	a	g	c	g	$\$_3$	$\$_4$	g	$\$_5$	c
$\pi_{x'_0}[i]$	1	0	1	0	0	1	0	1	0	1	0	0	1	0	0
$\pi_{x'_1}[i]$	1	0	1	0	0	1	0	1	0	1	0	0	1	0	0
$\pi_{x'_2}[i]$	1	0	1	0	0	5	0	1	0	1	0	0	1	0	0
$\pi_{x'_3}[i]$	1	0	1	0	0	1	0	1	0	1	0	0	1	0	0
$\pi_{x'_4}[i]$	0	0	0	0	0	0	0	0	1	0	0	0	0	0	1
$\mathsf{P}[i]$	15	5	6	5	2	9	0	7	0	5	5	4	2	2	0

of arrays $\pi_{x'_j}$. Let $i = 5$. We start by checking the lce between $x'[5 .. n - 1]$ and $x'_0 = \mathsf{g}$ which is given by $\pi_{x'_0}[5] = 1$. We proceed by skipping $x'[13]$ since $x'[13] = \$_5$, and terminate the check since $\pi_{x'_4}[9] = 0$; we thus set $\mathsf{P}[5] = 9$. We proceed by skipping $x'[8 .. 9]$ since $x'[3 .. 4] = \$_1\$_2$ and skipping $x'[10 .. 11]$ since $x'[10 .. 11] = \$_3\$_4$. We check the lce between $x'[7 .. n - 1]$ and $x'_3 = \mathsf{g}$ which is given by $\pi_{x'_3}[7] = 1$. We proceed by skipping $x'[13]$ since $x'[13] = \$_5$, and terminate the check since $\pi_{x'_4}[9] = 0$; we thus set $\mathsf{P}[5] = 9$.

Lemma 3. *Array* P *can be computed in time* $\mathcal{O}(\ell n)$.

Proof. The length of $x_j\$x'$, $0 \leq j \leq \ell$, is no more than $2n + 1$ and, by Lemma 1, there can be no more than $\ell + 1$ prefix tables to compute; using a standard algorithm [10], each takes no more than $2(2n + 1)$ letter comparisons, and no table entry is updated more than once, so $\mathcal{O}(\ell n)$ in total. For each position i, we may need to perform, by Lemma 1, at most 2ℓ letter comparisons for an entry in P, so $\mathcal{O}(\ell n)$ in total. Each entry in P is updated only once, and so we achieve the claim with a total of no more than $(6\ell + 4)n + 2\ell + 2$ letter comparisons. □

Lemma 4. $\mathsf{P}[i] \geq \mathsf{WP}[i]$, *for all* $0 \leq i < n$.

Proof. For those entries in P that are ∞ the claim is obvious. Those others entries in P are computed ignoring letters at black positions: should all those black positions match their corresponding positions and the probability threshold is not violated then $\mathsf{P}[i] = \mathsf{WP}[i]$; should any of those positions not match or the probability threshold is violated then $\mathsf{P}[i] > \mathsf{WP}[i]$. □

The method for computing table WP proceeds by determining $\mathsf{WP}[i]$ by increasing values of the position i on x. We introduce, the index i being fixed, two values g and f that constitute the key elements of our method. They satisfy the following relations

$$g = \max\{j + \mathsf{WP}[j] : 0 < j < i\} \tag{1}$$

and

$$f \in \{j : 0 < j < i \text{ and } j + \mathsf{WP}[j] = g\}. \tag{2}$$

We note that g and f are defined when $i > 1$. We note, moreover, that if $g < i$ we have then $g = i - 1$, and that on the contrary, by definition of f, we have $f < i \leq g$.

Lemma 5. *Let* $f < i < g$, u *be a factor of* x *with two valid occurrences at positions* 0 *and* f, $|u| = g - f$, *and* $\mathsf{WP}[i - f] < g - i$. *Then we can compute* $lcve(x, x[i .. g - 1])$ *in time* $\mathcal{O}(\ell z)$.

Proof. If $\mathsf{WP}[i - f] < g - i$ then there exist two factors v_1 and v_2, possibly $v_1 \neq v_2$, of u, $|v_1| = |v_2| = \mathsf{WP}[i - f]$, occurring at positions 0, f and $i - f$ and i, respectively. By Fact 1 factors v_1 and v_2 are valid. By the definition of table WP there does not exist another valid factor v, $|v| > |v_1|$, occurring at positions 0 and $i - f$. Let v_3 be the longest common valid prefix of x and $x[i .. g - 1]$, i.e. $lcve(x, x[i .. g - 1]) = |v_3|$. We have two cases: $v_1 = v_2$ and $v_1 \neq v_2$.

In case $v_1 = v_2$, then it holds that v_3, $|v_1| \leq |v_3| \leq g - i$, occurs at positions 0 and i. By Lemmas 1, 2, and 4, we can determine the length of v_3 using array P, $|v_3| \leq$ P$[i]$, and letter comparisons *only* at black positions (using array BP) in time $\mathcal{O}(\ell z)$.

In case $v_1 \neq v_2$ letter comparisons are required to determine the length of v_3. By Lemma 1 and triangle inequality, since v_1, v_2, and v_3 are valid factors, a prefix of v_1 may differ to a prefix of v_3 by at most 2ℓ (black) positions: v_1, occurring at position 0, differs to v_2, occurring at position $i - f$, by at most ℓ positions; and a prefix of v_2, occurring at position i, differs to a prefix of v_3, occurring at position i, by at most ℓ positions. Each position has *at most* z occurring letters. Therefore, by Lemmas 1 and 2, we can either determine the actual value of $\mathsf{lcve}(x, x[i \mathinner{.\,.} g - 1]) = |v_3| < |v_1|$ or determine that $|v_1| \leq |v_3| \leq g - i$ in time $\mathcal{O}(\ell z)$. In case $|v_1| \leq |v_3| \leq g - i$, by Lemmas 1, 2, and 4, we can determine the length of v_3 using array P, $|v_3| \leq$ P$[i]$, and letter comparisons *only* at black positions in time $\mathcal{O}(\ell z)$.

Example 4. Let the following string x and $z = 64$.

i	0	1	2	3	4	5	6	7	8	9	10	11	12	13 g	14
$x[i]$	g	(a, 0.5) (c, 0.5)	g	(g, 0.5) (t, 0.5)	(a, 0.5) (t, 0.5)	g	a	g	c	g	(g,0.5) (t,0.5)	(c,0.5) (t,0.5)	g	(a,0.5) (t,0.5)	c
P$[i]$	15	5	6	5	2	9	0	7	0	5	5	4	2	2	0

Further let $i = 10 < g = 14$, $u = $ gcgttga be the valid factor occurring at positions 0 and $f = 7$, $|u| = g - f = 7$, and WP$[i - f] = 3 < g - i = 4$. Factor $v_1 = $ gcg occurs at positions 0 and $f = 7$ and factor $v_2 = $ ttg occurs at positions $i - f = 3$ and $i = 10$. In this case $v_1 \neq v_2$. We apply Lemma 5.

Factor v_1, occurring at position 0, has a black position at index 1. Factor v_2, occurring at position 3, has two black positions at indices 3 and 4. Factors v_2 and v_3, starting at position 10, have two black positions at indices 10 and 11. Therefore we know that we have to compare the letters at positions 0 and 1 to the letters at positions 10 and 11, respectively, and that positions 2 and 12 are white and therefore $x[2] = x[12]$. There exist such letters that match and therefore the prefix of length $|v_1|$ of v_3 is gcg. And since P$[10] = 5$, by Lemma 4, we know that $|v_3| \leq 5$. We can determine $|v_3|$ via comparing black positions: 3 to 13. Therefore we determine that $\mathsf{lcve}(x, x[i \mathinner{.\,.} g - 1]) = 4$.

For a graphical illustration of the proof inspect Fig. 1.   □

Lemma 6. *Let $f < i < g$, u be a factor of x with two valid occurrences at positions 0 and f, $|u| = g - f$, and WP$[i - f] > g - i$. Then we can compute $\mathsf{lcve}(x, x[i \mathinner{.\,.} g - 1])$ in time $\mathcal{O}(\ell z)$.*

Proof. If WP$[i - f] > g - i$ then there exists a factor v, possibly $v \neq u$, $|v| = $ WP$[i - f]$, occurring at positions 0 and $i - f$. By the definition of WP factor v is valid and there does not exist another valid factor v', $|v'| > |v|$, occurring at positions 0 and $i - f$. Let v_1 be the longest common valid prefix of x and $x[i \mathinner{.\,.} g - 1]$, i.e. $\mathsf{lcve}(x, x[i \mathinner{.\,.} g - 1]) = |v_1|$. We have two cases: $v = u$ and $v \neq u$.

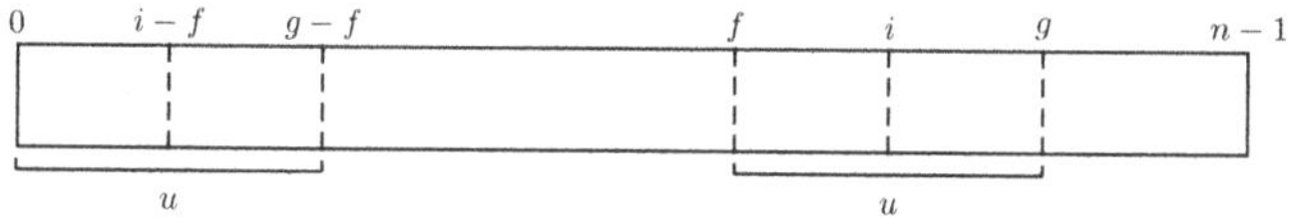

(a) String $x = x[0\,..\,n-1]$, u occurs at positions 0 and f, and $\mathsf{WP}[f] = |u|$

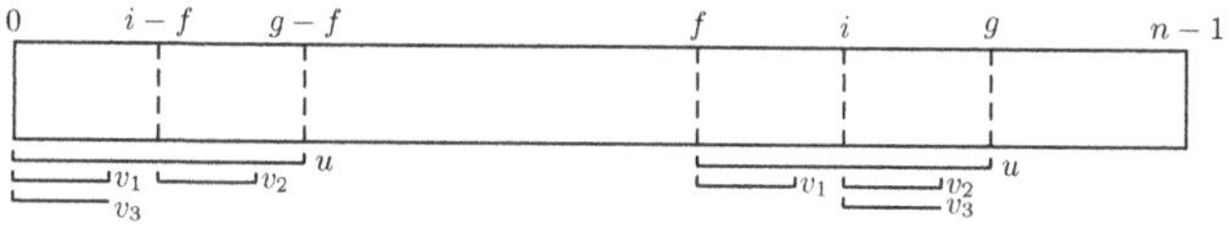

(b) If $v_1 = v_2$ then $|v_3| = \mathsf{lcve}(x, x[i\,..\,g-1]) \geq |v_1|$

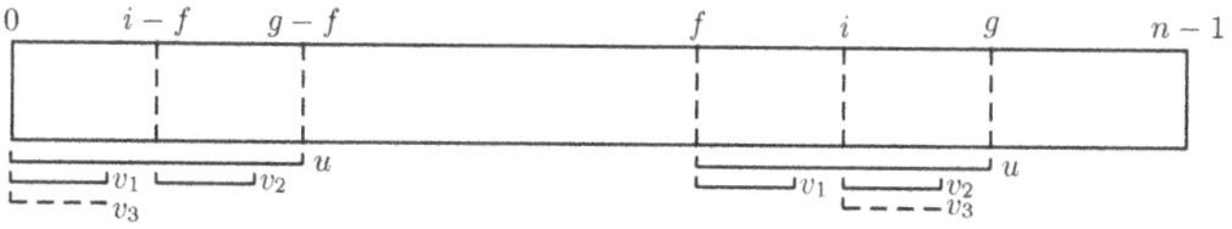

(c) If $v_1 \neq v_2$ then letter comparisons at black positions are required to determine the length of v_3

Fig. 1. Illustration of lemma 5

In case $v = u$, it holds that v_1, $|v_1| = g - i$, occurs at positions 0 and i, and hence $\mathsf{lcve}(x, x[i\,..\,g-1]) = |v_1|$.

In case $v \neq u$, letter comparisons are required to determine the length of v_1. By Lemma 1 and triangle inequality, since u, v, and v_1 are valid factors, v_1 may differ to some prefix of u by at most 2ℓ (black) positions: the prefix of length $g - i$ of u, occurring at position 0, differs to the suffix of length $g - i$ of u in at most ℓ positions; and the suffix of length $g - i$ of u, occurring at position i, differs to v_1 in at most ℓ positions. Each position has *at most* z occurring letters. Therefore, by Lemmas 1 and 2, we can either determine the actual value of $\mathsf{WP}[i] = \mathsf{lcve}(x, x[i\,..\,g-1]) < g - i$ or determine that $\mathsf{lcve}(x, x[i\,..\,g-1]) = g - i$ in time $\mathcal{O}(\ell z)$.

Example 5. Let the following string x and $z = 8$.

										f						g
i	0	1	2	3	4	5	6	7	8	9	10	11	12	13	14	15
$x[i]$	a	c	g	a	(c, 0.5) (t, 0.5)	(a, 0.5) (g, 0.5)	a	t	c	a	c	g	a	(c,0.5) (t,0.5)	(a,0.5) (g,0.5)	c

Further let $i = 12 < g = 15$, $u =$ `acgata` be the valid factor occurring at positions 0 and $f = 9$, $|u| = g - f = 6$. Factor $v =$ `acgat`, $|v| = \mathsf{WP}[i - f] = 5 > g - i = 3$, occurs at positions 0 and $i - f = 3$. In this case $u \neq v$. We apply Lemma 6.

The prefix of length $g - i = 3$ of $u =$ `acgata`, occurring at position 0, differs to the suffix of length $g - i = 3$ of u in two positions. The suffix of length $g - i = 3$ of u, occurring at position $i = 12$, differs to factor v_1, occurring at positions 0 and $i = 12$, in two positions. Therefore we know that we have to compare the letters at positions 1 and 2 to the letters at positions 13 and 14, respectively, and that positions 0 and 12 are white and therefore $x[0] = x[12]$. There exist such letters that match and therefore we determine that $\mathsf{lcve}(x, x[i \mathinner{.\,.} g - 1]) = g - i = 3$.

For a graphical illustration of the proof inspect Fig. 2.   □

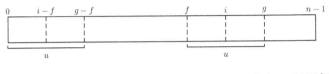

(a) String $x = x[0 \mathinner{.\,.} n - 1]$, u occurs at positions 0 and f, and $\mathsf{WP}[f] = |u|$

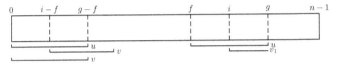

(b) If $u = v$ then $|v_1| = \mathsf{lcve}(x, x[i \mathinner{.\,.} g - 1])$

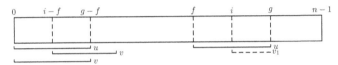

(c) If $u \neq v$ then letter comparisons at black positions are required to determine the length of v_1

Fig. 2. Illustration of lemma 6

Similar to Lemma 6 we can obtain the following.

Lemma 7. *Let* $f < i < g$, u *be a factor of* x *with two valid occurrences at positions* 0 *and* f, $|u| = g - f$, *and* $\mathsf{WP}[i - f] = g - i$. *Then we can compute* $\mathsf{lcve}(x, x[i \mathinner{.\,.} g - 1])$ *in time* $\mathcal{O}(\ell z)$.

3   Algorithm

We can now present Algorithm WeightedPrefixTable for computing table WP.

Lemma 8. *Algorithm* WeightedPrefixTable *correctly computes table* WP.

Proof. The computation of $\mathsf{WP}[0]$ is, by definition, correct. The variables f and g satisfy Eqs. 1 and 2 at each step of the execution of the loop. For i being fixed and satisfying the conditions $i < g$ *and* $\mathsf{lcve}(x, x[i \mathinner{.\,.} g - 1]) < g - i$, the algorithm applies the Lemmas 5–7 which produce a correct computation: $\mathsf{lcve}(x, x[i \mathinner{.\,.} n - 1]) = \mathsf{lcve}(x, x[i \mathinner{.\,.} g - 1])$. It remains thus to check that the computation is correct otherwise. But in this case, we compute $\mathsf{lcve}(x, x[i \mathinner{.\,.} n - 1]) = |x[f \mathinner{.\,.} g - 1]| = g - f$ which is, by definition, the value of $\mathsf{WP}[i]$. Therefore, Algorithm WeightedPrefixTable correctly computes table WP.   □

Lemma 9. *Given a weighted string x of length n and a constant cumulative weight threshold $1/z$ Algorithm WeightedPrefixTable requires time $\mathcal{O}(n)$.*

Proof. The computation of WP[0] can be trivially done in time $\mathcal{O}(n)$. Array P can be computed in time $\mathcal{O}(n)$, for $\ell = \mathcal{O}(1)$, by Lemma 3. As the value of g never decreases and that it varies from 0 to at most n, there will be, by Proposition 1, at most $\mathcal{O}(n)$ positive comparisons in the inner loop. Each negative comparison leads to the next step of the outer loop; and there are at most $n - 1$ of them. Thus $\mathcal{O}(n)$ comparisons on the overall. All other instructions, by Lemmas 5-7, take constant time for each value of i giving a total time of $\mathcal{O}(n)$. □

Algorithm *WeightedPrefixTable$(x, n, 1/z)$*
  WP$[0 \mathinner{..} n - 1] \leftarrow 0$;
  WP$[0] \leftarrow |u|$ where u is the longest valid prefix of x;
  Compute array P;
  $g \leftarrow 0$;
  foreach $i \in \{1, n - 1\}$ **do**
      if $i < g$ **and** $lcve(x, x[i \mathinner{..} g - 1]) < g - i$ **then**
          WP$[i] \leftarrow lcve(x, x[i \mathinner{..} g - 1])$;
      else
          $f \leftarrow i$; $g \leftarrow \max\{g, i\}$;
          while $g < n$ **and** *there exists $c \in \Sigma$ occurring at positions g
          and $g - f$ **and** there exists a common valid prefix, say v,
          $|v| = g - f$, of $x[0 \mathinner{..} n - 1]$ and $x[i \mathinner{..} n - 1]$, such that vc is a
          valid factor starting at position 0 and i* **do**
              $g \leftarrow g + 1$;
          WP$[i] \leftarrow g - f$;
      return WP;

Lemmas 8 and 9 imply the main result of this article (Theorem 1).

Corollary 1. *The number of letter comparisons performed by Algorithm Weight-edPrefixTable is no more than $(4z^2 \log z + 6z \log z + 4)n + 2z \log z + 2$.*

Proof. We perform no more than $(6\ell+4)n+2\ell+2$ letter comparisons to compute P (Lemma 3). Then each entry in WP is updated only once. The condition evaluation by the if statement requires at most $2\ell z$ letter comparisons (Lemmas 5–7). Each update in the else statement requires the same (Proposition 1). This gives at most $(4\ell z + 6\ell + 4)n + 2\ell + 2$ letter comparisons. For $z = 1$, the string is not essentially weighted so the claim clearly holds; for $z > 1$, we must show that:

$$\ell = \left\lceil \frac{\log z}{\log(\frac{z}{z-1})} \right\rceil = \left\lceil \frac{\log z}{\log(z) - \log(z - 1)} \right\rceil \le z \log z. \text{ Or equivalently that:}$$

$$\frac{\log z(z \log z - z \log(z - 1) - 1)}{\log z - \log(z - 1)} > 0.$$

Clearly the above is true if and only if: $z \log z - z \log(z-1) - 1 > 0$. There is a discontinuity at $z = 1$; after this it is always positive and the following holds:

$$\lim_{z \to \infty} z \log z - z \log(z-1) - 1 = 0.$$

$\square$

4   Final Remarks

In this article, we presented a linear-time algorithm for computing the prefix table for weighted strings with a very low constant factor (Corollary 1). This implies an $\mathcal{O}(nz^2 \log z)$-time algorithm for pattern matching on weighted strings for arbitrary z, which is simple and matches the best-known time complexity for this problem [2]. Furthermore, we anticipate that this structure will be of use in other problems on weighted strings such as computing the border table of a weighted string and computing the suffix/prefix overlap of two weighted strings.

References

1. Nomenclature Committee of the International Union of Biochemistry: (NC-IUB). Nomenclature for incompletely specified bases in nucleic acid sequences. Recommendations (1984). Eur. J. Biochem. **150**(1), 1–5 (1985)
2. Amir, A., Chencinski, E., Iliopoulos, C.S., Kopelowitz, T., Zhang, H.: Property matching and weighted matching. In: Lewenstein, M., Valiente, G. (eds.) CPM 2006. LNCS, vol. 4009, pp. 188–199. Springer, Heidelberg (2006)
3. Amir, A., Gotthilf, Z., Shalom, B.R.: Weighted LCS. J. Discrete Algorithms **8**(3), 273–281 (2010)
4. Amir, A., Iliopoulos, C.S., Kapah, O., Porat, E.: Approximate matching in weighted sequences. In: Lewenstein, M., Valiente, G. (eds.) CPM 2006. LNCS, vol. 4009, pp. 365–376. Springer, Heidelberg (2006)
5. Barton, C., Iliopoulos, C.S., Pissis, S.P.: Optimal computation of all tandem repeats in a weighted sequence. Algorithms Mol. Biol. **9**(21), 1–8 (2014)
6. Barton, C., Iliopoulos, C.S., Pissis, S.P., Smyth, W.F.: Fast and simple computations using prefix tables under hamming and edit distance. In: Jan, K., Miller, M., Froncek, D. (eds.) IWOCA 2014. LNCS, vol. 8986, pp. 49–61. Springer, Heidelberg (2015)
7. Christodoulakis, M., Iliopoulos, C.S., Mouchard, L., Perdikuri, K., Tsakalidis, A.K., Tsichlas, K.: Computation of repetitions and regularities of biologically weighted sequences. J. Comput. Biol. **13**(6), 1214–1231 (2006)
8. Iliopoulos, C.S., Makris, C., Panagis, Y., Perdikuri, K., Theodoridis, E., Tsakalidis, A.: The weighted suffix tree: an efficient data structure for handling molecular weighted sequences and its applications. Fundam. Inf. **71**(2–3), 259–277 (2006)
9. Iliopoulos, C.S., Mouchard, L., Perdikuri, K., Tsakalidis, A.K.: Computing the repetitions in a biological weighted sequence. J. Automata Lang. Comb. **10**(5/6), 687–696 (2005)
10. Smyth, W.F., Wang, S.: New perspectives on the prefix array. In: Amir, A., Turpin, A., Moffat, A. (eds.) SPIRE 2008. LNCS, vol. 5280, pp. 133–143. Springer, Heidelberg (2008)
11. Yan, T., Yoo, D., Berardini, T.Z., Mueller, L.A., Weems, D.C., Weng, S., Cherry, J.M., Rhee, S.Y.: PatMatch: a program for finding patterns in peptide and nucleotide sequences. Nucleic Acids Res. **33**(suppl. 2), W262–W266 (2005)

New Formulas for Dyck Paths in a Rectangle

José Eduardo Blažek[(✉)]

Laboratoire de Combinatoire Et D'Informatique Mathématique,
Université du Québec à Montréal, Montréal, Canada
jeblazek@lacim.ca

Abstract. We consider the problem of counting the set of $\mathscr{D}_{a,b}$ of Dyck paths inscribed in a rectangle of size $a \times b$. They are a natural generalization of the classical Dyck words enumerated by the Catalan numbers. By using Ferrers diagrams associated to Dyck paths, we derive formulas for the enumeration of $\mathscr{D}_{a,b}$ with a and b non relatively prime, in terms of Catalan numbers.

Keywords: Dyck paths · Ferrers diagrams · Catalan numbers · Bizley numbers · Christoffel words

1 Introduction

The study of Dyck paths is a central topic in combinatorics as they provide one of the many interpretations of Catalan numbers. A partial overview can be found for instance in Stanley's comprehensive presentation of enumerative combinatorics [1] (see also [2]). As a language generated by an algebraic grammar is characterized in terms of a Dyck language, they are important in theoretical computer science as well [3]. On a two-letter alphabet they correspond to well parenthesized expressions and can be interpreted in terms of paths in a square. Among the many possible generalizations, it is natural to consider paths in a rectangle, see for instance Labelle and Yeh [4], and more recently Duchon [5] or Fukukawa [6]. In algebraic combinatorics Dyck paths are related to parking functions and the representation theory of the symmetric group [7]. The motivation for studying these objects stems from this field in an attempt to better understand the links between these combinatorial objects.

In this work, we obtain a new formula for $|\mathscr{D}_{a,b}|$, when a and b are not relatively prime, in terms of the Catalan numbers using the notion of Christoffel path. More precisely, the main results of this article (**diagrams decomposition method** in Sect. 3, Theorems 1 and 2 in Sect. 4) are formulas for the case where $a = 2k$:

$$
|\mathscr{D}_{a,b}| =
\begin{cases}
\mathbf{Cat}_{(a,n)}^{(-)} - \displaystyle\sum_{j=1}^{k-1} \mathbf{Cat}_{(a-j,n)}^{(-)} \mathbf{Cat}_{(j,n)}^{(-)}, & \text{if } b = a(n+1) - 2, \\[2ex]
\mathbf{Cat}_{(a,n)}^{(+)} + \displaystyle\sum_{j=1}^{k} \mathbf{Cat}_{(a-j,n)}^{(+)} \mathbf{Cat}_{(j,n)}^{(+)}, & \text{if } b = an + 2,
\end{cases}
$$

© Springer International Publishing Switzerland 2015
F. Manea and D. Nowotka (Eds.): WORDS 2015, LNCS 9304, pp. 85–96, 2015.
DOI: 10.1007/978-3-319-23660-5_8

where $k, n \in \mathbb{N}$, $\mathbf{Cat}_{(a,n)}^{(-)} := \mathbf{Cat}_{(a,a(n+1)-1)}$, and $\mathbf{Cat}_{(a,n)}^{(+)} := \mathbf{Cat}_{(a,an+1)}$.

The paper is organized as follows. In Sect. 2 we fix the notation for Dyck and Christoffel paths, and present their encoding by Ferrers diagrams. Then, in Sect. 3, we develop the "Ferrers diagram comparison method" and "diagrams decomposition method". Section 4 contains several technical results in order to prove the main results, and in Sect. 5 we present the examples.

2   Definitions and Notation

We borrow the notation from Lothaire [8]. An *alphabet* is a finite set Σ, whose elements are called *letters*. The set of finite words over Σ is denoted Σ^* and $\Sigma^+ = \Sigma^* \setminus \{\varepsilon\}$ is the set of nonempty words where $\varepsilon \in \Sigma^*$ is the empty word. The number of occurrences of a given letter α in the word w is denoted $|w|_\alpha$ and $|w| = \sum_{\alpha \in \Sigma} |w|_\alpha$ is the length of the word. A *language* is a subset $L \subseteq \Sigma^*$. The *language* of a word w is $\mathcal{L}(w) = \{f \in \Sigma^* \mid w = pfs, \ p, s \in \Sigma^*\}$, and its elements are called the *factors* of w.

Dyck words and paths. It is well-known that the language of Dyck words on $\Sigma = \{\mathbf{0}, \mathbf{1}\}$ is the language generated by the algebraic grammar $D \to \mathbf{0}D\mathbf{1}D + \varepsilon$. They are enumerated by the Catalan numbers (see [9]),

$$\mathbf{Cat}_n = \frac{1}{n+1}\binom{2n}{n},$$

and can be interpreted as lattice paths inscribed in a square of size $n \times n$ using down and right unit steps (see Fig. 1 (a)).

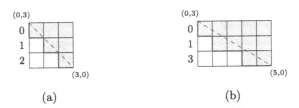

Fig. 1. Dyck path and Ferrers diagram.

More precisely an (a, b)-Dyck path is a south-east lattice path, going from $(0, a)$ to $(b, 0)$, which stays below the (a, b)-diagonal, that is the line segment joining $(0, a)$ to $(b, 0)$. In Fig. 1, the paths are respectively $\mathbf{010101}$ and $\mathbf{01011011}$.

Alternatively such word may be encoded as a Ferrers diagram corresponding to the set of boxes to left (under) the path. As usual, Ferrers diagrams are identified by the number of boxes on each line, thus corresponding to partitions:

$$\lambda = (\lambda_{a-1}, \lambda_{a-2}, \ldots, \lambda_1), \qquad \text{with } \lambda_{a-l} \leq \left\lfloor \frac{bl}{a} \right\rfloor \text{ where } 1 \leq l \leq a - 1. \qquad (1)$$

In the examples of Fig. 1, the paths are respectively encoded by the sequences $(2, 1, 0)$ and $(3, 1, 0)$. The cases where (a, b) are relatively prime, or $b = ak$ are of particular interest. For the case $b = ak$ with $k \geq 1$ we have the well-know formula of Fuss-Catalan (see [9]).

$$\mathbf{Cat}_{(a,k)} = \frac{1}{ak + 1} \binom{ak + a}{a}.$$

For $a \times b$ rectangles, with a and b are relatively prime, we also have the "classical" formula:

$$\mathbf{Cat}_{(a,b)} = \frac{1}{a + b + 1} \binom{a + b + 1}{a}.$$

In particular, when $a = p$ is prime, either b and p are relatively prime, or b is a multiple of p. Hence the relevant number of Dyck paths is:

$$|\mathscr{D}_{p,b}| = \begin{cases} \frac{1}{p+b} \binom{p+b}{p} & \text{if } \gcd(p, b) = 1, \\ \frac{1}{p+b+1} \binom{p+b+1}{p} & \text{if } b = kp. \end{cases}$$

The generalized ballot problem is related with the number of lattice paths form $(0, 0)$ to (a, b) that never go below the line $y = kx$ (see [10]):

$$\frac{b - ka + 1}{b} \binom{a + b}{a} \qquad \text{where } k \geq 1, \text{ and } b > ak \geq 0.$$

And the number of lattice paths of length $2(k + 1)n + 1$ that start at $(0, 0)$ and that avoid touching or crossing the line $y = kx$ (see [11]) has the formula:

$$\binom{2(k + 1)n}{2n} - (k - 1) \sum_{2n-1}^{i=0} \binom{2(k + 1)n}{i}, \qquad \text{where } n \geq 1 \text{ and } k \geq 0.$$

In the more general case we have a formula due to Bizley (see [12]) expressed as follows. Let $m = da$, $n = db$ and $d = \gcd(m, n)$, then:

$$B_k^{(a,b)} := \frac{1}{a + b} \binom{ka + kb}{ka} \qquad \text{for } k \in \mathbb{N},$$

$$B_\lambda^{a,b} := B_{\lambda_1}^{(a,b)} B_{\lambda_2}^{(a,b)} \cdots B_{\lambda_l}^{(a,b)} \qquad \text{if } \lambda = (\lambda_1, \lambda_2, \ldots, \lambda_l),$$

It is straightforward to show that the number of Dyck paths in $m \times n$ is:

$$|\mathscr{D}_{m,n}| := \sum_{\lambda \vdash d} \frac{1}{z_\lambda} B_\lambda^{(a,b)} \qquad \text{where } n \geq 1 \text{ and } k \geq 0.$$

Christoffel Paths and Words. A Christoffel path between two distinct points $P = (0, k)$ and $P' = (0, l)$ on a rectangular grid $a \times b$ is the closest lattice path that stays strictly below the segment PP' (see [13]). For instance, the Dyck path of Fig. 1(b) is also Christoffel, and the associated word is called a Christoffel word. The Christoffel path of a rectangular grid $a \times b$ is the Christoffel path associated to the line segment going from the north-west corner to the south-east corner of the rectangle of size $a \times b$. As in the case of Dyck paths, every Christoffel path in a fixed rectangular grid $a \times b$ is identified by a Ferrers diagram of shape $(\lambda_{a-1}, \lambda_{a-2}, \ldots, \lambda_1)$ given by Equation (1).

For later use, we define two functions associated to Ferrers diagram. Let $Q_{a,b}$ to be the total number of boxes in the Ferrers diagram associated to the Christoffel path of $a \times b$ (see [14]):

$$Q_{a,b} = \frac{(a-1)(b-1) + \gcd(a,b) - 1}{2}. \tag{2}$$

Also, let $\Delta_{a,b}(l)$ be the difference between the boxes of the Ferrers diagrams associated to the Christoffel paths of $a \times b$ and $a \times (b-1)$, respectively:

$$\Delta_{a,b}(l) := \left\lfloor \frac{bl}{a} \right\rfloor - \left\lfloor \frac{(b-1)l}{a} \right\rfloor,$$

where $a < b \in \mathbb{N}$ and $1 \leq l \leq a - 1$.

In the next section we give an alternate method to calculate the number of (a,b)-Dyck paths when a and b are not relatively prime, and satisfying certain conditions in terms of the Catalan numbers.

Isosceles Diagrams. An isosceles diagram $\mathscr{I}_n$ is a Ferrers diagram associated to a Christoffel path in a square having side length n. Given a Ferrers diagram $\mathscr{T}_{a,b}$, we call *maximum isosceles diagram* the largest isosceles diagram included in $\mathscr{T}_{a,b}$.

Ferrers Set. Let $\mathscr{T}_{a,b}$ be a Ferrers diagram. The Ferrers set of $\mathscr{T}_{a,b}$ is the set of all Dyck paths contained in $\mathscr{T}_{a,b}$.

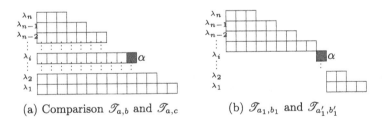

(a) Comparison $\mathscr{T}_{a,b}$ and $\mathscr{T}_{a,c}$ (b) $\mathscr{T}_{a_1,b_1}$ and $\mathscr{T}_{a_1',b_1'}$

Fig. 2. Rule 1.

3   Ferrers Diagrams Comparison Method

Let $\mathscr{T}_{a,b}$ be the Ferrers diagram associated to a Christoffel path of $a \times b$. In order to establish the main results we need to count the boxes in excess between the Christoffel paths in rectangles $a \times b$ and $a \times c$, for any $c > b$. We develop a method to do this by removing exceeding boxes between $\mathscr{T}_{a,b}$ and $\mathscr{T}_{a,c}$, for $c > b$. Using the functions $Q_{a,b}$ and $\Delta_{a,b}(l)$, our comparison method gives the following rules:

Rule 1: If $Q_{a,b} = 1$ and $\Delta_{a,b}(i) = 1$, there is only one corner in $\mathscr{T}_{a,c}$ which does not belong to the $\mathscr{T}_{a,b}$. Let $\mathscr{T}_{a_1,b_1}$ and $\mathscr{T}_{a_1',b_1'}$ be the Ferrers diagram obtained by erasing from $\mathscr{T}_{a,c}$ the row and the column that contain α (see Fig. 2(b)). These Ferrers diagrams are not associated to a Christoffel path in general. Let

$$\mathcal{J}_{a_1,b_1} \subseteq \mathscr{D}_{a_1,b_1}, \text{ and } \mathcal{J}_{a_1',b_1'} \subseteq \mathscr{D}_{a_1',b_1'}$$

be the sets of Dyck paths contained in the Ferrers diagrams $\mathscr{T}_{a_1,b_1}$ and $\mathscr{T}_{a_1',b_1'}$, respectively. We have:

$$|\mathscr{D}_{a,c}| - |\mathscr{D}_{a,b}| = -|\mathcal{J}_{a_1,b_1}| \cdot |\mathcal{J}_{a_1',b_1'}|,$$

It is clear that if the box α is located on the bottom line $(l = a - 1)$, the equation is reduced to:

$$|\mathscr{D}_{a,c}| - |\mathscr{D}_{a,b}| = -|\mathcal{J}_{a_1,b_1}|.$$

Rule 2: When $Q_{a,b} = k$ and there are exactly k rows with a difference of one box we need to calculate how many paths contain these boxes (see Fig. 3), so we construct a sequence of disjoint sets as follows. Let A_j be the set of all paths that do not contain the boxes α_i for each $i > j$, where $1 \leq j \leq k$. Also, let B_j be the set of all paths that do not contain the boxes α_i for each $i < j$, where $1 \leq j \leq k$. This strategy gives us disjoint sets that preserve the total union, so using **Rule 1** for every A_j or B_j we get:

$$|\mathscr{D}_{a,c}| - |\mathscr{D}_{a,b}| = -\sum_{j=1}^{k} (|\mathcal{J}_{a_j,b_j}| \cdot |\mathcal{J}_{a_j',b_j'}|),$$

where $\mathcal{J}_{a_j,b_j} \subseteq \mathscr{D}_{a_j,b_j}$, and $\mathcal{J}_{a_j',b_j'} \subseteq \mathscr{D}_{a_j',b_j'}$.

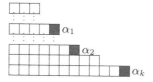

Fig. 3. More one box.

3.1   Diagrams Decomposition Method

Using the diagrams comparison method we make an iterative process erasing boxes in excess between the diagram $\mathscr{T}_{a,b}$ and its respective maximum isosceles diagram $\mathscr{I}_n$. It begins at the right upper box as shown in Fig. 5. The decomposition is give in sums and products of diagrams. The sum operation $+$ is given by the union of disjoint Ferrers sets. We can consider a red box in the border of

(a) $u10v$ is a path containing the red box. (b) $u01v$ is a path not containing the red box.

Fig. 4. Separation of diagrams.

the diagram. Any path contained in a diagrams having the red box is written as one of the two cases in Fig. 4.

The products of diagrams $\mathscr{T} \times \mathscr{T}'$ is a diagram containing all possible concatenation of a Dyck path of $\mathscr{T}$ with a Dyck path of $\mathscr{T}'$. For example, the diagram corresponding to $\mathscr{D}_{4,6}$ is $[4,3,1]$ ⊞ includes the isosceles diagram $[3,2,1]$ ⊞. When we remove the box the diagram splits into two pairs associated with operations that simplify the computation of paths (see Fig. 5).

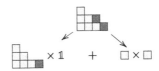

Fig. 5. First diagram decomposition.

In Fig. 5, $\mathbb{1}$ is an empty diagram. We repeat this method until all the diagrams are isosceles (the operation $\times$ distributes the operation $+$) (Fig 6).

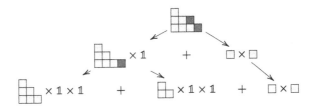

Fig. 6. Full diagram decomposition.

Clearly, we can count the Dyck paths in an isosceles diagram with a classical Catalan formula because there is a relation between the decomposition and the number of Dyck paths. This relation, denoted $\mathcal{H}$, between the isosceles diagrams and Catalan numbers is such as:

$$\mathcal{H}(\mathbb{1}) := 1, \quad \mathcal{H}(\mathscr{I}_n + \mathscr{I}_m) := \mathcal{H}(\mathscr{I}_n) + \mathcal{H}(\mathscr{I}_m),$$
$$\mathcal{H}(\mathscr{I}_n) := \mathbf{Cat}_n, \quad \mathcal{H}(\mathscr{I}_n \times \mathscr{I}_m) := \mathcal{H}(\mathscr{I}_n) \cdot \mathcal{H}(\mathscr{I}_m).$$

$$\mathcal{H}\left(\boxminus\right) = \mathcal{H}\left(\boxminus \times 1 \times 1 + \boxminus \times 1 \times 1 + \square \times \square\right)$$

$$= \mathcal{H}\left(\boxminus \times 1 \times 1\right) + \mathcal{H}\left(\boxminus \times 1 \times 1\right) + \mathcal{H}\left(\square \times \square\right)$$

$$= \mathcal{H}\left(\boxminus\right) + \mathcal{H}\left(\boxminus\right) + \mathcal{H}\left(\square\right) \times \mathcal{H}\left(\square\right)$$

$$= \mathbf{Cat_4} + \mathbf{Cat_3} + \mathbf{Cat_2} \times \mathbf{Cat_2} = 23$$

3.2 Technical Results

The following technical formulas are needed in the sequel (see [14]).

i) Let $a = 2k$, $b = 2k(n+1) - 1$, and $1 \leq l \leq 2k - 1$. Then,

$$\Delta_{2k,2k(n+1)-1}(l) = \begin{cases} 0 & \text{if } 1 \leq l \leq k, \\ 1 & \text{if } k+1 \leq l \leq 2k - 1. \end{cases} \tag{3}$$

ii) Let $a = 2k$, $b = 2kn + 2$, and $1 \leq l \leq 2k - 1$. Then,

$$\Delta_{2k,2kn+2}(l) = \begin{cases} 0 & \text{if } 1 \leq l \leq k - 1, \\ 1 & \text{if } k \leq l \leq 2k - 1. \end{cases} \tag{4}$$

iii) Let $a, k \in \mathbb{N}$, and $k < a$. There exists a unique $r \in \mathbb{N}$ such as for $k = 1, \dots, a - 1$:

$$\left\lfloor \frac{kr}{a} \right\rfloor = k - 1 \text{ or } \left\lfloor \frac{kr}{a} \right\rfloor = k, \tag{5}$$

and $\mathbf{gcd}(r, a) = 1$. The solution is given by $r = a - 1$

4 Theorems

Now we are ready to prove the two main results of this article. The formulas are obtained by studying the "Ferrers diagram comparison method" (see Sect. 3), in the cases where every Ferrers diagram obtained by subdivisions of $\mathcal{T}_{a,c}$ is associated to a Christoffel path inscribed in a rectangular box of co-prime dimension.

Theorem 1 (see [14]). Let $a = 2k$, $b = a(n+1) - 2$, and $k, n \in \mathbb{N}$, then the number of Dyck paths is:

$$|\mathcal{D}_{a,b}| = \mathbf{Cat}_{(a,n)}^{(-)} - \sum_{j=1}^{k-1} \mathbf{Cat}_{(a-j,n)}^{(-)} \mathbf{Cat}_{j,n}^{(-)},$$

where $\mathbf{Cat}_{a,n}^{(-)} := \mathbf{Cat}_{(a,a(n+1)-1)}$.

Proof. From Eqs. 2 and 3, we get that there are $k-1$ total difference between the Ferrers diagram associated to the Christoffel path of $a \times b$ and $a \times (b+1)$. We easily obtain that the Ferrers diagram associated to $\mathscr{D}_{c,cn+c-1}$ is $\lambda = ((c-1)n + c - 2, \ldots, 2n+1, n)$.

By definition of A_j, the rectangles $j \times b_1$ and $(2k-j) \times b_1'$ are such that their maximal underlying diagrams are:

$$\lambda = ((j-1)n + j - 2, \ldots, 3n + 2, 2n + 1, n),$$
$$\lambda' = ((2k - j - 1)n + 2k - j - 2, \ldots, 2n + 1, n),$$

respectively. So,

$$|A_j| = |\mathscr{D}_{2k-j,(2k-j)n+2k-j-1}||\mathscr{D}_{j,jn+j-1}|.$$

Since all rectangles are relatively prime, we have:

$$|\mathscr{D}_{a,b}| - |\mathscr{D}_{a,b+1}| = -\sum_{j=1}^{k-1}(|\mathscr{D}_{a-j,(a-j)(n+1)-1}| \cdot |\mathscr{D}_{j,j(n+1)-1}|).$$

then

$$|\mathscr{D}_{a,b}| = \mathbf{Cat}_{(a,n)}^{(-)} - \sum_{j=1}^{k-1} \mathbf{Cat}_{(a-j,n)}^{(-)} \mathbf{Cat}_{j,n}^{(-)}.$$

where $\mathbf{Cat}_{t,n}^{(-)} := \mathbf{Cat}_{(t,t(n+1)-1)}$, of course $\mathbf{Cat}_{(1,n)}^{(-)} = 1$. □

Theorem 2 (see [14]). *Let $a = 2k$, $b = an + 2$ and $k, n \in \mathbb{N}$, then the number of Dyck paths is:*

$$|\mathscr{D}_{a,b}| = \mathbf{Cat}_{(a,n)}^{(+)} + \sum_{j=1}^{k} \mathbf{Cat}_{(a-j,n)}^{(+)} \mathbf{Cat}_{j,n}^{(+)}.$$

where $\mathbf{Cat}_{a,n}^{(+)} := \mathbf{Cat}_{a,an+1}$.

Proof. From Eqs. 2 and 4, we get that there are k total difference between the Ferrers diagram associated to the Christoffel path of $a \times b$ and $a \times (b+1)$. We easily get that the Ferrers diagram associated to $\mathscr{D}_{c,cn+1}$ is $\lambda = ((c-1)n, \ldots, 2n, n)$. By definition of B_j, the rectangles $j \times b_1$ et $(2k-j) \times b_1'$ are such as that their maximal underlying diagram are:

$$\lambda = ((j-1)n, \ldots, 3n, 2n, n),$$
$$\lambda' = ((2k - j - 1)n, \ldots, 2n, n),$$

respectively. So,

$$|B_j| = |\mathscr{D}_{2k-j,(2k-j)n+1}||\mathscr{D}_{j,jn+1}|.$$

Since all rectangles are relatively prime, we have:

$$|\mathscr{D}_{a,b}| - |\mathscr{D}_{a,b-1}| = -\sum_{j=1}^{k-1}(|\mathscr{D}_{a-j,(a-j)n-1}| \cdot |\mathscr{D}_{j,jn+1}|).$$

then

$$|\mathscr{D}_{a,b}| = \mathbf{Cat}^{(+)}_{(a,n)} + \sum_{j=1}^{k} \mathbf{Cat}^{(+)}_{(a-j,n)} \mathbf{Cat}^{(+)}_{j,n}.$$

where $\mathbf{Cat}^{(+)}_{t,n} := \mathbf{Cat}_{(t,tn+1)}$, of course $\mathbf{Cat}^{(+)}_{(1,n)} = 1$.   □

5   Examples

In order to illustrate the main results, we consider the cases $\mathscr{D}_{8,8n+6}$, to generalize the case of discrepancies with a larger diagram. Then we study $\mathscr{D}_{6,6n+2}$ to generalize the case of discrepancies for shorter diagram. These cases corresponding to rectangles having relatively prime dimensions such that $\mathcal{J}_{a_j,b_j} = \mathscr{D}_{a_j,b_j}$ and $\mathcal{J}_{a'_j,b'_j} = \mathscr{D}_{a'_j,b'_j}$. Finally, we give some examples of the diagrams decomposition method.

5.1   Example $\mathscr{D}_{8,8n+6}$

We apply the comparison method to $\mathscr{D}_{8,8n+6}$ and $\mathscr{D}_{8,8n+7}$. Using Eq. 2, we get that the difference in total number of sub-diagonal boxes is $Q_{8,8n+7} - Q_{8,8n+6} = 3$. To find the lines where they are located we use the Eq. 3. In this cases $\Delta(l)$ is zero except for $l = 5, 6, 7$ (see Fig. 7).

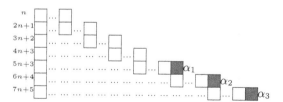

Fig. 7. $\mathscr{D}_{8,8n+6}$ and $\mathscr{D}_{8,8n+7}$

Applying Rule 2, we have that $A_1 = \{$ paths containing the box α_1 and not $(\alpha_2$ or $\alpha_3)\}$, $A_2 = \{$ paths containing the box α_2 and not $\alpha_3\}$, and $A_3 = \{$paths containing the box $\alpha_3\}$.

Applying the rule 1 to these sets, we have:

Case 1: for A_1, we must find the rectangles $5 \times b_1$ and $3 \times b'_1$ with underlying diagram $\lambda = (5n + 4, 4n + 3, 3n + 2, 2n + 1, n)$ and $\lambda' = (n)$, then

$$|A_1| = |\mathscr{D}_{5,5n+4}||\mathscr{D}_{3,3n+2}|.$$

Case 2: for A_2, we must find the rectangles $6 \times b_2$ and $2 \times b'_2$ with underlying diagram $\lambda = (5n + 4, 4n + 3, 3n + 2, 2n + 1, n)$ and $\lambda' = (n)$, then

$$|A_2| = |\mathscr{D}_{6,6n+5}||\mathscr{D}_{2,2n+1}|.$$

Case 3: for A_3, we must find the rectangle $7 \times b_3$ with underlying diagram $\lambda = (6n + 5, 5n + 4, 4n + 3, 3n + 2, 2n + 1, n)$, then,

$$|A_3| = |\mathscr{D}_{7,7n+6}|.$$

Finally, we obtain:

$$|\mathscr{D}_{8,8n+6}| = \mathbf{Cat}_{(8,8n+7)} - \mathbf{Cat}_{(7,7n+6)}$$
$$- \mathbf{Cat}_{(6,6n+5)}\mathbf{Cat}_{(2,2n+1)} - \mathbf{Cat}_{(5,5n+4)}\mathbf{Cat}_{(3,3n+2)}.$$

5.2   Example $\mathscr{D}_{6,6n+2}$

Similarly to the previous example, comparing $\mathscr{D}_{6,6n+2}$ and $\mathscr{D}_{6,6n+1}$ from Eq. 2, we get that the total difference is $Q_{6,6n+2} - Q_{6,6n+1} = 3$. From Eq. 4, in this cases $\Delta(l)$ is zero except for $l = 3, 4, 5$ (see Fig. 8).

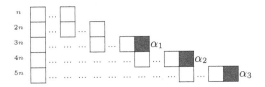

Fig. 8. $\mathscr{D}_{6,6n+1}$ et $\mathscr{D}_{6,6n+2}$

We consider the sets:

$$B_1 = \{\text{paths containing the box } \alpha_1\},$$
$$B_2 = \{\text{paths containing the box } \alpha_2 \text{ and not } \alpha_1\},$$
$$B_3 = \{\text{paths containing the box } \alpha_3 \text{ and not } (\alpha_1 \text{ or } \alpha_2)\}.$$

We have the following cases :

Case 1: For B_1, we find $3 \times b_1$ and $3 \times b_1'$ with underlying diagram $\lambda = (2n, n)$ and $\lambda' = (2n, n)$, respectively (see Eq. 5). Then,

$$|B_1| = |\mathscr{D}_{3,3n+1}||\mathscr{D}_{3,3n+1}|.$$

Case 2: For B_2, the rectangles $4 \times b_2$ and $2 \times b_2'$ such as $\lambda = (3n, 2n, n)$ et $\lambda' = (n)$ (see Eq. 5). Then,

$$|B_2| = |\mathscr{D}_{4,4n+1}||\mathscr{D}_{2,2n+1}|.$$

Case 3: for B_3, the rectangle $5 \times b_3$ such as $\lambda = (4n, 3n, 2n, n)$ (see Eq. 5). Then,

$$|B_3| = |\mathscr{D}_{5,5n+1}|.$$

Finally, we obtain:

$$|\mathscr{D}_{6,6n+2}| = \mathbf{Cat}_{(6,6n+1)} + \mathbf{Cat}_{(5,5n+1)}$$
$$+ \mathbf{Cat}_{(4,4n+1)}\mathbf{Cat}_{(2,2n+1)} + \mathbf{Cat}_{(3,3n+1)}\mathbf{Cat}_{(3,3n+1)}.$$

5.3   Example $\mathcal{D}_{6,9}$.

For $\mathcal{D}_{6,9}$ the Ferrers diagram is (Fig. 9):

Fig. 9. Ferrers diagram $\mathcal{D}_{6,9}$ and $\mathcal{D}_6$

and the diagrams decomposition method. After six iterations the decomposition is:

$$\text{[diagram decomposition]}$$

then,

$$\mathcal{H}\left(\text{[diagram]}\right) = \mathbf{Cat}_2^3 + 3\mathbf{Cat}_2\mathbf{Cat}_3 + \mathbf{Cat}_3^2 + 3\mathbf{Cat}_2\mathbf{Cat}_4 + \mathbf{Cat}_4 + 2\mathbf{Cat}_5 + \mathbf{Cat}_6$$
$$= 377.$$

We can also decompose (see Fig. 10), and after four iterations the decomposition is:

Fig. 10. Ferrers diagram $\mathcal{D}_{6,9}$ and $\mathcal{D}_{6,8}$

$$\text{[diagram decomposition]}$$

then,

$$\mathcal{H}\left(\text{[diagram]}\right) = \mathbf{Cat}_2|\mathcal{D}_{4,6}| + \mathbf{Cat}_2\mathbf{Cat}_3 + \mathbf{Cat}_2\mathbf{Cat}_4 + \mathbf{Cat}_{5,7} + |\mathcal{D}_{6,8}|$$
$$= 2 \cdot 23 + 2 \cdot 5 + 2 \cdot 14 + 66 + 227 = 377.$$

As they are many possible decomposition, finding the shortest one is an open problem.

Acknowledgements. The author would like to thank his advisor François Bergeron and Srečko Brlek for their advice and support during the preparation of this paper. The results presented here are part of J.E. Blažek's Master thesis. Algorithms in SAGE are available at http://thales.math.uqam.ca/~jeblazek/Sage_Combinatory.html.

References

1. Stanley, R.P.: Enumerative Combinatorics, 2nd edn. Cambridge University Press, New York (2011)
2. Bergeron, F., Labelle, G., Leroux, P., Readdy, M.: Combinatorial Species and Tree-like Structures, Encyclopedia of Mathematics and its Applications. Cambridge University Press, Cambridge (1998)
3. Eilenberg, S.: Automata, Languages, and Machines. Academic Press Inc, Orlando (1976)
4. Labelle, J., Yeh, Y.: Generalized dyck paths. Discrete Math. **82**(1), 1–6 (1990)
5. Duchon, P.: On the enumeration and generation of generalized dyck words. Discrete Math. **225**(1–3), 121–135 (2000)
6. Fukukawa, Y.: Counting generalized Dyck paths, April 2013. arXiv:1304.5595v1 [math.CO]
7. Gorky, E., Mazin, M., Vazirani, M.: Affine permutations and rational slope parking functions, March 2014. arXiv:1403.0303v1 [math.CO]
8. Lothaire, M.: Applied Combinatorics on Words. Cambridge University Press, Cambridge (2005)
9. Koshy, T.: Catalan Numbers with Applications. Oxford University Press, New York (2009)
10. Goulden, I., Serrano, L.: Maintaining the spirit of the reflection principle when the boundary has arbitrary integer slope. J. Comb. Theory Ser. A **104**, 317–326 (2003)
11. Chapman, R.J., Chow, T., Khetan, A., Moulton, D.P., Waters, R.J.: Simple formulas for lattice paths avoiding certain periodic staircase boundaries. J. Comb. Theory Ser. A **116**, 205–214 (2009)
12. Bizley, M.T.L.: Derivation of a new formula for the number of minimal lattice paths from $(0,0)$ to $(km, kn) \cdots$. JIA **80**, 55–62 (1954)
13. Melançon, G., Reutenauer, C.: On a class of Lyndon words extending Christoffel words and related to a multidimensional continued fraction algorithm. J. Integer Sequences **16**, 30 (2013)
14. Blazek, J.E.: Conbinatoire de ℕ-modules Catalan, Master thesis. Département de Mathématique, UQAM (2015)

Ambiguity of Morphisms in a Free Group

Joel D. Day$^{(\boxtimes)}$ and Daniel Reidenbach

Department of Computer Science, Loughborough University, Loughborough,
Leicestershire LE11 3TU, UK
{J.Day,D.Reidenbach}@lboro.ac.uk

Abstract. A morphism g is ambiguous with respect to a word u if there
exists a morphism $h \neq g$ such that $g(u) = h(u)$. The ambiguity of morphisms has so far been studied in a free monoid. In the present paper,
we consider the ambiguity of morphisms of the free group. Firstly, we
note that a direct generalisation results in a trivial problem. We provide a
natural reformulation of the problem along with a characterisation of elements of the free group which have an associated unambiguous injective
morphism. This characterisation matches an existing result known for
the free monoid. Secondly, we note a second formulation of the problem
which leads to a non-trivial situation: when terminal symbols are permit-
ted. In this context, we investigate the ambiguity of the morphism erasing
all non-terminal symbols. We provide, for any alphabet, a pattern which
can only be mapped to the empty word exactly by this morphism. We
then generalize this construction to give, for any morphism g, a pattern
α such that $h(\alpha)$ is the empty word if and only if $h = g$.

1 Introduction

A morphism g is *ambiguous* with respect to a word u if there exists a morphism
$h \neq g$ such that $g(u) = h(u)$. For example the morphism $g : \{a, b\}^* \to \{a, b\}^*$
such that $g(a) = a \cdot b \cdot a$ and $g(b) = b$ is ambiguous with respect to $a \cdot b \cdot b \cdot a$,
as the same result may be achieved with the morphism h given by $h(a) = a$ and
$h(b) = b \cdot a \cdot b$ (we have $g(a \cdot b \cdot b \cdot a) = h(a \cdot b \cdot b \cdot a) = a \cdot b \cdot a \cdot b \cdot b \cdot a \cdot b \cdot a$). On the
other hand, the (identity) morphism $g : \{a, b\}^* \to \{a, b\}^*$ given by $g(a) := a$ and
$g(b) = b$ is unambiguous with respect to $u := a \cdot b \cdot b \cdot a$, since no other morphism
$g : \{a, b\}^* \to \{a, b\}^*$ produces the same image when applied to u. However, if
instead of the monoid $\{a, b\}^*$, we consider the free group $\mathcal{F}_{\{a,b\}}$ generated by a
and b, then the identity morphism $g : \mathcal{F}_{\{a,b\}} \to \mathcal{F}_{\{a,b\}}$ becomes ambiguous with
respect to $u = a \cdot b \cdot b \cdot a$, as verified by, e.g., the morphism $h : \mathcal{F}_{\{a,b\}} \to \mathcal{F}_{\{a,b\}}$
given by $h(a) = a \cdot b \cdot b \cdot a \cdot b^{-2} \cdot a^{-1}$ and $h(b) = a \cdot b \cdot b \cdot a \cdot b \cdot a^{-1} \cdot b^{-2} \cdot a^{-1}$. Thus,
we see that by moving from the free monoid to the free group, the ambiguity of
morphisms can change.

The ambiguity of morphisms can be seen as both a property of the pair of
words $(u, g(u))$, and of the morphism itself. Put another way, it provides some
measure of (non-)determinism in the process of mapping u to $g(u)$. Similarly
to injectivity, the unambiguity of a morphism determines to some extent the

© Springer International Publishing Switzerland 2015
F. Manea and D. Nowotka (Eds.): WORDS 2015, LNCS 9304, pp. 97–108, 2015.
DOI: 10.1007/978-3-319-23660-5_9

information lost when the morphism is applied, and indeed unambiguity can be seen as a dual to injectivity. To substantiate this claim, consider two words u, v and a morphism g such that $g(u) = v$. It is possible to determine u from v and g if and only if g is injective. Similarly, if is possible to determine g from u and v if and only if g is unambiguous with respect to u. It is obvious, of course, that the final configuration, that v may be determined by u and g is the condition that g is a function.

All papers addressing the topic directly consider only morphisms of the free monoid, and are surprisingly recent (see e.g., Freydenberger et al. [1], Freydenberger, Reidenbach [2]), although earlier topics such as the Dual Post Correspondence Problem (Culik II, Karhumäki [3]) have addressed the question indirectly. Due to the nature of the question, it is clear that ambiguity of morphisms has implications on many areas such as equality sets (Salomaa [4], Engelfriet, Rozenberg [5]) or equalizers, the Post Correspondence Problem (Post [6]), and word equations (see, e.g., Makanin [7]); however, arguably the biggest achievements of research on the ambiguity of morphisms have come in the world of pattern languages (the set of all morphic images of a given word) (Reidenbach [8]).

In the current work, we extend the study of ambiguity to morphisms of the free group. While ambiguity and indeed pattern languages have generally been studied in the context of a free monoid, or semi-group, many of these ideas overlap with areas of study related to free groups, in which morphisms play a central role. Recent publications have addressed pattern languages in a group setting (Jain et al. [9]), and a group-equivalent of the Post Correspondence Problem (Bell, Potapov [10]), while more established areas include test words for automorphisms (Turner [11]), automorphisms themselves and their fixed subgroups (see e.g., Ventura [12]), equations of the free group (Makanin [13]), and equalizers – all of which are connected to the ambiguity of morphisms.

We begin in Sect. 3 with the observation is that, in its simplest form, a generalisation of the ambiguity of morphisms leads to a trivial problem: in the free group, all morphisms are ambiguous with respect to all words. However on closer inspection, it is possible to see that this evaluation does not reflect the richness and intricacies of the subject. We reformulate the problem in a natural way which results in a non-trivial notion of unambiguity in a free group. In Sect. 4 we provide a characterisation of when a word from the free group posesses an unambiguous morphism in terms of fixed points of non-trivial morphisms, providing an analogue to an existing result for the free monoid found in Freydenberger et al. [1].

Finally, in Sect. 5 we consider another setting in which the problem is non-trivial: when so-called terminal symbols are present. Many of the remaining open problems for ambiguity of morphisms in a free monoid are with regards to words which contain terminal symbols – symbols which are always preserved by morphisms. We provide some initial insights in the group setting, including a construction that, for any morphism g, provides a word which is uniquely erased by g.

2 Preliminaries

$\mathbb{N}$ denotes the natural numbers, and $\mathbb{N}_0 := \mathbb{N} \cup \{0\}$. $\mathbb{Z}$ denotes the integers. A set of symbols $\Sigma := \{a_1, a_2, \ldots\}$ is called an *alphabet*. The free semigroup generated by Σ is denoted Σ^+ and Σ^* is the free monoid. The identity element is denoted ε. For a given alphabet $\Sigma = \{a_1, a_2, \ldots\}$, its inverse Σ^{-1} is an alphabet $\{a_1^{-1}, a_2^{-1}, \ldots\}$ such that $\Sigma \cap \Sigma^{-1} = \emptyset$. The *free group* is the free monoid over $\Sigma \cup \Sigma^{-1}$ with additional relations $a_i a_i^{-1} = a_i^{-1} a_i = \varepsilon$ for each $a_i \in \Sigma$. For a free group $\mathcal{F}$ and a given word $u = u_1 u_2 \cdots u_n \in \mathcal{F}$, its inverse u^{-1} is the word $u_n^{-1} u_{n-1}^{-1} \cdots u_1^{-1}$. For an alphabet Σ, the free group generated by Σ is denoted $\mathcal{F}_\Sigma$. Generally, we will refer to generators as *letters* or *variables* (see pattern languages below).

For semigroups, monoids or groups $\mathcal{A}$ and $\mathcal{B}$, a *(homo)morphism* $\sigma : \mathcal{A} \to \mathcal{B}$ is a mapping such that, for every $u, v \in \mathcal{A}$, $\sigma(uv) = \sigma(u)\sigma(v)$. Note that for groups, this implies that $\sigma(u^{-1}) = \sigma(u)^{-1}$. By the *definition* of a morphism, we mean the set of images of individual generators (letters) of $\mathcal{A}$. A morphism $\sigma : \mathcal{A} \to \mathcal{B}$ is an *automorphism* if it is surjective, injective and $\mathcal{A} = \mathcal{B}$. A morphism $\sigma : \mathcal{A} \to \mathcal{B}$ is *periodic* if there exists some $u \in \mathcal{B}$ such that for every $a \in \mathcal{A}$, $\sigma(a) = u^n$ for some $n \in \mathbb{Z}$. A morphism $\sigma : \mathcal{A} \to \mathcal{B}$ is *ambiguous with respect to* $u \in \mathcal{A}$ if there exists a morphism $\tau : \mathcal{A} \to \mathcal{B}$ such that $\sigma(u) = \tau(u)$, and such that $\sigma(a) \neq \tau(a)$ for some letter a occurring in u. If, for some $u \in \mathcal{A}$, the identity morphism is ambiguous with respect to u, then u is said to be a *non-trivial fixed point*.

Given words w and u, u is a *factor* of w if there exist x, v such that $w = uxv$. A word w is *bordered* if there exist u, v such that $w = uvu$. A *contraction* is a factor or word with length greater than 0 which is equal to the empty word. A contraction is *primary* if it contains exactly one factor $x \cdot x^{-1}$ where x is a single letter. Words with no contractions are called *reduced*. It is assumed that all images of generators under morphisms are reduced.

Two words which are not necessarily reduced do not have to be graphically identical to be equal (within the structure of the group). Two reduced words must be graphically identical to be equal. If an equality of two words is considered, it should be assumed to be the least strict of the two types of equality (i.e., group equality over graphical equality), unless otherwise stated (sometimes this will be implied by assuming the words are reduced).

In line with notation from pattern languages, we will refer to pre-images of morphisms as *patterns*. The letters of a pattern are called *variables*, and the set of variables occurring in a pattern α is denoted var(α). Usually, we will use the infinitely generated group $\mathcal{F}_\mathbb{N}$ for patterns. We will use Σ to denote a finite (binary, unless otherwise specified) *terminal* alphabet. This consists of symbols which are not altered further by morphisms. Words over a terminal alphabet are referred to as *terminal words*. In Sect. 5, we consider patterns containing terminal symbols. Such patterns belong to the group $\mathcal{F}_{\mathbb{N} \cup \Sigma}$. A pattern language of a pattern α is the set $L_\Sigma(\alpha) := \{\sigma(\alpha) \mid \sigma : \mathcal{F}_{\mathbb{N} \cup \Sigma} \to \mathcal{F}_\Sigma$ is a terminal-preserving morphism$\}$ and for a terminal-free pattern is the set $L_\Sigma(\alpha) := \{\sigma(\alpha) \mid \sigma : \mathcal{F}_\mathbb{N} \to \mathcal{F}_\Sigma$ is a morphism$\}$.

For a pattern, or word $\alpha \in \Sigma^*$, we denote by $|\alpha|_x$ the number of occurrences of x in α. If $\alpha \in \mathcal{F}_\Sigma$, we denote by $|\alpha|_x$ the *balance* of occurrences of x, i.e., the number of occurrences of x minus the number of occurrences of x^{-1}.

3   Basic Ambiguity

We begin our investigation by considering the most basic generalisation of ambiguity from the monoid to the free group. The result is a trivial situation: that all morphisms are ambiguous with respect to all patterns.

Theorem 1. *All morphisms* $\sigma : \mathcal{F}_\mathbb{N} \to \mathcal{F}_\Sigma$ *are ambiguous with respect to every pattern* $\alpha \in \mathcal{F}_\mathbb{N}$ *with at least two variables.*

This triviality, which is not present in the free monoid, is primarily due to a 'trick' of composition with a particular class of inner automorphism which we demonstrate with an example below. In the case of the free monoid, all inner automorphisms degenerate into the identity. As a result the two morphisms τ and σ in the construction become identical.

Example 1. Let $\alpha := 1 \cdot 2 \cdot 2 \cdot 1$, and let σ be the morphism given by $\sigma(1) := \mathsf{ab}$ and $\sigma(2) = \mathsf{a}$. Then $\sigma(\alpha) = \mathsf{a} \cdot \mathsf{b} \cdot \mathsf{a} \cdot \mathsf{a} \cdot \mathsf{a} \cdot \mathsf{b}$. In order to apply the 'inner automorphism trick', we compose σ with the inner automorphism φ given by $\varphi(1) = \alpha \cdot 1 \cdot \alpha^{-1}$ and $\varphi(2) = \alpha \cdot 2 \cdot \alpha^{-1}$. The result is a morphism $\tau = \sigma \circ \varphi$ such that $\tau(1) = \mathsf{a} \cdot \mathsf{b} \cdot \mathsf{a}^3 \cdot \mathsf{b} \cdot \mathsf{a} \cdot \mathsf{b} \cdot \mathsf{b}^{-1} \cdot \mathsf{a}^{-3} \cdot \mathsf{b}^{-1} \cdot \mathsf{a}^{-1}$ and $\tau(2) = \mathsf{a} \cdot \mathsf{b} \cdot \mathsf{a}^3 \cdot \mathsf{b} \cdot \mathsf{a} \cdot \mathsf{b}^{-1} \cdot \mathsf{a}^{-3} \cdot \mathsf{b}^{-1} \cdot \mathsf{a}^{-1}$. Thus

$$\tau(\alpha) = \sigma(\alpha) \cdot \mathsf{a} \cdot \mathsf{b} \cdot \sigma(\alpha)^{-1} \cdot \sigma(\alpha) \cdot \mathsf{a} \cdot \sigma(1)^{-1} \cdot \sigma(\alpha) \cdot \mathsf{a} \cdot \sigma(\alpha)^{-1} \cdot \sigma(\alpha) \cdot \mathsf{a} \cdot \mathsf{b} \cdot \sigma(\alpha)^{-1}$$

$$= \sigma(\alpha) \cdot \mathsf{a} \cdot \mathsf{b} \cdot \mathsf{a} \cdot \mathsf{a} \cdot \mathsf{a} \cdot \mathsf{b} \cdot \sigma(\alpha)$$

$$= \sigma(\alpha) \cdot \sigma(\alpha) \cdot \sigma(\alpha)^{-1} = \sigma(\alpha).$$

Thus $\tau(\alpha) = \sigma(\alpha)$, and σ is ambiguous.

Of course, in order to complete the proof of Theorem 1, it is necessary to consider the possibility that $\tau = \sigma$ (i.e., when $\sigma(\alpha) \cdot \sigma(x) \cdot \sigma(\alpha)^{-1} = \sigma(x)$ for ever $x \in \mathrm{var}(\alpha)$). It turns out that this is true precisely when σ is periodic. Thus, we include the following observation, which can easily be produced with elementary number theory:

Proposition 1. *Let* $\alpha \in \mathcal{F}_\mathbb{N}$ *be a pattern with* $|\mathrm{var}(\alpha)| > 1$, *and let* $\sigma : \mathcal{F}_\mathbb{N} \to \mathcal{F}_\Sigma$ *be a periodic morphism. Then* σ *is ambiguous with respect to* α.

Proposition 1 is less surprising, as periodic morphisms can be seen to preserve the least amount of structural information. Indeed, in the free monoid, nearly all periodic morphisms are ambiguous. It is interesting, however, to note that the range of images obtainable by periodic morphisms is much larger in the free group than the free monoid. For example consider the pattern $\alpha := 1 \cdot 2 \cdot 1 \cdot 1 \cdot 2$. Any word

$w \in \mathcal{F}_\Sigma$ can be obtained as an image of α by applying the periodic morphism σ given by $\sigma(1) = w$ and $\sigma(2) = w^{-1}$. Thus all morphisms are ambiguous with respect to α, even without using the inner automorphism construction given in Example 1. By contrast, α has the largest possible set of unambiguous morphisms when considered in the free monoid.

As mentioned above, the fact that all periodic morphisms are ambiguous is intuitive. However, this is not true of those morphisms which are ambiguous only because of the composition with inner automorphisms. Since inner automorphisms are so closely related to the identity morphism, the composition τ in Example 1 can be seen as very closely related to σ. Hence, it is natural to consider a reformulation of the problem of ambiguity which disregards this particular phenomenon. From existing literature (Ivanonv [14]), we have the following proposition which shows this is sufficient to obtain a non-trivial problem.

Proposition 2 (Ivanov [14]). *Let* $\alpha := 1^{p_1} \cdot 2^{p_2} \cdots n^{p_n}$ *such that* $p_1, p_2, \ldots, p_n$ *are distinct even integers. Let* $\varphi : \mathcal{F}_{\mathrm{var}(\alpha)} \to \mathcal{F}_{\mathrm{var}(\alpha)}$ *be a morphism such that* $\varphi(\alpha) = \alpha$. *Then* φ *is an inner automorphism.*

Thus we can say that the identity morphism is unambiguous up to inner automorphism with respect to the pattern(s) α. We define this formally below.

Definition 1 (Ambiguity up to Inner Automorphism). *Let* $\alpha \in \mathcal{F}_\mathbb{N}$ *be a pattern. Let* $\sigma : \mathcal{F}_\mathbb{N} \to \mathcal{F}_\Sigma$ *be a morphism. Then* σ *is* unambiguous up to inner automorphism *with respect to* α *if, for every morphism* $\tau : \mathcal{F}_\mathbb{N} \to \mathcal{F}_\Sigma$ *with* $\tau(\alpha) = \sigma(\alpha)$, *there exists an inner automorphism* $\varphi : \mathcal{F}_\mathbb{N} \to \mathcal{F}_\mathbb{N}$ *such that* $\tau(x) = \sigma(\varphi(x))$ *for every* $x \in \mathrm{var}(\alpha)$. *Otherwise,* σ *is* ambiguous up to inner automorphism *with respect to* α.

Similarly, it is possible to consider ambiguity up to automorphism, injective morphism, or indeed any class of morphism. In the current work (in particular, Sect. 4) we will consider ambiguity up to inner automorphism, as this is, in a sense, the minimal restriction needed in order to obtain a rich theory.

It is worth mentioning here that there exists another variation of the problem which results in a non-trivial concept of ambiguity in the free group: namely when terminal symbols may occur in the patterns. This is due to the fact that since the terminal symbols must be preserved by morphisms, the inner-automorphism trick used in this section is not always possible. We consider this variation in Sect. 5.

4   Unambiguous Injective Morphisms

Referring to our discussion of Definition 1, the current section addresses the following question: Given a pattern α, does there exist a morphism $\sigma : \mathcal{F}_\mathbb{N} \to \mathcal{F}_\Sigma$ such that σ is unambiguous with respect to α (up to inner automorphism). More precisely, we wish to investigate the case that σ is also injective. For the free monoid, there exists a characterisation of this problem in terms of fixed points of morphism: a pattern $\alpha \in \mathbb{N}^+$ possesses an unambiguous injective morphism $\sigma : \mathbb{N}^* \to \Sigma^*$ if and only if it is not a fixed point of a morphism which is not the identity. We will firstly provide an equivalent characterisation for patterns in a free group, and will then use the remainder of the section to give some idea of the proof technique, which itself provides further useful insights.

4.1 Main Theorem

We saw in the previous section that in a free group, all patterns are fixed by morphisms which are not equal to the identity – namely inner automorphisms. Hence an analoguous result for free groups would rather be that a pattern possesses an injective morphism which is unambiguous up to inner automorphism if and only if it is not the fixed point of a morphism which is not an inner automorphism. This is precisely the statement we give in Theorem 2 below.

Theorem 2. *Let $\alpha \in \mathcal{F}_{\mathbb{N}}$ be a pattern. Then there exists an injective morphism $\sigma : \mathcal{F}_{\mathbb{N}} \to \mathcal{F}_{\Sigma}$ which is unambiguous with respect to α (up to inner automorphism) if and only if the identity morphism is unambiguous with respect to α (up to inner automorphism).*

It is worth noting that for the monoid, there exists a concise, combinatorial characterisation of patterns which are only fixed by the identity, namely the morphically primitive patterns (Reidenbach, Schneider [15]). It is even known that such patterns can be identified in polynomial time (Holub [16], Kociumaka et al. [17]). Unfortunately no such equivalent characterisation exists for free groups, although fixed points of morphisms is a wide area of study with many useful existing theorems. For example, one known class of patterns which are only fixed by inner automorphisms is given in Proposition 2 in the previous section.

4.2 Proof Outline

The main idea of the proof of Theorem 2 is to establish, for any pattern α, a morphic image $w \in \mathcal{F}_{\Sigma}$ of α, such that any morphism $\sigma : \mathcal{F}_{\mathbb{N}} \to \mathcal{F}_{\Sigma}$ mapping α to w must, in a specific sense, encode a morphism $\varphi_\sigma : \mathcal{F}_{\mathbb{N}} \to \mathcal{F}_{\mathbb{N}}$ which fixes α. The encoding works in such a way that if a morphism $\sigma : \mathcal{F}_{\mathbb{N}} \to \mathcal{F}_{\Sigma}$ mapping α to w is ambiguous up to inner automorphism, then either the encoded fixed point morphism φ_σ is ambiguous up to inner automorphism, or α belongs to a particular class of patterns, referred to as NCLR. A brief analysis shows that the identity morphism is ambiguous up to inner automorphism with respect to every NCLR pattern, and, thus, whenever σ is ambiguous up to inner automorphism, the identity morphism is ambiguous up to inner automorphism. Consequently if the identity is unambiguous up to inner automorphism with respect to α, then so is σ. Since the converse statement follows easily from the injectivity of σ, this proves the Theorem.

Constructing the Encoding Word w

The role of w in the proof is to encode the pre-image α in such a way that any morphism mapping α to w must be reducible to a morphism mapping α to α. We construct w as the image $\sigma(\alpha)$ for some morphism $\sigma : \mathcal{F}_\mathbb{N} \to \mathcal{F}_\Sigma$ so that it is guaranteed at least one morphism maps α to w. In fact, we generalise the situation slightly so that any morphism mapping any given pattern β to $w = \sigma(\alpha)$ must be reducible to a morphism mapping β to α. This allows us to give a characterisation of the inclusion problem for terminal free group pattern languages, generalising a result of Jiang et al. [18].

In the monoid, this encoding can be achieved by setting $\sigma(x)$ equal to k distinct *segments* $s_i := \mathsf{ab}^i\mathsf{a}$ which are unique to x, so that, e.g., $\sigma(1) = s_1 \cdot s_2 \cdots s_k$, $\sigma(2) = s_{k+1} \cdot s_{k+2} \cdots s_{2k}$ etc. By taking k large enough, it is guaranteed that for each $x \in \mathrm{var}(\alpha)$, there is a segment S_x which is not 'split' by τ (that is, if $\beta = b_1 \cdot b_2 \cdot b_3 \cdots b_{|\beta|}$, then every occurrence of S_x in $\tau(\beta)$ does not span two factors $\tau(b_j)$, $\tau(b_k)$), and such that every occurrence of S_x in the image $\sigma(\alpha)$ $(= \tau(\beta))$ corresponds to an occurrence of x in the pre-image α. These segments act as the encoding of α: the result of replacing each one in τ for the corresponding variable x, and removing all the surrounding letters a and b results in a morphism which exactly maps β to α.

However, if we wish to use the same technique within the free group, we have a more intricate task, due to the existence of contractions. The problem lies in the fact that these replacement operations are not compatible with reductions (the removal of contractions to produce a reduced word). This problem is demonstrated in more detail in Example 2, but first we introduce the following convenient notation. Note that by restricting ourselves to factors which are unbordered, we can ensure that occurrences of the factor do not overlap, and, thus, that our operation is a well-defined function. Let u, $v \in \mathcal{F}_\Sigma$ be words, and let w be an unbordered word. Denote by $R[w \to v](u)$ the word obtained by replacing all occurrences of w in u with v. For a morphism $\sigma : \mathcal{F}_\mathbb{N} \to \mathcal{F}_\Sigma$, denote by $R[w \to v](\sigma)$ the morphism defined by $\sigma(x) = R[w \to v](x)$ for each $x \in \mathbb{N}$.

Example 2. Let $\sigma : \mathcal{F}_\mathbb{N} \to \mathcal{F}_\Sigma$ be the morphism given by $\sigma(1) = \mathsf{a} \cdot \mathsf{b} \cdot \mathsf{b}$ and $\sigma(2) = \mathsf{a} \cdot \mathsf{b} \cdot \mathsf{a}^{-2}$. Let $\alpha := 1 \cdot 2 \cdot 2 \cdot 1$. Then

$$R[\mathsf{b} \cdot \mathsf{b} \to \mathsf{c}](\sigma(\alpha)) = R[\mathsf{b} \cdot \mathsf{b} \to \mathsf{c}](\mathsf{a} \cdot \mathsf{b} \cdot \mathsf{b} \cdot \mathsf{a} \cdot \mathsf{b} \cdot \mathsf{a}^{-2} \cdot \mathsf{a} \cdot \mathsf{b} \cdot \mathsf{a}^{-2} \cdot \mathsf{a} \cdot \mathsf{b} \cdot \mathsf{b})$$

$$= \mathsf{a} \cdot \mathsf{c} \cdot \mathsf{a} \cdot \mathsf{b} \cdot \mathsf{a}^{-1} \cdot \mathsf{b} \cdot \mathsf{a}^{-1} \cdot \mathsf{c} = R[\mathsf{b} \cdot \mathsf{b} \to \mathsf{c}](\sigma)(\alpha).$$

The equivalence holds because $\mathsf{b} \cdot \mathsf{b}$ is not split by a contraction, or the morphisms, so there is a one-to-one correspondence between the factors in the image $\sigma(\alpha)$ and factors in the definition of σ. Conversely, this does not hold for, e.g., the factor a^{-2}. In fact, all occurrences are contracted and thus do not fully appear in the image. Thus

$$R[\mathsf{a}^{-2} \to \mathsf{c}](\sigma(\alpha)) = \sigma(\alpha) = \mathsf{a} \cdot \mathsf{b} \cdot \mathsf{a} \cdot \mathsf{b} \cdot \mathsf{a}^{-1} \cdot \mathsf{b} \cdot \mathsf{a}^{-1} \cdot \mathsf{b} \cdot \mathsf{b}$$

and

$$R[\mathsf{a}^{-2} \to \mathsf{c}](\sigma)(\alpha) = \mathsf{a} \cdot \mathsf{b} \cdot \mathsf{b} \cdot \mathsf{a} \cdot \mathsf{b} \cdot \mathsf{c} \cdot \mathsf{a} \cdot \mathsf{b} \cdot \mathsf{c} \cdot \mathsf{a} \cdot \mathsf{b} \cdot \mathsf{b} \neq \sigma(\alpha).$$

so we have $R[\mathsf{a}^{-2} \to \mathsf{c}](\sigma(\alpha)) \neq R[\mathsf{a}^{-2} \to \mathsf{c}](\sigma)(\alpha)$.

It is clear from the above example that if we wish to extend the reasoning from Jiang et al. [18] to use for free groups, we must guarantee not only that the encoding segments S_x are not split by the morphism τ, but are also not split by any contraction in the image $\tau(\beta)$. We can achieve this by increasing the number of segments in the definition of σ, but first we need the following bound for the number of contractions which may occur in the image $\tau(\beta)$.

Proposition 3. *Let $\alpha \in \mathcal{F}_{\mathbb{N}}$ and let $\sigma : \mathcal{F}_{\mathbb{N}} \to \mathcal{F}_{\Sigma}$ be a morphism. Then $\sigma(\alpha)$ can be fully reduced by removing at most $\frac{|\alpha|(|\alpha|-1)}{2}$ primary contractions.*

We are now ready to construct our 'encoding morphism' σ. It has a similar structure to the one in Jiang et al. [18]. However, it must contain a much higher number of segments as segments can be split not only between the variables, but also between contractions. We also alter our segments so that they are unbordered, although this is purely for convenience when formally considering replacements.

Definition 2. *Let $s_i := \mathsf{ab}^i$. For any k, $p \in \mathbb{N}$, let $\sigma_{k,p}$ be the morphism given by $\sigma_{k,p}(x) := \gamma_x \cdot s_p \cdot s_{p+(x-1)k} \cdots s_{p+xk-1} \cdot \gamma_x$ for every $x \in \mathbb{N}$, where $\gamma_x := s_x \cdot \mathsf{a}$.*

The factors γ_x ensure that no contractions occur within the main segments s_x, meaning that each segment is guaranteed to appear in the (reduced) image $\sigma(\alpha)$. By giving the segments a minimal length p, we can guarantee that no segment can be obtained 'accidentally' by contracting γ_x factors. Thus, we can be sure that at least k segments appear, uniquely associated to each pre-image variable. The encoding is achieved by making k large enough that not all segments can be split and, hence, for every variable x there is at least one which can be used to successfully encode (and later decode) x in the image.

It is worth noting at this stage that we can apply our reasoning so far to produce a result for group pattern languages. If $L_\Sigma(\alpha) \subseteq L_\Sigma(\beta)$, then $\sigma(\alpha) \in L(\beta)$, and thus our construction yields the following characterisation of the inclusion problem for terminal-free group pattern languages, extending a known equivalent result for terminal free pattern languages in the free monoid (Jiang et al. [18]).

Theorem 3. *Let α, $\beta \in \mathcal{F}_{\mathbb{N}}$ be patterns. Then $L_\Sigma(\alpha) \subseteq L_\Sigma(\beta)$ if and only if there exists a morphism $\varphi : \mathcal{F}_{\mathbb{N}} \to \mathcal{F}_{\mathbb{N}}$ with $\varphi(\beta) = \alpha$.*

Returning to the case that $\beta = \alpha$, we are able to infer that if $\sigma \; (= \sigma_{k,p}$ for some large k, p) is ambiguous up to inner automorphism with respect to α, then we have two distinct morphisms σ and τ mapping α to $w = \sigma(\alpha)$. As a result we have two encoded morphisms φ_σ and φ_τ both fixing α. Due to the construction of σ, it is not difficult to see that φ_σ is the identity.

Consequently, if φ_τ is not an inner automorphism, then the identity morphism is ambiguous up to inner automorphism with respect to α, and the claim of Theorem 2 holds for that case. Assuming that φ_τ is an inner automorphism, we can reverse the process of obtaining φ_τ from τ, to learn the structure of τ.

In particular we know from the positions of the variables $x \in \text{var}(\alpha)$ in the definition of φ the relative positions of certain segments S_x in the definition of τ (recall that the segments S_x are replaced by the variables x, and then all other letters a and b removed to obtain φ_τ from τ). Thus in order to completely reconstruct τ it remains to find out the factors between the segments S_x. It turns out that these factors are determined by the following system of equations Π:

$$u_1 = w_1$$

$$v_i u_j = w_k$$

$$v_{|\alpha|} = w_{|\alpha|+1}$$

for every k such that $i \cdot j$ or $j^{-1} \cdot i^{-1}$ is a factor of α at positions $k-1$, k. If Π has only one solution (it must have at least one, corresponding to the case that $\tau = \sigma \circ \varphi_\tau$), then σ is unambiguous up to inner automorphism with respect to α. In the final part of the proof, we consider the case that Π has multiple solutions by investigating a particular class of patterns, which we call NCLR.

NCLR Patterns

We define the set of patterns NCLR as follows. It is worth noting that our definition generalises an existing class known for the monoid, described by Freydenberger, Reidenbach in [2].

Definition 3 (NCLR). *Let* N_0, C_0, L_0, R_0 *be pairwise-disjoint sets of variables with* $L_0 \cup R_0 \neq \emptyset$. *Let* $N := N_0 \cup N_0^{-1}$, $C := C_0 \cup C_0^{-1}$, $L := L_0 \cup R_0^{-1}$ *and* $R := R_0 \cup L_0^{-1}$. *A pattern* $\alpha \in \mathcal{F}_\mathbb{N}$ *is in the set* NCLR *if and only if it has the form*

$$N^*(L \cdot C^* \cdot R \cdot N^*)^+.$$

We motivate the above definition with the following remark.

Remark 1. The system of equations Π (above) has more than one solution exactly when $\alpha \in NCLR$.

In order to complete the proof (sketch) of Theorem 2 it is necessary to assert that the identity morphism is ambiguous up to inner automorphism with respect to α whenever $\alpha \in NCLR$. Fortunately this is not difficult, and in fact we can even prove this statement for all morphisms.

Proposition 4. *Every morphism* $\sigma : \mathcal{F}_\mathbb{N} \to \mathcal{F}_\Sigma$ *is ambiguous up to inner automorphism with respect to every pattern* $\alpha \in NCLR$.

Example 3. Consider the pattern $\alpha := 1 \cdot 2 \cdot 3 \cdot 4 \cdot 2 \cdot 3 \cdot 1 \cdot 2 \cdot 3 \cdot 4 \cdot 2 \cdot 4^{-1}$. Let $N_0 = \emptyset$, $C_0 = 2$, $L_0 = \{1, 4\}$, $R_0 = \{3\}$, and note that this shows $\alpha \in NCLR$. Consider the morphism $\varphi : \mathcal{F}_\mathbb{N} \to \mathcal{F}_\mathbb{N}$ given by $\varphi(1) = 1 \cdot 5$, $\varphi(2) = 5^{-1} \cdot 2 \cdot 5$, $\varphi(3) = 5^{-1} \cdot 3$ and $\varphi(4) = 4 \cdot 5$. Clearly φ is not an inner automorphism. Moreover

$$\varphi(\alpha) = 1 \cdot 5 \cdot 5^{-1} \cdot 2 \cdot 5 \cdot 5^{-1} \cdot 3 \cdot 4 \cdot 5 \cdot 5^{-1} \cdot 2 \cdot 5 \cdot 5^{-1} \cdot 3 \cdot 1 \cdot 3 \cdot 5^{-1} 2 \cdot 5 \cdot 5^{-1} \cdot 3 \cdot 4 \cdot 5 \cdot 5^{-1} \cdot 2 \cdot 5 \cdot 5^{-1} \cdot 4^{-1}$$

$$= \alpha.$$

Thus, the identity is ambiguous up to inner automorphism with respect to α. Furthermore, we can compose any morphism $\sigma : \mathcal{F}_\mathbb{N} \to \mathcal{F}_\Sigma$ with φ to get $\sigma \circ \varphi(1) = \sigma(1) \cdot \sigma(7)$, $\sigma \circ \varphi(2) = \sigma(7)^{-1} \cdot \sigma(2) \cdot \sigma(7)$... etc. Clearly, σ and $\sigma \circ \varphi$ are distinct whenever $\sigma(7) \neq \varepsilon$. Consequently, σ is unambiguous with respect to α.

Thus, if α is only fixed by inner automorphisms, there exists an (injective) morphism $\sigma : \mathcal{F}_\mathbb{N} \to \mathcal{F}_\Sigma$ which is unambiguous up to inner automorphism with respect to α. Conversely, if α is fixed by a morphism which is not an inner automorphism it is clear that all injective morphisms are ambiguous up to inner automorphism with respect to α.

5   Patterns with Terminal Symbols

In the previous section, we saw that by ignoring a particular construction involving inner automorphisms, we return to a rich, non-trivial theory, and we are able to partition patterns according to whether they possess an unambiguous morphism up to inner automorphism. In the present section, we discuss an alternative situation in which the question of whether a morphism is ambiguous is again non-trivial, this time in the stronger, traditional sense, where composition with an inner automorphism is allowed. More precisely, we look at patterns containing terminal symbols – symbols which must be preserved by morphisms. This condition greatly reduces the number of inner automorphisms which can be constructed. For example, if we consider patterns with at least two terminal symbols, the only inner automorphism we can apply to that pattern is the identity morphism. It is also worth noting that the same restriction applies to periodic morphisms – if two terminal symbols are present in a pattern, then any morphism preserving those terminal symbols cannot be periodic. Thus, the constructions given in Sect. 3 to show that ambiguity of morphisms is trivial in the free group cannot be reproduced in general for patterns with terminal symbols.

We begin with the following proposition, supporting our claim that the question of ambiguity is non-trivial for patterns with terminal symbols. In fact, we produce a far stronger construction. Using so-called C-test words (Lee [19]), we construct a pattern for which *all* morphisms are unambiguous.

Proposition 5. *For any finite $\Delta \subset \mathbb{N}$, there exists a pattern $\alpha \in \mathcal{F}_{\mathbb{N} \cup \Sigma}$ with* $\mathrm{var}(\alpha) = \Delta$ *such that all terminal-preserving morphisms $\sigma : \mathcal{F}_{\mathbb{N} \cup \Sigma} \to \mathcal{F}_\Sigma$ are unambiguous with respect to α.*

Corollary 1. *For any finite $\Delta \subset \mathbb{N}$, every terminal-preserving morphism $\sigma : \mathcal{F}_{\Delta \cup \Sigma} \to \mathcal{F}_\Sigma$ is unambiguous with respect to some pattern α with $\mathrm{var}(\alpha) = \Delta$.*

Since we are unable to provide a characterisation for when an unambiguous morphism exists with respect to a pattern, we will instead attempt to categorise patterns according to how many of their morphisms are unambiguous, and also focus on some specific examples – in particular morphisms which erase the pattern completely. The next proposition serves several purposes in this respect. Firstly, it provides a concise example of an unambiguous morphism (the C-test words on which the previous proposition relies are defined recursively, and thus the construction becomes exceedingly long). Secondly, we establish the unambiguity of a morphism erasing a pattern α, and thirdly we note that since we can also find a morphism which is ambiguous with respect to α, we have patterns for which there are both unambiguous and ambiguous morphisms.

Proposition 6. *Let* $\alpha := 1 \cdot a \cdot 1^{-1} \cdot b \cdot 1 \cdot 2 \cdot a \cdot b \cdot 2$. *Then the terminal-preserving morphism mapping* α *to* ε *is unambiguous, and any terminal-preserving morphism mapping* α *to* $a \cdot b \cdot a \cdot a \cdot b^{-1} \cdot a^{-1}$ *is ambiguous.*

It is clear that there exist patterns for which all morphisms are ambiguous – take for example $1 \cdot 2 \cdot a$. Thus we can complete the following classification of patterns:

Theorem 4. *There exist patterns* $\alpha_1, \alpha_2, \alpha_3 \in \mathcal{F}_{\mathbb{N} \cup \Sigma}$ *such that*

- *All terminal-preserving morphisms are ambiguous with respect to* α_1,
- *there exist both ambiguous and unambigous terminal-preserving morphisms with respect to* α_2, *and*
- *all terminal-preserving morphisms are unambiguous with respect to* α_3.

Another short example of a pattern with an unambiguous morphism is given below. In particular, we show that the pattern has the particularly interesting property that the only morphism which completely erases the pattern is the one which erases each individual variable. It also forms the basis for Theorem 5 below.

Proposition 7. *Let* $\alpha := 1 \cdot a \cdot 1^{-1} \cdot b \cdot 1 \cdot b^{-1} \cdot a^{-1} \cdot 2 \cdot 2 \cdot a \cdot 2 \cdot 2 \cdot a^{-1}$ *and let* $\sigma : \mathcal{F}_{\mathbb{N} \cup \Sigma} \to \mathcal{F}_{\Sigma}$ *be the terminal-preserving morphism such that* $\sigma(x) = \varepsilon$ *for every* $x \in \operatorname{var}(\alpha)$. *Then* σ *is unambiguous with respect to* α, *and* $\sigma(\alpha) = \varepsilon$.

The above proposition is easily generalised for larger alphabets.

Proposition 8. *Let* Δ *be a finite alphabet and let* $\sigma : \mathcal{F}_{\Delta \cup \Sigma} \to \mathcal{F}_{\Sigma}$ *be the terminal-preserving morphism given by* $\sigma(x) = \varepsilon$ *for every* $x \in \Delta$. *There exists a pattern* $\alpha \in \mathcal{F}_{\Delta \cup \Sigma}$ *with* $\operatorname{var}(\alpha) = \Delta$, *such that* σ *is unambiguous with respect to* α, *and* $\sigma(\alpha) = \varepsilon$.

Finally, as a further generalisation, we can obtain a construction, for any morphism σ, of a pattern α which is only erased by σ.

Theorem 5. *Let* Δ *be a finite subset of* $\mathbb{N}$. *For every terminal-preserving morphism* $\sigma : \mathcal{F}_{\Delta \cup \Sigma} \to \mathcal{F}_{\Sigma}$, *there exists an* $\alpha \in \mathcal{F}_{\Delta \cup \Sigma}$ *such that* $\sigma(\alpha) = \varepsilon$, *and* σ *is unambiguous with respect to* α.

Acknowledgements. The authors wish to thank Alexey Bolsinov for his helpful input on the subject of inner automorphisms.

References

1. Freydenberger, D., Reidenbach, D., Schneider, J.: Unambiguous morphic images of strings. Int. J. Found. Comput. Sci. **17**, 601–628 (2006)
2. Freydenberger, D., Reidenbach, D.: The unambiguity of segmented morphisms. Discrete Appl. Math. **157**, 3055–3068 (2009)
3. Culik II, K., Karhumäki, J.: On the equality sets for homomorphisms on free monoids with two generators. Theor. Inform. Appl. (RAIRO) **14**, 349–369 (1980)
4. Salomaa, A.: Equality sets for homomorphisms of free monoids. Acta Cybern. **4**, 127–239 (1978)
5. Engelfriet, J., Rozenberg, G.: Fixed point languages, equality languages and representation of recursively enumerable languages. J. ACM **27**, 499–518 (1980)
6. Post, E.: A variant of a recursively unsolvable problem. Bull. Am. Math. Soc. **52**, 264–268 (1946)
7. Makanin, G.: Equations in a free semigroup. Am. Math. Soc. Transl. Ser. **2**(117), 1–6 (1981)
8. Reidenbach, D.: Discontinuities in pattern inference. Theor. Comput. Sci. **397**, 166–193 (2008)
9. Jain, S., Miasnikov, A., Stephan, F.: The complexity of verbal languages over groups. In: 27th Annual IEEE Symposium on Logic in Computer Science (LICS), pp. 405–414 (2012)
10. Bell, P., Potapov, I.: On the undecidability of the identity correspondence problem and its applications for word and matrix semigroups. Int. J. Found. Comput. Sci. **21**, 963–978 (2010)
11. Turner, E.: Test words for automorphisms of free groups. Bull. Lond. Math. Soc. **28**, 255–263 (1996)
12. Ventura, E.: Fixed subgroups in free groups: a survey. Contemp. Math. **296**, 231–256 (2002)
13. Makanin, G.: Equations in a free group. Math. USSR Izvestiya **21**, 483–546 (1983)
14. Ivanov, S.: On certain elements of free groups. J. Algebra **204**, 394–405 (1998)
15. Reidenbach, D., Schneider, J.: Morphically primitive words. Theor. Comput. Sci. **410**, 2148–2161 (2009)
16. Holub, Š.: Polynomial-time algorithm for fixed points of nontrivial morphisms. Discrete Math. **309**, 5069–5076 (2009)
17. Kociumaka, T., Radoszewski, J., Rytter, W., Waleń, T.: Linear-time version of holub's algorithm for morphic imprimitivity testing. In: Dediu, A.-H., Martín-Vide, C., Truthe, B. (eds.) LATA 2013. LNCS, vol. 7810, pp. 383–394. Springer, Heidelberg (2013)
18. Jiang, T., Salomaa, A., Salomaa, K., Yu, S.: Decision problems for patterns. J. Comput. Syst. Sci. **50**, 53–63 (1995)
19. Lee, D.: On certain c-test words for free groups. J. Algebra **247**, 509–540 (2002)

The Degree of Squares is an Atom

Jörg Endrullis[1], Clemens Grabmayer[1], Dimitri Hendriks[1](✉),
and Hans Zantema[2,3]

[1] Department of Computer Science, VU University Amsterdam,
Amsterdam, Netherlands
diem@cs.vu.nl
[2] Department of Computer Science, Eindhoven University of Technology,
Eindhoven, Netherlands
[3] Institute for Computing and Information Science,
Radboud University Nijmegen, Nijmegen, Netherlands

Abstract. We answer an open question in the theory of degrees of infinite sequences with respect to transducibility by finite-state transducers. An initial study of this partial order of degrees was carried out in [1], but many basic questions remain unanswered. One of the central questions concerns the existence of atom degrees, other than the degree of the 'identity sequence' $10^0 10^1 10^2 10^3 \cdots$. A degree is called an 'atom' if below it there is only the bottom degree $\mathbf{0}$, which consists of the ultimately periodic sequences. We show that also the degree of the 'squares sequence' $10^0 10^1 10^4 10^9 10^{16} \cdots$ is an atom.

As the main tool for this result we characterise the transducts of 'spiralling' sequences and their degrees. We use this to show that every transduct of a 'polynomial sequence' either is in $\mathbf{0}$ or can be transduced back to a polynomial sequence for a polynomial of the same order.

1 Introduction

Finite-state transducers are ubiquitous in computer science, but little is known about the transducibility relation they induce on infinite sequences. A finite-state transducer (FST) is a deterministic finite automaton which reads the input sequence letter by letter, in each step producing an output word and changing its state. An example of an FST is depicted in Fig. 1, where we write '$a|w$' along the transitions to indicate that the input letter is a and the output word is w. For example, it transduces the Thue-Morse sequence $\mathsf{T} = 0110100110010110 \cdots$ to the period doubling sequence $\mathsf{P} = 1011101010111011 \cdots$.

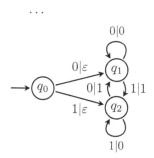

Fig. 1. A finite-state transducer realising the sum of consecutive bits modulo 2.

We are interested in transductions of infinite sequences. We say that a sequence σ is *transducible* to a sequence τ, $\sigma \geq \tau$, if there exists an FST that

© Springer International Publishing Switzerland 2015
F. Manea and D. Nowotka (Eds.): WORDS 2015, LNCS 9304, pp. 109–121, 2015.
DOI: 10.1007/978-3-319-23660-5_10

transforms σ into τ. The relation $\geq$ is a preorder on the set $\Sigma^{\mathbb{N}}$ of infinite sequences, which induces an equivalence relation $\equiv$ on $\Sigma^{\mathbb{N}}$, and a partial order on the set of *(FST) degrees*, that is, the equivalence classes with respect to $\equiv$.

So we have $\mathsf{T} \geq \mathsf{P}$. Also the back transformation can be realised by an FST, $\mathsf{P} \geq \mathsf{T}$. Hence the sequences are equivalent, $\mathsf{T} \equiv \mathsf{P}$, and are in the same degree.

The bottom degree $\mathbf{0}$ is formed by the ultimately periodic sequences, that is, all sequences of the form $uvvv \cdots$ for finite words u, v with v non-empty. Every infinite sequence can be transduced to any ultimately periodic sequence.

There is a clear analogy between degrees induced by transducibility and the recursion-theoretic degrees of unsolvability (Turing degrees). Hence many of the problems settled for Turing degrees, predominantly in the 1940s, 50s and 60s, can be asked again for FST degrees.

Some initial results on FST degrees have been obtained in [1]: the partial order of degrees is not dense, not well-founded, there exist no maximal degrees, and a set of degrees has an upper bound if and only if the set is countable. The morphic degrees, and the computable degrees form subhierarchies. Also shown in [1] is the existence of an 'atom' degree, namely the degree of $10^0 10^1 10^2 10^3 \cdots$.[1] A degree $D \neq \mathbf{0}$ is an atom if there exists no degree strictly between D and the bottom degree $\mathbf{0}$. Thus every trans-

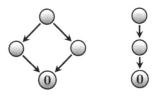

Fig. 2. Possible structures in the hierarchy (no intermediate points on the arrows).

duct of a sequence in an atom degree D is either in $\mathbf{0}$ or still in D. The following questions, which have been answered for Turing degrees, remain open for the FST degrees:

(i) How many atom degrees exist?
(ii) When do two degrees have a supremum (or infimum)? In particular, are there pairs of degrees without a supremum (infimum)?
(iii) Do the configurations displayed in Fig. 2 exist?
(iv) Can every finite distributive lattice be embedded in the hierarchy?

At the British Colloquium for Theoretical Computer Science 2014, Jeffrey Shallit offered £20 for each of the following questions, see [2]:

(a) Is the degree of the Thue-Morse sequence an atom?
(b) Are there are atom degrees other than that of $10^0 10^1 10^2 10^3 \cdots$?

We answer (b) by showing that the degree of the 'squares' $10^0 10^1 10^4 10^9 \cdots$ is an atom. The main tool that we use in the proof is a characterisation of transducts of 'spiralling' sequences (Theorem 4.22), which has the following consequence (Proposition 5.1): For all $k \geq 1$ it holds that every transduct $\sigma \not\equiv \mathbf{0}$ of the sequence $\langle n^k \rangle = 10^{0^k} 10^{1^k} 10^{2^k} 10^{3^k} \cdots$ has, in its turn, a transduct of the form $\langle p(n) \rangle = 10^{p(0)} 10^{p(1)} 10^{p(2)} 10^{p(3)} \cdots$ where $p(n)$ is a polynomial also of degree k.

---

[1] In [1] atom degrees were called 'prime degrees'. We prefer the more general notion of 'atom' because prime factorisation does not hold, see Theorem 4.24.

This fact not only enables us to show that the degree of the squares sequence is an atom, by using it for $k = 2$, but it also suggests that the analogous result for the degree of $\langle n^k \rangle$, for arbitrary $k \geq 1$, has come within reach. For this it would namely suffice to show, for polynomials $p(n)$ of degree k, that $\langle p(n) \rangle$ can be transduced back to $\langle n^k \rangle$.

We obtain that there is a pair of non-zero degrees whose infimum is $\mathbf{0}$, namely the pair of atom degrees of $\langle n \rangle$ and $\langle n^2 \rangle$. We moreover use Theorem 4.22 to show that there is a non-atom degree that has no atom degree below it (Theorem 4.24).

2 Preliminaries

We use standard terminology and notation, see, e.g., [3]. Let Σ be an alphabet, i.e., a finite non-empty set of symbols. We denote by Σ^* the set of all finite words over Σ, and by ε the empty word. We let $\Sigma^+ = \Sigma^* \setminus \{\varepsilon\}$. The set of infinite sequences over Σ is $\Sigma^{\mathbb{N}} = \{\sigma \mid \sigma : \mathbb{N} \to \Sigma\}$ with $\mathbb{N} = \{0, 1, 2, \ldots\}$, the set of natural numbers. For $u \in \Sigma^*$, we let $u^\omega = uuu \cdots$. We define $\mathbb{N}_{<k} = \{0, 1, \ldots, k - 1\}$. We let $\Sigma^\infty = \Sigma^* \cup \Sigma^{\mathbb{N}}$ denote the set of all words over Σ. For $u \in \Sigma^*$ and $v \in \Sigma^\infty$ we write $u \sqsubseteq v$ to denote that u is a prefix of v, that is, when $v = uv'$ for some $v' \in \Sigma^\infty$. For $f : \mathbb{N} \to A$ and $k \in \mathbb{N}$, the k-th shift of f is defined by $\mathcal{S}^k(f)(n) = f(n + k)$, for all $n \in \mathbb{N}$.

We use the notation $\boldsymbol{a}$ to denote a tuple $\boldsymbol{a} = \langle a_0, a_1, \ldots, a_{k-1} \rangle$ where the a_i are elements from some set A; the length of $\boldsymbol{a}$ is k. Given a tuple $\boldsymbol{a}$ we use a_i for the element indexed i, that is, we start indexing from 0 onward. We use the notation $\boldsymbol{a}'$ to denote the rotated tuple $\boldsymbol{a}' = \langle a_1, \ldots, a_{k-1}, a_0 \rangle$.

3 Finite-State Transducers and Degrees

For a thorough introduction to finite-state transducers, we refer to [3]. Here we only consider *complete pure sequential* finite-state transducers, where for every state q and input letter a there is precisely one successor state $\delta(q, a)$, and the functions realised by these transducers preserve prefixes. We use $\mathbf{2} = \{0, 1\}$ both for the input and the output alphabet.

Definition 3.1. A *finite-state transducer (FST)* is a tuple $T = \langle Q, q_0, \delta, \lambda \rangle$ where Q is a finite set of states, $q_0 \in Q$ is the initial state, $\delta : Q \times \mathbf{2} \to Q$ is the transition function, and $\lambda : Q \times \mathbf{2} \to \mathbf{2}^*$ is the output function.

We homomorphically extend the transition function δ to $Q \times \mathbf{2}^* \to Q$ and the output function λ to $Q \times \mathbf{2}^\infty \to \mathbf{2}^\infty$ as follows:

$$\delta(q, \varepsilon) = q \quad \delta(q, au) = \delta(\delta(q, a), u) \qquad (q \in Q, a \in \mathbf{2}, u \in \mathbf{2}^*)$$
$$\lambda(q, \varepsilon) = \varepsilon \quad \lambda(q, au) = \lambda(q, a) \cdot \lambda(\delta(q, a), u) \quad (q \in Q, a \in \mathbf{2}, u \in \mathbf{2}^\infty).$$

The function $T : \mathbf{2}^\infty \to \mathbf{2}^\infty$ *realised* by the FST T is defined by $T(u) = \lambda(q_0, u)$, for all $u \in \mathbf{2}^\infty$.

Definition 3.2. Let $T = \langle Q, q_0, \delta, \lambda \rangle$ be an FST. A *zero-loop* in T is a sequence of states $q_1, \ldots, q_n$ with $n > 1$ such that $q_1 = q_n$ and $q_i \neq q_j$ for all i, j with $1 \leq i < j < n$ and $q_{i+1} = \delta(q_i, 0)$ for all $1 \leq i < n$. The *length* of the zero-loop is $n - 1$. (Note that there can only be finitely many zero-loops in an FST.) Let $T = \langle Q, q_0, \delta, \lambda \rangle$ be an FST. We define $\mathsf{Z}(T)$ as the least common multiple of the lengths of all zero-loops of T.

Let T be an FST with states Q. From any state $q \in Q$, after reading the word $0^{|Q|}$, the automaton must have entered a zero-loop (by the pigeonhole principle there must be a state repetition). By definition of $\mathsf{Z}(T)$, the length ℓ of this loop divides $\mathsf{Z}(T)$; say $\mathsf{Z}(T) = d\ell$ for some $d \geq 1$. As a consequence, reading $0^{|Q|+i \cdot \mathsf{Z}(T)}$ yields the output $\lambda(q, 0^{|Q|})$ followed by di copies of the output of the zero-loop. This yields a pumping lemma for FSTs, see also [1, Lemma 29].

Lemma 3.3. *Let $T = \langle Q, q_0, \delta, \lambda \rangle$ be an FST. For every $q \in Q$ and $n \geq |Q|$ there exist words $p, c \in \mathbf{2}^*$ such that for all $i \in \mathbb{N}$, $\delta(q, 10^{n+i \cdot \mathsf{Z}(T)}) = \delta(q, 10^n)$ and $\lambda(q, 10^{n+i \cdot \mathsf{Z}(T)}) = pc^i$.*

Definition 3.4. Let T be an FST, and let $\sigma, \tau \in \mathbf{2}^{\mathbb{N}}$ be infinite sequences. We say that T *transduces* σ to τ, and that τ is the T-*transduct* of σ, which we denote by $\sigma \geq_T \tau$, whenever $T(\sigma) = \tau$. We write $\sigma \geq \tau$, and call τ a *transduct* of σ, if there exists an FST T such that $\sigma \geq_T \tau$.

Clearly, the relation $\geq$ is reflexive. By composition of FSTs (the so-called 'wreath product'), the relation $\geq$ is also transitive, see [1, Remark 9]. We write $\sigma > \tau$ when $\sigma \geq \tau$ but not $\tau \geq \sigma$. Whenever we have $\sigma \geq \tau$ as well as a back-transduction $\tau \geq \sigma$, we consider σ and τ to be equivalent.

Definition 3.5. We define the relation $\equiv \subseteq \mathbf{2}^{\mathbb{N}} \times \mathbf{2}^{\mathbb{N}}$ by $\equiv = \geq \cap \geq^{-1}$. For a sequence $\sigma \in \mathbf{2}^{\mathbb{N}}$ the equivalence class $[\sigma] = \{\tau \mid \sigma \equiv \tau\}$ is called the *degree* of σ.

A degree $[\sigma]$ is an *atom* if $[\sigma] \neq \mathbf{0}$ and there is no degree $[\tau]$ such that $[\sigma] > [\tau] > \mathbf{0}$.

4   Characterising Transducts of Spiralling Sequences

In this section we characterize the transducts of 'spiralling' sequences. Proofs omitted in the text can be found in the extended version [4].

Definition 4.1. For a function $f : \mathbb{N} \to \mathbb{N}$ we define the sequence $\langle f \rangle \in \mathbf{2}^{\mathbb{N}}$ by

$$\langle f \rangle = \prod_{i=0}^{\infty} 10^{f(i)} = 10^{f(0)}\, 10^{f(1)}\, 10^{f(2)} \cdots .$$

For a sequence $\langle f \rangle$, we often speak of the n-th *block* of $\langle f \rangle$ to refer to the occurrence of the word $10^{f(n)}$ in $\langle f \rangle$.

In the sequel we often write $\langle f(n)]\rangle$ to denote the sequence $\langle n \mapsto f(n)\rangle$. We note that there is a one-to-one correspondence between functions $f : \mathbb{N} \to \mathbb{N}$, and infinite sequences over the alphabet $\mathbf{2}$ that start with the letter 1 and that contain infinitely many occurrences of the letter 1. Every degree is of the form $[\langle f\rangle]$ for some $f : \mathbb{N} \to \mathbb{N}$.

The following lemma is concerned with some basic operations on functions that have no effect on the degree of $\langle f\rangle$ (multiplication with a constant, and x- and y-shifts), and others by which we may go to a lower degree (taking subsequences, merging blocks).

Lemma 4.2. *Let $f : \mathbb{N} \to \mathbb{N}$, and $a,b \in \mathbb{N}$. It holds that:*

(i) $\langle af(n)\rangle \equiv \langle f(n)\rangle$, for $a > 0$,
(ii) $\langle f(n+a)\rangle \equiv \langle f(n)\rangle$,
(iii) $\langle f(n)+a\rangle \equiv \langle f(n)\rangle$,
(iv) $\langle f(n)\rangle \geq \langle f(an)\rangle$, for $a > 0$,
(v) $\langle f(n)\rangle \geq \langle af(2n)+bf(2n+1)\rangle$.

Definition 4.3. *Let A be a set. A function $f : \mathbb{N} \to A$ is* ultimately periodic *if for some integers $n_0 \geq 0$, $p > 0$ we have $f(n+p) = f(n)$ for all $n \geq n_0$.*

Definition 4.4. *A function $f : \mathbb{N} \to \mathbb{N}$ is called* spiralling *if*

(i) $\lim_{n\to\infty} f(n) = \infty$, and
(ii) for every $m \geq 1$, the function $n \mapsto f(n) \bmod m$ is ultimately periodic.

Functions with the property (ii) in Definition 4.4 have been called 'ultimately periodic reducible' by Siefkes [5] (quotation from [6]). Note that the identity function is spiralling. Furthermore scalar products, and pointwise sums and products of spiralling functions are again spiralling. As a consequence also polynomials $a_k n^k + a_{k-1}n^{k-1} + \cdots + a_0$ are spiralling.

In the remainder of this section, we will characterise the transducts σ of $\langle f\rangle$ for spiralling f. We will show that if such a transduct σ is not ultimately periodic, then it is equivalent to a sequence $\langle g\rangle$ for a spiralling function g, and moreover, g can be obtained from f by a 'weighted product'.

Lemma 4.5. *Let $f : \mathbb{N} \to \mathbb{N}$ be a spiralling function. We have $\langle f\rangle \geq \sigma$ if and only if σ is of the form*

$$\sigma = w \cdot \prod_{i=0}^{\infty}\prod_{j=0}^{m-1} p_j\, c_j^{\varphi(i,j)} \qquad \text{where} \qquad \varphi(i,j) = \frac{f(n_0 + mi + j) - a_j}{z}, \qquad (1)$$

for some integers $n_0, m, a_j \geq 0$ and $z > 0$, and finite words $w, p_j, c_j \in \mathbf{2}^*$ $(0 \leq j < m)$ such that $\varphi(i,j) \in \mathbb{N}$ for all $i \in \mathbb{N}$ and $j \in \mathbb{N}_{<m}$.

Proof. Assume $\langle f\rangle \geq \sigma$ and let $T = \langle Q, q_0, \delta, \lambda\rangle$ be an FST that transduces $\langle f\rangle$ to σ. As f is spiralling, there exist $\ell_0, p \in \mathbb{N}$, $p > 0$ such that $f(n) \equiv f(n+p)$ (mod $Z(T)$) for every $n \geq \ell_0$. Moreover, as a consequence of $\lim_{n\to\infty} f(n) = \infty$, there exists $\ell_1 \in \mathbb{N}$ such that $f(n) \geq |Q|$ for every $n \geq \ell_1$.

For $n \in \mathbb{N}$, let $q_n \in Q$ be the state that the automaton T is in, before reading the n-th occurrence of 1 in the sequence σ (i.e., the start of the block $10^{f(n)}$). By the pigeonhole principle there exist $n_0, m \in \mathbb{N}$ with $\max\{\ell_0, \ell_1\} < n_0$ and $m > 0$ such that $m \equiv 0 \pmod{p}$ and $q_{n_0} = q_{n_0+m}$. Then, for every $i \in \mathbb{N}$ and $j \in \mathbb{N}_{<m}$, we have $f(n_0 + mi + j) \geq |Q|$ and $f(n_0 + mi + j) \equiv f(n_0 + j) \pmod{\mathsf{Z}(t)}$. For $j \in \mathbb{N}_{<m}$, we define $a_j = \min\{f(n_0 + mi + j) \mid i \in \mathbb{N}\}$. Note that $a_j \geq |Q|$. Then for all $i \in \mathbb{N}$ and $j \in \mathbb{N}_{<m}$ we have: $f(n_0 + mi + j) = a_j + \varphi(i,j) \cdot z$ where $\varphi(i,j) = (f(n_0 + mi + j) - a_j)/z$ and $z = \mathsf{Z}(T)$. Using Lemma 3.3 it follows by induction that $q_{n_0+n} = q_{n_0+m+n}$ for every $n \in \mathbb{N}$. Hence $q_{n_0+j} = q_{n_0+mi+j}$ for every $i \in \mathbb{N}$ and $j \in \mathbb{N}_{<m}$. Then, again by Lemma 3.3, for every $j \in \mathbb{N}_{<m}$ there exist words $p_j, c_j \in \mathbf{2}^*$ such that $\lambda(q_{n_0+mi+j}, 10^{f(n_0+mi+j)}) = \lambda(q_{n_0+j}, 10^{a_j+\varphi(i,j)\cdot\mathsf{Z}(T)}) = p_j c_j^{\varphi(i,j)}$ for all $i \in \mathbb{N}$. We conclude that

$$\sigma = \prod_{n=0}^{\infty} \lambda(q_n, 10^{f(n)}) = w \cdot \prod_{i=0}^{\infty} \prod_{j=0}^{m-1} p_j \, c_j^{\varphi(i,j)} \,,$$

where $w = \prod_{n=0}^{n_0-1} \lambda(q_n, 10^{f(n)})$.

For the other direction, we refer to the extended version [4].   □

Example 4.6. We illustrate the influence of both the zero-loops of the automaton as well as the growth of the block lengths of the input sequence, on the size m of the innermost product of Lemma 4.5. Consider the FST $T = \langle\{q_0, q_1, q_2\}, q_0, \delta, \lambda\rangle$ in Fig. 3 on the left, and the sequence $\langle\lfloor\frac{n}{2}\rfloor\rangle = 11101010^2 10^2 10^3 10^3 10^4 10^4 \cdots$, where $\lfloor x \rfloor = \max\{n \in \mathbb{N} \mid n \leq x\}$. We investigate the sequence $r_0 a_0 r_1 a_1 r_2 \cdots$ of states r of T alternated with letters a from the input sequence, in such a way that T is in state r_n after having read the word $a_0 a_1 \cdots a_{n-1}$,

$$\underline{q_0} 1 q_2 1 q_1 1 q_0 0 q_1 1 q_0 0 q_1 1 q_0 0 q_1 0 q_2 1 q_1 0 q_2 0 \underline{q_0} 1 q_2 0 q_0 0 q_1 0 q_2 1 q_1 0 q_2 0 q_0 0 \cdots$$

The two underlined occurrences of q_0 indicate a repetition of states in combination with a repetition of the block size modulo $\mathsf{Z}(T) = 3$. The number of blocks between these occurrences, forming the repetition, is $m = 6$. Actually, following the proof of Lemma 4.5 precisely, the algorithm would instead select the repetition starting from the second underlined occurrence of q_0. The reason is that

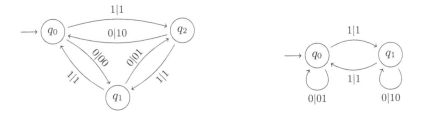

Fig. 3. Transducers used in Examples 4.6 and 4.8.

in general, when reading a block $100\cdots0$, only after reading $|Q|$ zeros, we are guaranteed to be in a zero-loop. For this FST T, all states are on a zero-loop, and so we enter the loop immediately.

From Lemma 4.5 it follows that FSTs can only transform the length of blocks by linear functions (and merge blocks). As a consequence, FSTs typically cannot slow down the growth rate of blocks in spiralling sequences by more than a linear factor. This yields the following simple criterion for non-reducibility.

Lemma 4.7. *Let* $f, g : \mathbb{N} \to \mathbb{N}$ *be such that* f *is spiralling,* g *is not ultimately periodic and* $g \in o(f)$, *i.e., for every* $a \in \mathbb{N}$ *there is* $n_0 \in \mathbb{N}$ *such that for all* $n \geq n_0$ *it holds that* $f(n) \geq ag(n)$. *Then* $\langle f \rangle \not\geq \langle g \rangle$.

Note that the reverse $\langle g \rangle \not\geq \langle f \rangle$ does *not* follow. For example, we have $\langle 4^n \rangle \not\geq \langle 2^n \rangle$, but $\langle 2^n \rangle \geq \langle 4^n \rangle$.

There can be several ways to factor the transduct σ in the statement of Lemma 4.5, as shown by the following example.

Example 4.8. Consider the FST $T = \langle \{q_0, q_1\}, q_0, \delta, \lambda \rangle$ in Fig. 3 on the right, and the sequence $\langle n \rangle = 11010^2 10^3 10^4 \cdots$. The T-transduct of $\langle n \rangle$ is $T(\langle n \rangle) = 1\,1(01)\,1(10)^2\,1(01)^3\,1(10)^4\cdots$. Using the double product (1) of Lemma 4.5 we have $w = \varepsilon$, $n_0 = 0$, $m = 2$, $p_0 = p_1 = 1$, $c_0 = 01$, $c_1 = 10$, $a_0 = a_1 = 0$, and $z = Z(T) = 1$. Thus, $\varphi(i, j) = 2i + j$. Note that $c_1^\omega = p_0 c_0^\omega$, and so we can merge the factors of the innermost product, decreasing its size to $m = 1$. Then $T(\langle n \rangle) = 1101\,1101(0101)\,1101(0101)^2\,1101(0101)^3\cdots$, that is, in the double product (1), $T(\langle n \rangle)$ can be factored by choosing $m = 1$, $p_0 = 1101$, and $c_0 = 0101$.

Whenever we have the situation $c_j^\omega = p_{j+1} c_{j+1}^\omega$ in the representation (1) for some $j \in \mathbb{N}_{<m}$ (addition is modulo m), we speak of a 'transition ambiguity'. We will eliminate such ambiguities by merging factors of the innermost product, as in Lemma 4.10. In general, this merging may involve weighted sums in the exponents. The following lemma is folklore, see for instance [7].

Lemma 4.9. *If a sequence* σ *is periodic with period lengths* k *and* ℓ, *then it is periodic with period length* $\gcd(k, \ell)$.

Lemma 4.10. *Let* $u, v, w \in 2^*$ *be finite words with* $u, w \neq \varepsilon$ *such that* $u^\omega = vw^\omega$. *Then there exists a word* $x \in 2^*$ *and* $a, b \in \mathbb{N}$ *such that* $u^m v w^n = v x^{am + bn}$ *for all* $m, n \in \mathbb{N}$.

Our goal is to obtain a simple characterisation of the degrees of the transducts σ of spiralling sequences $\langle f \rangle$. To this end, we transform the transduct σ into a sequence σ' by replacing in the double product of Lemma 4.5 the displayed occurrence of p_j by 1 and of c_j by 0 for every $j \in \mathbb{N}_{<m}$. To guarantee that this transformation does not change the degree, that is $\sigma' \equiv \sigma$, we first have to resolve transition ambiguities.

For the back transformation blocks $10^{\varphi(i,j)}$ have to be replaced by $p_j c_j^{\varphi(i,j)}$, an operation that is easily realised by an FST. If the product does not contain transition ambiguities, then also the transformation from σ into σ' can be realised by an FST and thus does not change the degree of the sequence, hence $\sigma \equiv \sigma'$. If there exists $j \in \mathbb{N}_{<m}$ with $c_j^\omega = p_{j+1}c_{j+1}^\omega$, then, by this transformation of σ into σ', one possibly leaves the degree of σ, i.e., $\sigma > \sigma'$. This is because for large enough $i \in \mathbb{N}$, an FST cannot recognise where a block $p_j c_j^{\varphi(i,j)}$ ends and where the next block $p_{j+1}c_{j+1}^{\varphi(i,j+1)}$ starts. This might make it impossible to realise the transformation by an FST, as then the FST cannot replace p_{j+1} by 1.

Definition 4.11. A *weight* is a tuple $\langle a_0, \ldots, a_{k-1}, b \rangle \in \mathbb{Q}^{k+1}$ of rational numbers such that $a_0, \ldots, a_{k-1} \geq 0$. Given a weight $\alpha = \langle a_0, \ldots, a_{k-1}, b \rangle$ and a function $f : \mathbb{N} \to \mathbb{N}$ we define $\alpha \cdot f \in \mathbb{Q}$ by

$$\alpha \cdot f \ = \ a_0 f(0) + a_1 f(1) + \cdots + a_{k-1} f(k-1) + b \,.$$

The weight α is called *constant* when $a_j = 0$ for all $j \in \mathbb{N}_{<k}$. For a tuple of weights $\boldsymbol{\alpha} = \langle \alpha_0, \ldots, \alpha_{m-1} \rangle$ we define its *rotation* by $\boldsymbol{\alpha}' = \langle \alpha_1, \ldots, \alpha_{m-1}, \alpha_0 \rangle$.

For functions $f : \mathbb{N} \to \mathbb{N}$, and tuples $\boldsymbol{\alpha} = \langle \alpha_0, \alpha_1, \ldots, \alpha_{m-1} \rangle$ of weights, the *weighted product* of $\boldsymbol{\alpha}$ and f is a function $\boldsymbol{\alpha} \otimes f : \mathbb{N} \to \mathbb{Q}$ that is defined by induction on n through the following scheme of equations:

$$(\boldsymbol{\alpha} \otimes f)(0) = \alpha_0 \cdot f$$
$$(\boldsymbol{\alpha} \otimes f)(n+1) = (\boldsymbol{\alpha}' \otimes \mathcal{S}^{|\alpha_0|-1}(f))(n) \qquad\qquad (n \in \mathbb{N})$$

where $|\alpha_i|$ is the length of the tuple α_i, and $\mathcal{S}^k(f)$ is the k-th shift of f. A weighted product $\boldsymbol{\alpha} \otimes f$ is called *natural* if $(\boldsymbol{\alpha} \otimes f)(n) \in \mathbb{N}$ for all $n \in \mathbb{N}$.

In what follows, all weighted products $\boldsymbol{\alpha} \otimes f$ that we consider are assumed to be natural.

Example 4.12. Let $f(n) = n$ for all $n \in \mathbb{N}$, and $\boldsymbol{\alpha} = \langle \alpha_1, \alpha_2 \rangle$ with $\alpha_1 = \langle 1, 2, 3, 4 \rangle$, $\alpha_2 = \langle 0, 1, 1 \rangle$. Interpreting the functions f and $\boldsymbol{\alpha} \otimes f$ as sequences, the computation of $\boldsymbol{\alpha} \otimes f$ can be visualised as follows:

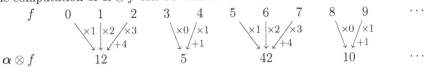

Thus, for $n = 0, 1, 2, 3 \ldots$, $(\boldsymbol{\alpha} \otimes f)(n)$ takes the values $12, 5, 42, 10, \ldots$.

Lemma 4.13. *Let $\boldsymbol{\alpha}$ be an m-tuple of weights ($m > 0$), and let $f : \mathbb{N} \to \mathbb{N}$. For all $n \in \mathbb{N}$ we have $(\boldsymbol{\alpha} \otimes f)(n) = \alpha_r \cdot \mathcal{S}^t(f)$ where $q, r \in \mathbb{N}$ with $r < m$ are such that $n = qm + r$, and $t = q \cdot \sum_{j=0}^{m-1}(|\alpha_j| - 1) + \sum_{j=0}^{r-1}(|\alpha_j| - 1)$.*

Lemma 4.14. *Let $f : \mathbb{N} \to \mathbb{N}$. If $\langle \boldsymbol{\alpha} \otimes f \rangle \not\equiv 0$, then there exists $i \in \mathbb{N}_{<|\boldsymbol{\alpha}|}$ such that α_i is a non-constant weight.*

Lemma 4.15. *Let $f : \mathbb{N} \to \mathbb{N}$ be spiralling, and let $\boldsymbol{\alpha}$ be a tuple of non-constant weights. Then $\boldsymbol{\alpha} \otimes f$ is spiralling.*

We will show that weighted products give rise to a characterisation, up to equivalence $\equiv$, of functions realised by FSTs on the set of spiralling sequences, see Theorem 4.22.

Lemma 4.16. *Let* $f : \mathbb{N} \to \mathbb{N}$, *and* $\boldsymbol{\alpha}$ *a tuple of weights. If* $\boldsymbol{\alpha} \otimes f$ *is a natural weighted product, then we have* $\langle f \rangle \geq \langle \boldsymbol{\alpha} \otimes f \rangle$.

Proof. Let $m = |\boldsymbol{\alpha}|$, $k_i = |\alpha_i| - 1$, and $\alpha_i = \langle a_{i,0}/d_i, a_{i,1}/d_i, \ldots, a_{i,k_i-1}/d_i, b_i/d_i \rangle$ with $a_{i,j}, d_i \in \mathbb{N}$ and $b_i \in \mathbb{Z}$ for all $i \in \mathbb{N}_{<m}$ and $j \in \mathbb{N}_{<k_i}$ (clearly weights can always be brought into this form).

We define $T = \langle Q, q^0_{m-1,k_{m-1}-1}, \delta, \lambda \rangle$ consisting of states $q^h_{i,j}$ for every $i \in \mathbb{N}_{<m}$, $j \in \mathbb{N}_{<k_i}$, and h such that $\min(0, b_i) \leq h < d_i$. The superscript h ($\min(0, b_i) \leq h < d_i$) in a state $q^h_{i,j}$ indicates the amount h/d_i of zeros that still has to be consumed/produced. The transition and output functions $\langle \delta, \lambda \rangle : Q \times \mathbf{2} \to Q \times \mathbf{2}^*$ of T are defined as follows; let $\mathrm{id}_{\geq 0} : \mathbb{Z} \to \mathbb{N}$ be defined by $\mathrm{id}_{\geq 0}(z) = z$ if $z \geq 0$ and $\mathrm{id}_{\geq 0}(z) = 0$ otherwise.

$$\langle \delta, \lambda \rangle(q^h_{i,j}, 0) = \langle q^{h'}_{i,j}, 0^e \rangle \quad \text{where } e = \lfloor \mathrm{id}_{\geq 0}(h + a_{i,j})/d_i \rfloor, h' = h + a_{i,j} - ed_i$$

$$\langle \delta, \lambda \rangle(q^h_{i,j}, 1) = \langle q^h_{i,j+1}, \varepsilon \rangle \quad (j < k_i - 1)$$

$$\langle \delta, \lambda \rangle(q^h_{i,k_i-1}, 1) = \langle q^{h'}_{i',0}, 10^e \rangle \quad \text{where } e = \lfloor \mathrm{id}_{\geq 0}(b_{i'})/d_{i'} \rfloor, h' = b_{i'} - ed_{i'},$$

where $i' = i + 1 \bmod m$. See [4] for the proof of $\langle f \rangle \geq_T \langle \boldsymbol{\alpha} \otimes f \rangle$. $\square$

Definition 4.17. Let $f : \mathbb{N} \to \mathbb{N}$ be a function, and, for some $m > 0$, let $\boldsymbol{\alpha}$ be an m-tuple of weights such that $\boldsymbol{\alpha} \otimes f$ is a natural weighted product. Let $\boldsymbol{p}$ and $\boldsymbol{c}$ be m-tuples of finite words. We define the sequence $\Phi(f, \boldsymbol{\alpha}, \boldsymbol{p}, \boldsymbol{c}) \in \mathbf{2}^{\mathbb{N}}$ by

$$\Phi(f, \boldsymbol{\alpha}, \boldsymbol{p}, \boldsymbol{c}) = \prod_{i=0}^{\infty} \prod_{j=0}^{m-1} p_j \, c_j^{\varphi(i,j)} \quad \text{where} \quad \varphi(i,j) = (\boldsymbol{\alpha} \otimes f)(mi + j).$$

Note that $\langle f \rangle$ can also be cast into this notation: $\langle f \rangle = \Phi(f, \langle\langle 1, 0 \rangle\rangle, \langle 1 \rangle, \langle 0 \rangle)$. For the following lemma we recall that for a tuple $\boldsymbol{a} = \langle a_0, a_1, \ldots, a_{k-1} \rangle$, we write $\boldsymbol{a}'$ for the rotation $\langle a_1, \ldots, a_{k-1}, a_0 \rangle$.

Lemma 4.18. *Let* $f, \boldsymbol{\alpha}, \boldsymbol{p}, \boldsymbol{c}$ *be as in Definition 4.17. We have* $\Phi(f, \boldsymbol{\alpha}, \boldsymbol{p}, \boldsymbol{c}) = p_0 c_0^{\alpha_0 \cdot f} \cdot \Phi(\mathcal{S}^{|\alpha_0|-1}(f), \boldsymbol{\alpha}', \boldsymbol{p}', \boldsymbol{c}')$.

Lemma 4.19. *Let* $f : \mathbb{N} \to \mathbb{N}$ *be a spiralling function, and let* $\sigma \in \mathbf{2}^{\mathbb{N}}$ *be such that* $\langle f \rangle \geq \sigma$ *and* $\sigma \not\equiv \mathbf{0}$. *Then there exist* $n_0, m \in \mathbb{N}$, *a word* $w \in \mathbf{2}^*$, *a tuple of weights* $\boldsymbol{\alpha}$, *and tuples of words* $\boldsymbol{p}$ *and* $\boldsymbol{c}$ *with* $|\boldsymbol{\alpha}| = |\boldsymbol{p}| = |\boldsymbol{c}| = m > 0$ *such that:*

(i) $\sigma = w \cdot \Phi(\mathcal{S}^{n_0}(f), \boldsymbol{\alpha}, \boldsymbol{p}, \boldsymbol{c})$,
(ii) $c_j^{\omega} \neq p_{j+1} c_{j+1}^{\omega}$ *for every* j *with* $0 \leq j < m - 1$, *and* $c_{m-1}^{\omega} \neq p_0 c_0^{\omega}$, *and*
(iii) $c_j \neq \varepsilon$, *and* α_j *is non-constant, for all* $j \in \mathbb{N}_{<m}$.

Proof. By Lemma 4.5, there exist $n_0, m, a_j, z \in \mathbb{N}$ ($j \in \mathbb{N}_{<m}$), $w \in \mathbf{2}^*$, and $\boldsymbol{p}, \boldsymbol{c} \in (\mathbf{2}^*)^m$ such that $\sigma = w \cdot \Phi(\mathcal{S}^{n_0}(f), \boldsymbol{\alpha}, \boldsymbol{p}, \boldsymbol{c})$, where, for $j \in \mathbb{N}_{<m}$, α_j is defined by $\alpha_j = \langle \frac{1}{z}, -\frac{a_j}{z} \rangle$.

We now repeatedly alter the tuples $\boldsymbol{\alpha}$, $\boldsymbol{p}$, $\boldsymbol{c}$ until conditions (ii) and (iii) are fulfilled while condition (i) is upheld. For this we let $n_0 \in \mathbb{N}$, $w \in \mathbf{2}^*$, $m \in \mathbb{N}$, $\boldsymbol{\alpha}$, $\boldsymbol{p}$, and $\boldsymbol{c}$ with $|\boldsymbol{\alpha}| = |\boldsymbol{p}| = |\boldsymbol{c}| = m$ be arbitrary such that (i) holds.

First note that, if $m = 1$ and condition (ii) or (iii) are violated, then $\sigma \equiv \mathbf{0}$, contradicting the assumption.

In case (ii) does not hold, consider the smallest $h \in \mathbb{N}_{<m}$ such that $c_h^\omega = p_{h+1}c_{h+1}^\omega$ where addition in the subscripts is computed modulo m. We assume $h < m - 1$; the case $h = m - 1$ proceeds analogously, using Lemma 4.18. For $i \in \mathbb{N}$ and $j \in \mathbb{N}_{<m}$, we let $\varphi(i, j) = (\boldsymbol{\alpha} \otimes \mathcal{S}^{n_0}(f))(mi + j)$. By Lemma 4.10 there are integers $a, b \geq 0$ and a word $x \in \mathbf{2}^*$ such that $c_h^{\varphi(i,h)} p_{h+1} c_{h+1}^{\varphi(i,h+1)} = p_{h+1} x^a \varphi(i,h) + b\varphi(i,h+1)$ $(\star)$. We now define a tuple of weights $\boldsymbol{\beta}$, and tuples of words $\boldsymbol{q}$ and $\boldsymbol{d}$, with $|\boldsymbol{\beta}| = |\boldsymbol{q}| = |\boldsymbol{d}| = m - 1$, as follows: Let $j \in \mathbb{N}_{<m-1}$. If $j < h$, then we define $q_j = p_j$, $d_j = c_j$, and $\beta_j = \alpha_j$. If $j > h$, we define $q_j = p_{j+1}$, $d_j = c_{j+1}$, and $\beta_j = \alpha_{j+1}$. If $j = h$, we define $q_j = p_h p_{h+1}$, $d_j = x$, and we let the weight β_j be defined as follows: For $\alpha_h = \langle r_0, r_1, \ldots, r_{k-1}, e \rangle$ and $\alpha_{h+1} = \langle r_0', r_1', \ldots, r_{\ell-1}', e' \rangle$, let $\beta_h = \langle ar_0, ar_1, \ldots, ar_{k-1}, br_0', br_1', \ldots, br_{\ell-1}', ae + be' \rangle$. By definition of $\boldsymbol{q}$, $\boldsymbol{d}$, and $\boldsymbol{\beta}$, to verify $\Phi(\mathcal{S}^{n_0}(f), \boldsymbol{\alpha}, \boldsymbol{p}, \boldsymbol{c}) = \Phi(\mathcal{S}^{n_0}(f), \boldsymbol{\beta}, \boldsymbol{q}, \boldsymbol{d})$, it suffices to check, for all $i \in \mathbb{N}$, $p_h c_h^{\varphi(i,h)} p_{h+1} c_{h+1}^{\varphi(i,h+1)} = q_h d_h^{\varphi'(i,h)}$; here, for $i \in \mathbb{N}$ and $j \in \{0, \ldots, m-2\}$, $\varphi'(i, j)$ is defined by $\varphi'(i, j) = (\boldsymbol{\beta} \otimes \mathcal{S}^{n_0}(f))((m-1)i + j)$. Fix $i \in \mathbb{N}$. By Lemma 4.13 we have $\varphi(i, h) = \alpha_h \cdot \mathcal{S}^t(f)$ and $\varphi(i, h+1) = \alpha_{h+1} \cdot \mathcal{S}^{t+k}(f)$, for some $t \in \mathbb{N}$. Also we have $\varphi'(i, h) = \beta_h \cdot \mathcal{S}^{t'}(f)$ for some $t' \in \mathbb{N}$. By definition of $\boldsymbol{\beta}$ we obtain $t' = t$. It follows that $\varphi'(i, h) = a \cdot \varphi(i, h) + b \cdot \varphi(i, h+1)$, and we conclude by $(\star)$. Repeat the procedure with $\boldsymbol{\beta}$, $\boldsymbol{q}$, $\boldsymbol{d}$.

For the case (iii) that does not hold we refer to the extended version [4]. □

For the proof of the following theorem we allow for a more liberal version of transducers. Instead of input letters along the edges we now allow input *words*. Transitions of these transducers are of the form $q \xrightarrow{\langle u, v \rangle | w} q'$. The idea is that this transition is taken if the automaton is in state q and the input word is of the form $uv\tau$. Then the automaton produces output w and switches to state q', consuming u and continuing with $v\tau$.

Definition 4.20. An FST with *look-ahead* (FST*) is a tuple $T = \langle Q, q_0, D, \delta, \lambda \rangle$ where Q is a finite set of states, $q_0 \in Q$ is the initial state, the finite set $D \subseteq Q \times \mathbf{2}^+ \times \mathbf{2}^*$ is the input domain of the transition function $\delta : D \to Q$, and the output function $\lambda : D \to \mathbf{2}^*$, satisfying the following condition: for all $q \in Q$, $u_1, u_2, v_1, v_2 \in \mathbf{2}^*$ if $u_1 u_2$ is a prefix of $v_1 v_2$ and $\langle q, u_1, u_2 \rangle \in D$ and $\langle q, v_1, v_2 \rangle \in D$, then $u_1 = v_1$ and $u_2 = v_2$.

We lift δ to a partial function $\delta^\star : Q \times \mathbf{2}^* \rightharpoonup Q$ by $\delta^\star(q, \varepsilon) = q$ and

$$\delta^\star(q, u_1 u_2 v) = \delta^\star(\delta(q, u_1, u_2), u_2 v) \qquad (\langle q, u_1, u_2 \rangle \in D, v \in \mathbf{2}^*).$$

Similarly, we lift λ to a partial function $\lambda^\star : Q \times \mathbf{2}^\infty \rightharpoonup \mathbf{2}^\infty$ by $\lambda^\star(q, \varepsilon) = \varepsilon$ and

$$\lambda^\star(q, u_1 u_2 v) = \lambda(q, u_1, u_2) \cdot \lambda^\star(\delta(q, u_1, u_2), u_2 v) \qquad (\langle q, u_1, u_2 \rangle \in D, v \in \mathbf{2}^\infty).$$

The partial function $T : \mathbf{2}^\infty \rightharpoonup \mathbf{2}^\infty$ *realised* by the FST* T is defined by $T(u) = \lambda^\star(q_0, u)$, for all $u \in \mathbf{2}^\infty$.

These transducers can be simulated by FSTs.

Lemma 4.21. *For every FST* T there is an FST T' such that for all $u \in \mathbf{2}^\infty$, $T'(u) = T(u)$ whenever $T(u)$ is defined.*

Theorem 4.22. *Let $f : \mathbb{N} \to \mathbb{N}$ be spiralling, and $\sigma \in \mathbf{2}^\mathbb{N}$. Then $\langle f \rangle \geq \sigma$ if and only if $\sigma \equiv \langle \alpha \otimes \mathcal{S}^{n_0}(f) \rangle$ for some integer $n_0 \geq 0$, and a tuple of weights α.*

Proof. One direction is by Lemma 4.16. For the other, assume $\langle f \rangle \geq \sigma$. If $\sigma \in \mathbf{0}$, then $\sigma \equiv \langle \langle \langle 0, 0 \rangle \rangle \otimes \mathcal{S}^0(f) \rangle = \langle n \mapsto 0 \rangle = 1^\omega$. Thus let $\sigma \notin \mathbf{0}$. By Lemma 4.19 there exist $n_1, m \in \mathbb{N}$, $w \in \mathbf{2}^*$, α, p and c with $|\alpha| = |p| = |c| = m > 0$ such that $\sigma = w \cdot \Phi(\mathcal{S}^{n_1}(f), \alpha, p, c)$, and fulfilling the conditions (ii) and (iii) of Lemma 4.19. We abbreviate $g = \alpha \otimes \mathcal{S}^{n_1}(f)$. We will show that $\sigma \equiv \langle g \rangle$.

By Lemma 4.15 we have that the function g is spiralling too.

By conditions (ii) and (iii), for every $j \in \mathbb{N}_{<m}$, there exists $t_j \in \mathbb{N}$ such that $c_j^\omega(t_j) \neq (p_{j+1} c_{j+1}^\omega)(t_j)$ (where addition is modulo m); let t_j be minimal with this property.

For $j \in \mathbb{N}_{<m}$, let $\ell_j, \ell_j' \in \mathbb{N}$ be minimal such that $|c_j c_j^{\ell_j}| > t_j$ and $|p_{j+1} c_{j+1}^{\ell_j'}| > t_j$. Then by minimality of t_j and ℓ_j, we obtain

(i) $c_j c_j^{\ell_j - 1} \sqsubseteq p_{j+1} c_{j+1}^{\ell_j'} \tau$ for every $\tau \in \mathbf{2}^\mathbb{N}$, and

(ii) $c_j c_j^{\ell_j} \not\sqsubseteq p_{j+1} c_{j+1}^{\ell_j'} \tau$ for every $\tau \in \mathbf{2}^\mathbb{N}$

(with again addition computed modulo m). From (i) we moreover obtain

(iii) $c_j c_j^{\ell_j} \sqsubseteq c_j^n p_{j+1} c_{j+1}^{\ell_j'} \tau$ for every $n > 0$ and $\tau \in \mathbf{2}^\mathbb{N}$.

Next, we take a suffix σ' of σ such that every occurrence of a block $p_{j+1} c_{j+1}^{\varphi(i,j)}$ has as a prefix $p_{j+1} c_{j+1}^{\ell_j}$. Let $n_2 \in \mathbb{N}$ be such that for $g' = \mathcal{S}^{n_2 \cdot m}(g)$ we have that $g'(n) > \max\{t_j \mid j \in \mathbb{N}_{<m}\}$ for all $n \in \mathbb{N}$; the existence of such an n_2 follows from g being spiralling. To prove $\sigma \equiv \langle g \rangle$ it suffices to show $\sigma' \equiv \langle g' \rangle$ where

$$\sigma' = \prod_{i=n_2}^\infty \prod_{j=0}^{m-1} p_j c_j^{\varphi(i,j)} = \prod_{i=0}^\infty \prod_{j=0}^{m-1} p_j c_j^{\varphi'(i,j)} \qquad \langle g' \rangle = \prod_{i=0}^\infty \prod_{j=0}^{m-1} 10^{\varphi'(i,j)}$$

where $\varphi(i,j) = g(mi + j)$, and $\varphi'(i,j) = g'(mi + j)$. Note that by the choice of n_2, we have $\varphi'(i, j+1) \geq \ell_j'$ for all $i \in \mathbb{N}$ and $j \in \mathbb{N}_{<m}$.

It is clear how to construct an FST that transduces $\langle g' \rangle$ to σ'. For $\sigma' \geq \langle g' \rangle$, we define a FST* $T = \langle Q, q_{m-1}, D, \delta, \lambda \rangle$, as follows, and apply Lemma 4.21. Let

$Q = \{q_j \mid j \in \mathbb{N}_{<m}\}$ and $D = \{\langle q_j, c_j, c_j^{\ell_j}\rangle \mid j \in \mathbb{N}_{<m}\} \cup \{\langle q_j, p_{j+1}, c_{j+1}^{\ell'_j}\rangle \mid j \in \mathbb{N}_{<m}\}$, and define δ, λ by

$$\langle \delta, \lambda \rangle (q_j, c_j, c_j^{\ell_j}) = \langle q_j, 0 \rangle \qquad \langle \delta, \lambda \rangle (q_j, p_{j+1}, c_{j+1}^{\ell'_j}) = \langle q_{j+1}, 1 \rangle.$$

We now argue that $\sigma' \geq_T \langle g' \rangle$. This follows from the following facts:

(a) $\lambda^*(q_j, p_{j+1}c^{\varphi'(i,j+1)}\tau) = 1 \cdot \lambda^*(q_{j+1}, c^{\varphi'(i,j+1)}\tau)$,
(b) $\lambda^*(q_j, c_j^n p_{j+1}c^{\varphi'(i,j+1)}\tau) = 0 \cdot \lambda^*(q_{j+1}, c^{\varphi'(i,j+1)}\tau)$ for all $n > 0$, since by item (iii) we have $c_j c_j^{\ell_j} \sqsubseteq c_j^n p_{j+1}c^{\varphi'(i,j+1)}\tau$.   □

Lemma 4.23. *Let $f : \mathbb{N} \to \mathbb{N}$ be spiralling, and $\sigma \not\equiv \mathbf{0}$ with $\langle f \rangle \geq \sigma$. Then we have $\sigma \geq \langle \langle \beta \rangle \otimes \mathcal{S}^{n_0}(f) \rangle$ for some integer $n_0 \geq 0$, and a non-constant weight β.*

Theorem 4.24. *There is a non-atom, non-zero degree $[\sigma]$ that has no atom degree below it. Hence, non-zero transducts of σ start an infinite descending chain.*

Proof. We define the function $f : \mathbb{N} \to \mathbb{N}$ by $f(n) = 2^n$. We show that the degree $[\langle f \rangle]$ has no atom degree below it. Let $\sigma \not\equiv \mathbf{0}$ with $\langle f \rangle \geq \sigma$. By Lemma 4.23 there is a non-constant weight $\beta = \langle a_0, a_1, \ldots, a_{k-1}, b \rangle$ such that $\sigma \geq \langle g \rangle$ where $g = \langle \beta \rangle \otimes \mathcal{S}^{n_0}(f)$. Since $f(n) = 2^n$ it follows that

$$g(n) = b + \sum_{i=0}^{k-1} a_i 2^{n_0 + nk + i} = b + 2^{nk} \sum_{i=0}^{k-1} a_i 2^{n_0 + i} = (g(0) - b) \cdot 2^{nk} + b.$$

By Lemma 4.16 we have that $\langle g \rangle = \langle (g(0) - b) \cdot 2^{nk} + b \rangle \equiv \langle 2^{nk} \rangle$. Thus we have $\sigma \geq \langle 2^{nk} \rangle$. Also $\langle 2^{nk} \rangle \geq \langle 2^{2nk} \rangle$ holds by Lemma 4.2 (iv), and by Lemma 4.7 we conclude $\langle 2^{2nk} \rangle \not\geq \langle 2^{nk} \rangle$.   □

5   Squares

In [1] it is shown that $[\langle n \rangle]$ is an atom degree. One of the main questions of [1] is whether there exist other atom degrees. Here we show that also $[\langle n^2 \rangle]$ is an atom degree. The main tool is Theorem 4.22, the characterisation of transducts of spiralling sequences, which implies the following proposition.

Proposition 5.1. *Let $p(n)$ be a polynomial of degree k with non-negative integer coefficients, and let σ be a transduct of $\langle p(n) \rangle$ with $\sigma \not\equiv \mathbf{0}$. Then $\sigma \geq \langle q(n) \rangle$ for some polynomial $q(n)$ of degree k with non-negative integer coefficients.*

Proof. By Lemma 4.23 it follows that $\sigma \geq \langle \langle \alpha \rangle \otimes \mathcal{S}^{n_0}(p) \rangle$ for some integer $n_0 \geq 0$, and a non-constant weight $\alpha = \langle a_0, \ldots, a_{k-1}, b \rangle$. Let $h \in \mathbb{N}_{<k}$ be such that $a_h \neq 0$. Then we find $((\langle \alpha \rangle \otimes \mathcal{S}^{n_0}(p))(n) = b + \sum_{j=0}^{k-1} a_j \cdot p(n_0 + nk + j)$, which can easily be recognised to be a polynomial $q(n)$ of degree k.   □

Theorem 5.2. *The degree $[\langle n^2 \rangle]$ is an atom.*

Proof. Let $\sigma \in \mathbf{2}^{\mathbb{N}}$ be a transduct of $\langle n^2 \rangle$ such that σ is not ultimately periodic. By Proposition 5.1 there are integers $a > 0$, $b, c \geq 0$ such that $\sigma \geq \langle an^2 + bn + c \rangle$. We first assume $2a \geq b$. Abbreviate $f(n) = an^2 + (2a+b)n$. The roman numbers below refer to Lemma 4.2. We derive

$$
\begin{aligned}
\langle an^2 + bn + c \rangle &\equiv \langle an^2 + bn \rangle && \text{by (iii)} \\
&\equiv \langle a(n+1)^2 + b(n+1) \rangle && \text{by (ii)} \\
&\equiv \langle f(n) \rangle && \text{by (iii)} \\
&\geq \langle b(f(2n)) + (2a - b)(f(2n+1)) \rangle && \text{by (v)} \\
&\equiv \langle 8a^2 n^2 + 16a^2 n \rangle \equiv \langle 8a^2 (n+1)^2 \rangle && \text{by (iii)} \\
&\equiv \langle (n+1)^2 \rangle && \text{by (i)} \\
&\equiv \langle n^2 \rangle && \text{by (ii)} .
\end{aligned}
$$

If $2a < b$, choose d such that $2ad \geq b$. Then we have $\langle an^2 + bn \rangle \geq \langle ad^2 n^2 + bdn \rangle$ by (iv), and we reason as above for $\langle a'n^2 + b'n \rangle$ with $a' = ad^2$ and $b' = bd$.

This shows that every non-ultimately periodic transduct of $\langle n^2 \rangle$ can be transduced back to $\langle n^2 \rangle$. Hence, the degree of $\langle n^2 \rangle$ is an atom. □

References

1. Endrullis, J., Hendriks, D., Klop, J.: Degrees of streams. J. Integers **11B**(A6), 1–40 (2011). Proceedings of the Leiden Numeration Conference 2010
2. Shallit, J.: Open problems in automata theory: an idiosyncratic view. LMS Keynote Talk in Discrete Mathematics, BCTCS (2014). https://cs.uwaterloo.ca/shallit/Talks/bc4.pdf
3. Sakarovitch, J.: Elements Of Automata Theory. Cambridge University Press, Cambridge (2003)
4. Endrullis, J., Grabmayer, C., Hendriks, D., Zantema, H.: The Degree of Squares is an Atom (Extended Version). Technical report 1506.00884, arxiv.org, June 2015
5. Siefkes, D.: Undecidable extensions of monadic second order successor arithmetic. Math. Logic Q. **17**(1), 385–394 (1971)
6. Seiferas, J., McNaughton, R.: Regularity-preserving relations. Theor. Comput. Sci. **2**(2), 147–154 (1976)
7. Cautis, S., Mignosi, F., Shallit, J., Wang, M., Yazdani, S.: Periodicity, morphisms, and matrices. Theor. Comput. Sci. **295**, 107–121 (2003)

Words with the Maximum Number of Abelian Squares

Gabriele Fici[1](✉) and Filippo Mignosi[2]

[1] Dipartimento di Matematica e Informatica, Università di Palermo, Palermo, Italy
`Gabriele.Fici@unipa.it`
[2] Dipartimento di Ingegneria e Scienze dell'Informazione e Matematica,
Università dell'Aquila, L'Aquila, Italy
`Filippo.Mignosi@di.univaq.it`

Abstract. An abelian square is the concatenation of two words that are anagrams of one another. A word of length n can contain $\Theta(n^2)$ distinct factors that are abelian squares. We study infinite words such that the number of abelian square factors of length n grows quadratically with n.

1 Introduction

A fundamental topic in Combinatorics on Words is the study of repetitions. A repetition in a word is a factor that is formed by the concatenation of two or more identical blocks. The simplest kind of repetition is a square, that is the concatenation of two copies of the same block, like *sciascia*. A famous conjecture of Fraenkel and Simpson [1] states that a word of length n contains less than n distinct square factors. Experiments strongly suggest that the conjecture is true, but a theoretical proof of the conjecture seems difficult. In [1], the authors proved a bound of $2n$. In [2], Ilie improved this bound to $2n - \Theta(\log n)$, but the conjectured bound is still far away.

Among the different generalizations of the notion of repetition, a prominent one is that of an abelian repetition. An abelian repetition in a word is a factor that is formed by the concatenation of two or more blocks that have the same number of occurrences of each letter in the alphabet. Of course, the simplest kind of abelian repetition is an abelian square, that is therefore the concatenation of a word with an anagram of itself, like *viavai*. Abelian squares were considered in 1961 by Erdös [3], who conjectured that there exist infinite words avoiding abelian squares (this conjecture has later been proved to be true, and the smallest possible size of an alphabet for which it holds has been proved to be 4 [4]).

We focus on the maximum number of abelian squares that a word can contain. Opposite to case of ordinary squares, a word of length n can contain $\Theta(n^2)$ distinct abelian square factors (see [5]). Since the total number of factors in a word of length n is quadratic in n, this means that there exist words in which a fixed proportion of all factors are abelian squares. So we turn our attention to infinite words, and we wonder whether there exist infinite words such that for every n any factor of length n contains, on average, a number of abelian

F. Manea and D. Nowotka (Eds.): WORDS 2015, LNCS 9304, pp. 122–134, 2015.
DOI: 10.1007/978-3-319-23660-5_11

squares that is quadratic in n. We call such an infinite word *abelian-square rich*. Since a random binary word of length n contains $\Theta(n\sqrt{n})$ distinct abelian square factors [6], the existence of abelian-square rich words is not immediate. We also introduce *uniformly abelian-square rich* words, that are infinite words such that for every n, every factor of length n contains a quadratic number of abelian squares.

As a first result, we prove that the famous Thue-Morse word is uniformly abelian-square rich. Then we look at the class of Sturmian words, that are aperiodic infinite words with the lowest factor complexity. In this case, we prove that if a Sturmian word is β-power free for some $\beta \geq 2$ (that is, does not contain repetitions of order β or higher), then it is uniformly abelian-square rich.

2 Notation and Background

Let $\Sigma = \{a_1, a_2, \ldots, a_\sigma\}$ be an ordered σ-letter alphabet. Let Σ^* stand for the free monoid generated by Σ, whose elements are called *words* over Σ. The *length* of a word w is denoted by $|w|$. The *empty word*, denoted by ε, is the unique word of length zero and is the neutral element of Σ^*. We also define $\Sigma^+ = \Sigma^* \setminus \{\varepsilon\}$.

A *prefix* (resp. a *suffix*) of a word w is any word u such that $w = uz$ (resp. $w = zu$) for some word z. A *factor* of w is a prefix of a suffix (or, equivalently, a suffix of a prefix) of w. The set of prefixes, suffixes and factors of the word w are denoted by $Pref(w)$, $Suff(w)$ and $Fact(w)$, respectively. From the definitions, we have that ε is a prefix, a suffix and a factor of any word.

For a word w and a letter $a_i \in \Sigma$, we let $|w|_{a_i}$ denote the number of occurrences of a_i in w. The *Parikh vector* (sometimes called *composition vector*) of a word w over $\Sigma = \{a_1, a_2, \ldots, a_\sigma\}$ is the vector $P(w) = (|w|_{a_1}, |w|_{a_2}, \ldots, |w|_{a_\sigma})$. An *abelian k-power* is a word of the form $v_1 v_2 \cdots v_k$ where all the v_i's have the same Parikh vector. An abelian 2-power is called an *abelian square*.

An *infinite word* w over Σ is an infinite sequence of letters from Σ, that is, a function $w : \mathbb{N} \mapsto \Sigma$. Given an infinite word w, the *recurrence index* $R_w(n)$ of w is the least integer m (if any exists) such that every factor of w of length m contains all factors of w of length n. If the recurrence index is defined for every n, the infinite word w is called *uniformly recurrent* and the function $R_w(n)$ the *recurrence function* of w. A uniformly recurrent word w is called *linearly recurrent* if the ratio $R_w(n)/n$ is bounded. Given a linearly recurrent word w, the real number $r_w = \limsup_{n \to \infty} R_w(n)/n$ is called the *recurrence quotient* of w.

The *factor complexity function* of an infinite word w is the integer function $p_w(n)$ defined by $p_w(n) = |Fact(w) \cap \Sigma^n|$. An infinite word w has *linear complexity* if $p_w(n) = O(n)$.

A *substitution* over the alphabet Σ is a map $\tau : \Sigma \mapsto \Sigma^+$. Using the extension to words by concatenation, a substitution can be iterated. Note that for every substitution τ and every $n > 0$, τ^n is again a substitution. Moreover, a substitution τ over Σ can be naturally extended to a morphism from Σ^* to Σ^*, since for every $u, v \in \Sigma^*$, one has $\tau(uv) = \tau(u)\tau(v)$, provided that one defines

$\tau(\varepsilon) = \varepsilon$. A substitution τ is *k-uniform* if there exists an integer $k \geq 1$ such that for all $a \in \Sigma$, $|\tau(a)| = k$. We say that a substitution is *uniform* if it is *k*-uniform for some $k \geq 1$. A substitution τ is *primitive* if there exists an integer $n \geq 1$ such that for every $a \in \Sigma$, $\tau^n(a)$ contains every letter of Σ at least once. In this paper, we will only consider primitive substitutions such that $\tau(a_1) = a_1 v$ for some non-empty word v. These substitutions always have a fixed point, which is the infinite word $w = \lim_{n \to \infty} \tau^n(a_1)$. Moreover, this fixed point is linearly recurrent (see for example [7]) and therefore has linear complexity.

3   Abelian-square Rich Words

Kociumaka et al. [5] showed that a word of length n can contain a number of distinct abelian square factors that is quadratic in n. We give here a proof of this fact for the sake of completeness.

Proposition 1. *A word of length n can contain $\Theta(n^2)$ distinct abelian square factors.*

Proof. Consider the word $w_n = a^n b a^n b a^n$, of length $3n + 2$. For every $0 \leq i, j \leq n$ such that $i + j + n$ is even, the factor $a^i b a^n b a^j$ of w is an abelian square. Since the number of possible choices for the pair (i, j) is quadratic in n, we are done. □

Motivated by the previous result, we wonder whether there exist infinite words such that all their factors contain a number of abelian squares that is quadratic in their length. But first, we relax this condition and consider words in which, for every sufficiently large n, a factor of length n contains, on average, a number of distinct abelian square factors that is quadratic in n.

Definition 1. *An infinite word w is abelian-square rich if and only if there exists a positive constant C such that for every n sufficiently large one has*

$$\frac{1}{p_w(n)} \sum_{v \in Fact(w) \cap \Sigma^n} \{\# \ abelian \ square \ factors \ of \ v\} \geq Cn^2.$$

Notice that Christodoulakis et al. [6] proved that a binary word of length n contains $\Theta(n\sqrt{n})$ distinct abelian square factors on average, hence an infinite binary random word is almost surely not abelian-square rich.

Given a finite or infinite word w, we let $ASF_w(n)$ denote the number of abelian square factors of w of length n. Of course, $ASF_w(n) = 0$ if n is odd, so this quantity is significant only for even values of n.

The following lemma is a consequence of the definition of linearly recurrent word.

Lemma 1. *Let w be a linearly recurrent word. If there exists a constant C such that for every n sufficiently large one has $\sum_{m \leq n} ASF_w(m) \geq Cn^2$, then w is abelian-square rich.*

In an abelian-square rich word the average number of abelian squares in a factor is quadratic in the length of the factor. A stronger condition is that *every* factor contains a quadratic number of abelian squares. We thus introduce uniformly abelian-square rich words.

Definition 2. *An infinite word w is uniformly abelian-square rich if and only if there exists a positive constant C such that for every n sufficiently large one has*

$$\inf_{v \in Fact(w) \cap \Sigma^n} \{\# \ abelian \ square \ factors \ of \ v\} \geq Cn^2.$$

Clearly, if a word is uniformly abelian-square rich, then it is also abelian-square rich, but the converse is not always true. However, in the case of linearly recurrent words, the two definitions are equivalent, as shown in the next lemma.

Lemma 2. *If w is abelian-square rich and linearly recurrent, then it is uniformly abelian-square rich.*

Proof. Since w is linearly recurrent, there exists a positive integer K such that every factor of w of length Kn contains all the factors of w of length n. Let v be a factor of w of length n containing the largest number of abelian squares among the factors of w of length n. Hence the number of abelian squares in v is at least the average number of abelian squares in a factor of w of length n. Since w is abelian square rich, the number of abelian squares in v is greater than or equal to Cn^2, for a positive constant C and n sufficiently large. Since v is contained in any factor of w of length Kn, the number of abelian squares in any factor of w of length Kn is greater than or equal to Cn^2, whence the statement follows. $\square$

The rest of this section is devoted to prove that the Thue-Morse word and the Sturmian words that do not contain arbitrarily large repetitions are uniformly abelian-square rich.

3.1 The Thue-Morse Word

Let

$$t = 0110100110010110100110110 \cdots$$

be the Thue-Morse word, i.e., the fixed point of the uniform substitution $\mu : 0 \mapsto 01, 1 \mapsto 10$. For every $n \geq 4$, the factors of length n of t belong to two disjoint sets: those that start only at even positions in t, and those that start only at odd positions in t. This is a consequence of the fact that t is overlap-free, hence 0101 cannot be preceded by 1 nor followed by 0, and that 00 and 11 are not images of letters, so they cannot appear at even positions.

Let $p(n)$ be the factor complexity function of t. It is known [8, Proposition 4.3], that for every $n \geq 1$ one has $p(2n) = p(n) + p(n+1)$ and $p(2n+1) = 2p(n+1)$.

The next lemma (proved in [9]) shows that the Thue-Morse word has the property that for every length there are at least one third of the factors that begin and end with the same letter, and at least one third of the factors that begin and end with different letters. We define $f_{aa}(n)$ (resp. $f_{ab}(n)$) as the number of factors of t of length n that begin and end with the same letter (resp. with different letters).

Lemma 3 ([9]). *For every* $n \geq 2$, *one has* $f_{aa}(n) \geq p(n)/3$ *and* $f_{ab}(n) \geq p(n)/3$.

Since $p(n) \geq 3(n-1)$ for every n [10, Corollary 4.5], we get the following result.

Corollary 1. *For every* $n \geq 2$, *one has* $f_{aa}(n) \geq n-1$ *and* $f_{ab}(n) \geq n-1$.

Proposition 2. *The Thue-Morse word t is uniformly abelian-square rich.*

Proof. Let u be a factor of length $n > 1$ of t that begins and ends with the same letter. Since the image of any even-length word under μ is an abelian square, we have that $\mu^2(u)$ is an abelian square factor of t of length $4n$ that begins and ends with the same letter. Moreover, the word obtained from $\mu^2(u)$ by removing the first and the last letter is an abelian square factor of t of length $4n - 2$. So, by Corollary 1, t contains at least $n - 1$ abelian square factors of length $4n$ and at least $n - 1$ abelian square factors of length $4n - 2$. This implies that for every even n the number of abelian square factors of t of length n is linear in n. Hence, for every n the number of abelian square factors of t of length at most n is quadratic in n. The statement then follows from Lemmas 1 and 2.   □

3.2 Sturmian Words

In this section we fix the alphabet $\Sigma = \{\mathbf{a}, \mathbf{b}\}$.

Recall that a (finite or infinite) word w over Σ is *balanced* if and only if for any u, v factors of w of the same length, one has $||u|_{\mathbf{a}} - |v|_{\mathbf{a}}| \leq 1$.

We start with a simple lemma.

Lemma 4. *Let w be a finite balanced word over Σ. Then for any $k > 0$, $P(w) = (0, 0) \mod k$ if and only if w is an abelian k-power.*

Proof. Let w be balanced and $P(w) = (ks, kt)$, for a positive integer k and some $s, t \geq 0$. Then we can write $w = v_1 v_2 \cdots v_k$ where each v_i has length $s + t$. Now, each v_i must have Parikh vector equal to (s, t) otherwise w would not be balanced, whence the only if part of the statement follows. The if part is straightforward.   □

A binary infinite word is *Sturmian* if and only if it is balanced and aperiodic. Sturmian words are precisely the infinite words having $n + 1$ distinct factors of length n for every $n \geq 0$. There is a lot of other equivalent definitions of Sturmian words. A classical reference on Sturmian words is [11, Chap. 2]. Let us recall here the definition of Sturmian words as codings of a rotation.

We fix the torus $I = \mathbb{R}/\mathbb{Z} = [0,1)$. Given α, β in I, if $\alpha > \beta$, we use the notation $[\alpha, \beta)$ for the interval $[\alpha, 1) \cup [0, \beta)$. Recall that given a real number α, $\lfloor \alpha \rfloor$ is the greatest integer smaller than or equal to α, $\lceil \alpha \rceil$ is the smallest integer greater than or equal to α, and $\{\alpha\} = \alpha - \lfloor \alpha \rfloor$ is the fractional part of α. Notice that $\{-\alpha\} = 1 - \{\alpha\}$.

Let $\alpha \in I$ be irrational, and $\rho \in I$. The Sturmian word $s_{\alpha,\rho}$ (resp. $s'_{\alpha,\rho}$) of *angle* α and *initial point* ρ is the infinite word $s_{\alpha,\rho} = a_0 a_1 a_2 \cdots$ defined by

$$a_n = \begin{cases} \mathbf{b} \text{ if } \{\rho + n\alpha\} \in I_\mathbf{b}, \\ \mathbf{a} \text{ if } \{\rho + n\alpha\} \in I_\mathbf{a}, \end{cases}$$

where $I_\mathbf{b} = [0, 1-\alpha)$ and $I_\mathbf{a} = [1-\alpha, 1)$ (resp. $I_\mathbf{b} = (0, 1-\alpha]$ and $I_\mathbf{a} = (1-\alpha, 1]$).

In other words, take the unitary circle and consider a point initially in position ρ. Then start rotating this point on the circle (clockwise) of an angle α, 2α, 3α, etc. For each rotation, take the letter $\mathbf{a}$ or $\mathbf{b}$ associated with the interval within which the point falls. The infinite sequence obtained in this way is the Sturmian word $s_{\alpha,\rho}$ (or $s'_{\alpha,\rho}$, depending on the choice of the two intervals). See Fig. 1 for an illustration.

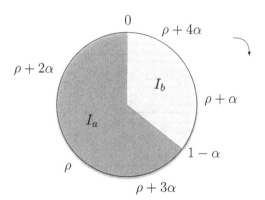

Fig. 1. The rotation of angle $\alpha = \varphi - 1 \approx 0.618$ and initial point $\rho = \alpha$ generating the Fibonacci word $F = s_{\varphi-1,\varphi-1} = \mathbf{abaababaabaabab} \cdots$.

For example, if $\varphi = (1 + \sqrt{5})/2 \approx 1.618$ is the golden ratio, the Sturmian word

$$F = s_{\varphi-1,\varphi-1} = \mathbf{abaababaabaababaabababaabaababaabaab} \cdots$$

is called the *Fibonacci word*:

A Sturmian word for which $\rho = \alpha$, like the Fibonacci word, is called *characteristic*. Note that for every α one has $s_{\alpha,0} = \mathbf{b}s_{\alpha,\alpha}$ and $s'_{\alpha,0} = \mathbf{a}s_{\alpha,\alpha}$.

An equivalent way to see the coding of a rotation consists in fixing the point and rotating the intervals. In this representation, the interval $I_\mathbf{b} = I_\mathbf{b}^0$ is rotated at each step, so that after i rotations it is transformed into the interval $I_{\mathbf{b}}^{-i} = [\{-i\alpha\}, \{-(i+1)\alpha\})$, while $I_{\mathbf{a}}^{-i} = I \setminus I_{\mathbf{b}}^{-i}$.

This representation is convenient since one can read within it not only a Sturmian word but also any of its factors. More precisely, for every positive integer n, the factor of length n of $s_{\alpha,\rho}$ starting at position $j \geq 0$ is determined by the value of $\{\rho + j\alpha\}$ only. Indeed, for every j and i, we have:

$$a_{j+i} = \begin{cases} \mathbf{b} & \text{if } \{\rho + j\alpha\} \in I_{\mathbf{b}}^{-i}; \\ \mathbf{a} & \text{if } \{\rho + j\alpha\} \in I_{\mathbf{a}}^{-i}. \end{cases}$$

As a consequence, we have that given a Sturmian word $s_{\alpha,\rho}$ and a positive integer n, the $n + 1$ different factors of $s_{\alpha,\rho}$ of length n are completely determined by the intervals $I_{\mathbf{b}}^0, I_{\mathbf{b}}^{-1}, \dots, I_{\mathbf{b}}^{-(n-1)}$, that is, only by the points $\{-i\alpha\}$, $0 \leq i < n$. In particular, they do not depend on ρ, so that the set of factors of $s_{\alpha,\rho}$ is the same as the set of factors of $s_{\alpha,\rho'}$ for any ρ and ρ'. Hence, from now on, we let s_α denote any Sturmian word of angle α.

If we arrange the $n + 2$ points $0, 1, \{-\alpha\}, \{-2\alpha\}, \dots, \{-n\alpha\}$ in increasing order, we determine a partition of I in $n+1$ subintervals, $L_0(n), L_1(n), \dots, L_n(n)$. Each of these subintervals is in bijection with a different factor of length n of any Sturmian word of angle α (see Fig. 2).

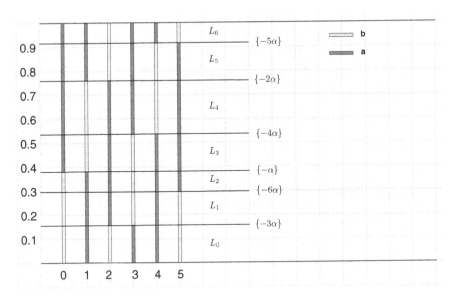

Fig. 2. The points 0, 1 and $\{-\alpha\}, \{-2\alpha\}, \{-3\alpha\}, \{-4\alpha\}, \{-5\alpha\}, \{-6\alpha\}$, arranged in increasing order, define the intervals $L_0(6) \approx [0, 0.146)$, $L_1(6) \approx [0.146, 0.292)$, $L_2(6) \approx [0.292, 0.382)$, $L_3(6) \approx [0.382, 0.528)$, $L_4(6) \approx [0.528, 0.764)$, $L_5(6) \approx [0.764, 0.910)$, $L_6(6) \approx [0.910, 1)$. Each interval is associated with one of the factors of length 6 of the Fibonacci word, respectively **babaab**, **baabab**, **baabaa**, **ababaa**, **abaaba**, **aababa**, **aabaab**.

Recall that a factor of length n of a Sturmian word s_α has a Parikh vector equal either to $(\lfloor n\alpha \rfloor, n - \lfloor n\alpha \rfloor)$ (in which case it is called *light*) or to $(\lceil n\alpha \rceil, n - \lceil n\alpha \rceil)$ (in which case it is called *heavy*). The following proposition relates the intervals $L_i(n)$ to the Parikh vectors of the factors of length n (see [12,13]).

Proposition 3. *Let s_α be a Sturmian word of angle α, and n a positive integer. Let t_i be the factor of length n associated with the interval $L_i(n)$. Then t_i is heavy if $L_i(n) \subset [\{-n\alpha\}, 1)$, while it is light if $L_i(n) \subset [0, \{-n\alpha\})$.*

Example 1. Let $\alpha = \varphi - 1 \approx 0.618$ and $n = 6$. We have $6\alpha \approx 3.708$, so that $\{-6\alpha\} \approx 0.292$. The reader can see in Fig. 2 that the factors of length 6 corresponding to intervals above (resp. below) $\{-6\alpha\} \approx 0.292$ all have Parikh vector $(4, 2)$ (resp. $(3, 3)$). That is, the intervals L_0 and L_1 are associated with light factors (**babaab, baabab**), while the intervals L_2 to L_6 are associated with heavy factors (**baabaa, ababaa, abaaba, aababa, aabaab**).

Observe that, by Lemma 4, every factor of a Sturmian word having even length and containing an even number of **a**'s (or, equivalently, of **b**'s) is an abelian square. The following proposition relates the abelian square factors of a Sturmian word of angle α with the arithmetic properties of α.

Proposition 4. *Let s_α be a Sturmian word of angle α, and n a positive even integer. Let t_i be the factor of length n associated with the interval $L_i(n)$. Then t_i is an abelian square if and only if $L_i(n) \subset [\{-n\alpha\}, 1)$ if $\lfloor n\alpha \rfloor$ is even, or $L_i(n) \subset [0, \{-n\alpha\})$ if $\lfloor n\alpha \rfloor$ is odd.*

Proof. By Proposition 3, t_i is heavy if and only if $L_i(n) \subset [\{-n\alpha\}, 1)$, while it is light if and only if $L_i(n) \subset [0, \{-n\alpha\})$. If $\lfloor n\alpha \rfloor$ is even, then every light factor of length n contains an even number of **a**'s and hence is an abelian square, while if $\lfloor n\alpha \rfloor$ is odd, then every heavy factor of length n contains an even number of **a**'s and hence is an abelian square, whence the statement follows.   □

Recall that given a finite or infinite word w, $ASF_w(n)$ denotes the number of abelian square factors of w of length n.

Corollary 2. *Let s_α be a Sturmian word of angle α. For every positive even n, let $I_n = \{\{-i\alpha\} \mid 1 \le i \le n\}$. Then*

$$ASF_{s_\alpha}(n) = \begin{cases} \#\{x \in I_n \mid x \le \{-n\alpha\}\} \text{ if } \lfloor n\alpha \rfloor \text{ is even}; \\ \#\{x \in I_n \mid x \ge \{-n\alpha\}\} \text{ if } \lfloor n\alpha \rfloor \text{ is odd}. \end{cases}$$

Example 2. The factors of length 6 of the Fibonacci word F are, lexicographically ordered: **aabaab, aababa, abaaba, ababaa, baabaa** (heavy factors), **baabab, babaab** (light factors). The light factors, whose number of **a**'s is $\lfloor 6\alpha \rfloor = 3$, are not abelian squares; the heavy factors, whose number of **a**'s is $\lceil 6\alpha \rceil = 4$, are all abelian squares.

We have $I_6 = \{0.382, 0.764, 0.146, 0.528, 0.910, 0.292\}$ (values are approximated) and $6\alpha \simeq 3.708$, so $\lfloor 6\alpha \rfloor$ is odd. Thus, there are 5 elements in I_6 that are $\geq \{-6\alpha\}$, so by Corollary 2 there are 5 abelian square factors of length 6.

The factors of length 8 of the Fibonacci word are, lexicographically ordered: **aabaabab, aababaab, abaabaab, abaababa, ababaaba, baabaaba, baababaa, babaabaa** (heavy factors), **babaabab** (light factor). The light factor, whose number of **a**'s is $\lfloor 8\alpha \rfloor = 4$, is an abelian square; the heavy factors, whose number of **a**'s is $\lceil 8\alpha \rceil = 5$, are not abelian squares. We have $I_8 = \{0.382, 0.764, 0.146, 0.528, 0.910, 0.292, 0.674, 0.056\}$ (values are approximated) and $8\alpha \simeq 4.944$, so $\lfloor 8\alpha \rfloor$ is even. Thus, there is only one element in I_8 that is $\leq \{8\alpha\}$, so by Corollary 2 there is only one abelian square factor of length 8.

In Table 1 we report the first values of the sequence $ASF_F(n)$ for the Fibonacci word F.

Table 1. The first values of the sequence $ASF_F(n)$ of the number of abelian square factors of length n in the Fibonacci word $F = s_{\varphi-1, \varphi-1}$.

n	0	2	4	6	8	10	12	14	16	18	20	22	24	26	28	30	32	34	36
$ASF_F(n)$	0	1	3	5	1	9	5	5	15	3	13	13	5	25	9	15	25	21	27

Recall that every irrational number α can be uniquely written as a (simple) continued fraction as follows:

$$\alpha = a_0 + \cfrac{1}{a_1 + \cfrac{1}{a_2 + \cdots}} \tag{1}$$

where $a_0 = \lfloor \alpha \rfloor$, and the infinite sequence $(a_i)_{i \geq 0}$ is called the sequence of partial quotients of α. The continued fraction expansion of α is usually denoted by its sequence of partial quotients as follows: $\alpha = [a_0; a_1, a_2, \ldots]$, and each its finite truncation $[a_0; a_1, a_2, \ldots, a_k]$ is a rational number n_k/m_k called the kth convergent to α. We say that an irrational $\alpha = [a_0; a_1, a_2, \ldots]$ has bounded partial quotients if and only if the sequence $(a_i)_{i \geq 0}$ is bounded.

The development in continued fraction of α is deeply related to the exponent of the factors of the Sturmian word s_α. Recall that an infinite word w is said to be β-power free, for some $\beta \geq 2$, if for every factor v of w, the ratio between the length of v and its minimal period is smaller than β. The second author [14] proved that a Sturmian word of angle α is β-power free for some $\beta \geq 2$ if and only if α has bounded partial quotients.

Since the golden ratio φ is defined by the equation $\varphi = 1 + 1/\varphi$, we have from Eq. 1 that $\varphi = [1; 1, 1, 1, 1, \ldots]$ and therefore $\varphi - 1 = [0; 1, 1, 1, 1, \ldots]$, so the Fibonacci word is an example of β-power free Sturmian word (actually, it is $(2 + \varphi)$-power free [15]).

We are now proving that if α has bounded partial quotients, then the Sturmian word s_α is abelian-square rich. For this, we will use a result on the discrepancy of uniformly distributed modulo 1 sequences from [16]. To the best of our knowledge, this is the first application of this result to the theory of Sturmian words, and we think that the correspondence we are now showing might be useful for deriving other results on Sturmian words.

Let $\omega = (x_n)_{n \geq 0}$ be a given sequence of real numbers. For a positive integer N and a subset E of the torus I, we define $A(E; N; \omega)$ as the number of terms x_n, $0 \leq n \leq N$, for which $\{x_n\} \in E$. If there is no risk of confusion, we will write $A(E; N)$ instead of $A(E; N; \omega)$.

Definition 3. *The sequence $\omega = (x_n)_{n \geq 0}$ of real numbers is said to be uniformly distributed modulo 1 if and only if for every pair a, b of real numbers with $0 \leq a < b \leq 1$ we have*

$$\lim_{N \to \infty} \frac{A([a,b); N; \omega)}{N} = b - a.$$

Definition 4. *Let $x_0, x_1, \ldots, x_N$ be a finite sequence of real numbers. The number*

$$D_N = D_N(x_0, x_1, \ldots, x_N) = \sup_{0 \leq \gamma < \delta \leq 1} \left| \frac{A([\gamma, \delta); N)}{N} - (\delta - \gamma) \right|$$

is called the discrepancy of the given sequence. For an infinite sequence ω of real numbers the discrepancy $D_N(\omega)$ is the discrepancy of the initial segment formed by the first $N + 1$ terms of ω.

The two previous definitions are related by the following result.

Theorem 1 ([16]). *The sequence ω is uniformly distributed modulo 1 if and only if $\lim_{N \to \infty} D_N(\omega) = 0$.*

An important class of uniformly distributed modulo 1 sequences is given by the sequence $(n\alpha)_{n \geq 0}$ with α a given irrational number and $n \in \mathbb{N}$. The discrepancy of the sequence $(n\alpha)$ will depend on the finer arithmetical properties of α. In particular, we have the following theorem, stating that if α has bounded partial quotients, then its discrepancy has the least order of magnitude possible.

Theorem 2 ([16]). *Suppose the irrational $\alpha = [a_0; a_1, \ldots]$ has partial quotients bounded by K. Then the discrepancy $D_N(\omega)$ of $\omega = (n\alpha)$ satisfies $ND_N(\omega) = O(\log N)$. More exactly, we have*

$$ND_N(\omega) \leq 3 + \left(\frac{1}{\log \varphi} + \frac{K}{\log(K+1)} \right) \log N. \tag{2}$$

We are now using previous definitions and results to prove that β-power free Sturmian words are abelian-square rich.

Theorem 3. *Let s_α be a Sturmian word of angle α such that α has bounded partial quotients. Then there exists a positive constant C such that for every n sufficiently large one has $\sum_{m \leq n} ASF_{s_\alpha}(m) \geq Cn^2$.*

Proof. For every even n, let $I'_n = \{\{i\alpha\} \mid 1 \leq i \leq n\}$. By Corollary 2 and basic arithmetical properties of the fractional part, we have:

$$ASF_{s_\alpha}(n) = \begin{cases} \#\{x \in I'_n \mid x \geq \{n\alpha\}\} \text{ if } \lfloor n\alpha \rfloor \text{ is even;} \\ \#\{x \in I'_n \mid x \leq \{n\alpha\}\} \text{ if } \lfloor n\alpha \rfloor \text{ is odd.} \end{cases}$$

So:

$$\sum_{m \leq n} ASF_{s_\alpha}(m) \tag{3}$$

$$\geq \sum_{m \leq n} \#\{\{i\alpha\} \mid \{i\alpha\} \leq 1/2, i \leq m, \text{ and } \{m\alpha\} \leq 1/2, \lfloor m\alpha \rfloor \text{ even }\} \tag{4}$$

$$\geq \sum_{m \leq n} \#\{\{i\alpha/2\} \mid \{i\alpha/2\} \in [1/4, 1/2), i \leq m, \text{ and } \{m\alpha/2\} \leq 1/4\} \tag{5}$$

$$\geq \sum_{n/2 \leq m \leq n} \#\{\{i\alpha/2\} \mid \{i\alpha/2\} \in [1/4, 1/2), i \leq n/2, \text{ and } \{m\alpha/2\} \leq 1/4\} \tag{6}$$

$$= \#\{\{i\alpha/2\} \mid \{i\alpha/2\} \in [1/4, 1/2), i \leq n/2\} \times \sum_{n/2 \leq m \leq n} \{m \mid \{m\alpha/2\} \leq 1/4\} \tag{7}$$

where: (4) follows from (3) by Corollary 2; (5) follows from (4) because $\{m\alpha/2\} \leq 1/4$ implies $\{m\alpha\} \leq 1/2$ and $\lfloor m\alpha \rfloor$ is even if and only if $\{m\alpha/2\} \leq 1/2$; (6) follows from (5) is obvious; finally (7) follows from (6) because the cardinality of the first set is independent from the sum.

Now, $\alpha/2$ has bounded partial quotients (since α has) and we can apply Theorem 2 to evaluate the two factors of (7). So we have:

$$\#\{\{i\alpha/2\} \mid \{i\alpha/2\} \in [1/4, 1/2), i \leq n/2\}$$
$$= A([1/4, 1/2); n/2; (n\alpha/2))$$
$$\geq (1/2 - 1/4)n/2 - C_1 \log n$$
$$= n/8 - C_1 \log n,$$

for n sufficiently large and a positive constant C_1. We also have:

$$\sum_{n/2 \leq m \leq n} \{m \mid \{m\alpha/2\} \leq 1/4\}$$
$$= A([0, 1/4); n; (n\alpha/2)) - A([0, 1/4); n/2; (n\alpha/2))$$
$$\geq n/4 - C_2 \log n - n/8 - C_3 \log n$$
$$= n/8 - C_4 \log n,$$

for n sufficiently large and positive constants C_2, C_3, C_4. The product of the two factors of (6) is therefore greater than a constant times n^2, as required.   □

The recurrence quotient r_α of a Sturmian word of angle $\alpha = [0; a_1, a_2, \ldots]$ such that α has bounded partial quotients verifies $2 + r_\alpha < \limsup a_i < 3 + r_\alpha$ [17, Proposition 5]. Moreover, Durand [18] proved that a Sturmian word of angle α is linearly recurrent if and only if α has bounded partial quotients. Thus, we have the following:

Corollary 3. *Let s_α be a Sturmian word of angle α. If s_α is β-power free, then s_α is uniformly abelian-square rich.*

Proof. We known that s_α is β-power free for some $\beta \geq 2$ if and only if α has bounded partial quotients if and only if s_α is linearly recurrent. The statement then follows from Theorem 3 and Lemmas 1 and 2.   □

4   Conclusions and Future Work

We proved that the Thue-Morse is uniformly abelian-square rich. We think that the technique we used for the proof can be generalized to some extent, and could be used, for example, to prove that a class of fixed points of uniform substitutions are uniformly abelian-square rich.

We also proved that Sturmian words that are β-power free for some $\beta \geq 2$ are uniformly abelian-square rich. The proof we gave is based on a classical result on the discrepancy of the uniformly distributed modulo 1 sequence $(n\alpha)_{n \geq 0}$, where α is the slope of the Sturmian word. To the best of our knowledge, this is the first application of this result to the theory of Sturmian words, and we think that the correspondence we have shown might be useful for deriving other results on Sturmian words.

The natural question that then arises is whether the hypothesis of power freeness is necessary for a Sturmian word being (uniformly) abelian-square rich. We leave open the question to determine whether s_α is not uniformly abelian-square rich nor abelian-square rich in the case when α has unbounded partial quotients.

We mostly investigated binary words in this paper. We conjecture that binary words have the largest number of abelian square factors. More precisely, we propose the following conjecture.

Conjecture 1. If a word of length n contains k many distinct abelian square factors, then there exists a binary word of length n containing at least k many distinct abelian square factors.

A slightly different point of view from the one we considered in this paper consists in identifying two abelian squares if they have the same Parikh vector. Two abelian squares are therefore called *inequivalent* if they have different Parikh vectors [19]. Sturmian words only have a linear number of inequivalent abelian squares. Nevertheless, a word of length n can contain $\Theta(n\sqrt{n})$ inequivalent abelian squares [20]. Computations support the following conjecture:

Conjecture 2 (see [5]). A word of length n contains $O(n\sqrt{n})$ inequivalent abelian squares.

Acknowledgements. The authors acknowledge the support of the PRIN 2010/2011 project "Automi e Linguaggi Formali: Aspetti Matematici e Applicativi" of the Italian Ministry of Education (MIUR).

References

1. Fraenkel, A.S., Simpson, J.: How many squares can a string contain? J. Comb. Theory, Ser. A **82**(1), 112–120 (1998)
2. Ilie, L.: A note on the number of squares in a word. Theoret. Comput. Sci. **380**(3), 373–376 (2007)
3. Erdös, P.: Some unsolved problems. Magyar Tud. Akad. Mat. Kutato. Int. Kozl. **6**, 221–254 (1961)
4. Keränen, V.: Abelian squares are avoidable on 4 letters. In: Kuich, W. (ed.) ICALP 1992. LNCS, vol. 623, pp. 41–52. Springer, Heidelberg (1992)
5. Kociumaka, T., Radoszewski, J., Rytter, W., Waleń, T.: Maximum number of distinct and nonequivalent nonstandard squares in a word. In: Shur, A.M., Volkov, M.V. (eds.) DLT 2014. LNCS, vol. 8633, pp. 215–226. Springer, Heidelberg (2014)
6. Christodoulakis, M., Christou, M., Crochemore, M., Iliopoulos, C.S.: On the average number of regularities in a word. Theoret. Comput. Sci. **525**, 3–9 (2014)
7. Damanik, D., Zare, D.: Palindrome complexity bounds for primitive substitution sequences. Discrete Math. **222**(1–3), 259–267 (2000)
8. Brlek, S.: Enumeration of factors in the Thue-Morse word. Discr. Appl. Math. **24**(1–3), 83–96 (1989)
9. Cassaigne, J., Fici, G., Sciortino, M., Zamboni, L.: Cyclic Complexity of Words, Submitted (2015). http://arxiv.org/abs/1402.5843
10. de Luca, A., Varricchio, S.: Some combinatorial properties of the Thue-Morse sequence and a problem in semigroups. Theoret. Comput. Sci. **63**(3), 333–348 (1989)
11. Lothaire, M.: Algebraic Combinatorics on Words. Cambridge University Press, Cambridge (2002)
12. Fici, G., Langiu, A., Lecroq, T., Lefebvre, A., Mignosi, F., Prieur-Gaston, É.: Abelian repetitions in Sturmian words. In: Béal, M.-P., Carton, O. (eds.) DLT 2013. LNCS, vol. 7907, pp. 227–238. Springer, Heidelberg (2013)
13. Rigo, M., Salimov, P., Vandomme, E.: Some properties of abelian return words. J. Integer Seq. **16**, 13.2.5 (2013)
14. Mignosi, F.: Infinite words with linear subword complexity. Theoret. Comput. Sci. **65**(2), 221–242 (1989)
15. Mignosi, F., Pirillo, G.: Repetitions in the Fibonacci infinite word. RAIRO Theor. Inform. Appl. **26**, 199–204 (1992)
16. Kuipers, L., Niederreiter, H.: Uniform Distribution of Sequences. Wiley, New York (1974)
17. Cassaigne, J.: Limit values of the recurrence quotient of sturmian sequences. Theoret. Comput. Sci. **218**(1), 3–12 (1999)
18. Durand, F.: Corrigendum and addendum to: "Linearly recurrent subshifts have a finite number of non-periodic subshift factors" [Ergodic Theory Dynam. Systems **20**(4), 1061–1078 (2000)]. Ergodic Theory Dynam. Systems **23**(2), pp. 663–669 (2003)
19. Fraenkel, A.S., Simpson, J., Paterson, M.: On weak circular squares in binary words. In: Hein, J., Apostolico, A. (eds.) CPM 1997. LNCS, vol. 1264, pp. 76–82. Springer, Heidelberg (1997)
20. Kociumaka, T., Radoszewski, J., Rytter, W., Walen, T.: Personal communication (2015)

Arithmetics on Suffix Arrays of Fibonacci Words

Dominik Köppl$^{(\boxtimes)}$ and Tomohiro I

Department of Computer Science, TU Dortmund, Dortmund, Germany
dominik.koeppl@tu-dortmund.de, tomohiro.i@cs.tu-dortmund.de

Abstract. We study the sequence of Fibonacci words and some of its derivatives with respect to their suffix array, inverse suffix array and Burrows-Wheeler transform based on the respective suffix array. We show that the suffix array is a rotation of its inverse under certain conditions, and that the factors of the LZ77 factorization of any Fibonacci word yield again similar characteristics.

1 Introduction

The sequence of Fibonacci words is one of the best studied set of strings in the field of combinatorics. A Fibonacci word, composed by the concatenation of its predecessor with its pre-predecessor, excels at many interesting properties regarding factorizations [18], powers [16], fractals [15], entropy [7] and palindromes [4]. The sequence is often used as a testbed for algorithms, since they are a representative for some worst case scenarios [8]. Regarding benchmarks, it is beneficial to know the shape of the considered data structures when they are applied to a Fibonacci word; studying the combinatorial properties may help understanding experimental results, or may even lead to designing new algorithms.

We study properties of Fibonacci words and of some derivatives with respect to its suffix array (SA) [13]. The SA induces the structure of its inverse and of the SA-based Burrows-Wheeler transform (BWT) [2,10]. Under certain conditions, the SA is a rotation, a reversed rotation, or the copy of its inverse.

With this insight, it is easy to reason about complex data structures like compressed suffix trees [19], the FM-Index [5] or LZ77-based self-indexes [6].

2 Related Work

Regarding suffix data structures, Rytter [17] considered building a directed acyclic word graph and a suffix tree on the j-th Fibonacci word. By some properties of the Fibonacci words, Rytter showed an easy way to modify both data structures to match with the $(j + 1)$-th word. Considering BWT based on rotations, Mantaci et al. [14] discovered that the BWT rearranges any Fibonacci word in a block of consecutive b's, followed by a block of consecutive a's. More generally, Mantaci et al. [14] and Simpson and Puglisi [20] gave a general theorem regarding this special shape of the BWT applied to a class of binary strings,

© Springer International Publishing Switzerland 2015
F. Manea and D. Nowotka (Eds.): WORDS 2015, LNCS 9304, pp. 135–146, 2015.
DOI: 10.1007/978-3-319-23660-5_12

so called standard words, to which the Fibonacci words belong. Christodoulakis et al. [3] proposed a constant time algorithm for querying different properties on the BWT's rotations. Without any delimiting, unique character at the string's end, the BWT defined on rotations and the BWT based on the SA (may) differ. In fact, the BWT based on the SA of F_n (for any odd n) does not transform the string into two homogenous blocks. Another result with respect to rotations is done by Droubay [4]: they showed for every F_n with $n \bmod 3 \neq 0$ that there exists exactly one k such that the k-th rotation of F_n is a palindrome.

Since some current, popular indexing strategies perform compression on the text (e.g., [6]), we further consider the LZ77 factorization [23]. In the special case of the Fibonacci words, Berstel and Savelli [1] pointed out that the LZ77 factorization coincides with the palindromic factorization studied by Wen and Wen [22]. We will show that our results apply analogously to the LZ77 factors.

3   Preliminaries

Let Σ denote an ordered alphabet. An element in Σ^* is called a **string**. For any string T, let $|T|$ denote the length of T. The string of length zero is denoted by ϵ. Σ^* forms with ϵ and the concatenation $\Sigma^* \times \Sigma^* \to \Sigma^*, (u, v) \mapsto uv$ a free monoid. For any $1 \leq i \leq |T|$, $T[i]$ denotes the i-th character of T. When $T \in \Sigma^*$ is represented by the concatenation of $x, y, z \in \Sigma^*$, i.e., $T = xyz$, then x, y and z are called a **prefix**, **substring** and **suffix** of T, respectively. For any $1 \leq i \leq j \leq |T|$, a substring of T starting at i and ending at j is denoted by $T[i..j]$. Especially, a suffix starting at position i of T is denoted by $T[i..]$. For any $x, y \in \Sigma^*$, let $\mathsf{lcp}(x, y)$ denote the length of the longest common prefix of x and y.

The **lexicographical order** is denoted by $<\subset \Sigma^* \times \Sigma^*$, i.e., $x < y$ iff (either) x is a proper prefix of y, or $l := \mathsf{lcp}(x, y)$ is less than $\min(|x|, |y|)$ and $x[l+1] < y[l+1]$. We use another ordering $\lessdot$ for that $x \lessdot y$ iff the latter condition holds. The ordering $\lessdot$ is finer than $<$. If $x \lessdot y$, then $xu \lessdot yv$ holds for any $u, v \in \Sigma^*$. For instance, $a < aa$ and $a \not\lessdot aa$ since a is a prefix of aa. Appending b to both strings flips the lexicographic order to $ab > aab$. Taking $aa \lessdot ab$ as an example, appending characters to both strings does not affect their ordering.

The **inverse** R^{-1} of an array R is an array with the same length for that $R^{-1}[R[i]] = i$ holds for every $1 \leq i \leq |R|$; the inverse of R exists iff R is a **permutation** over $\{1, \ldots, |R|\}$, i.e., $\lambda.j \mapsto R[j]$ is an injective endomorphism. The **suffix array** SA_T of a string T is an array of length $|T|$ such that $T[\mathsf{SA}_T[i]..] < T[\mathsf{SA}_T[i+1]..]$ for every $1 \leq i < |T|$. Since SA_T is a permutation, its inverse (i.e., the **inverse suffix array**) exists, and is denoted by ISA_T.

Like in common literature, we let arrays start at position one (not zero); therefore we will modify the modulo operator not to map any value to zero. For this purpose, we define the modulo operator on the natural numbers by $\bmod n$: $\mathbb{N} \to \{1, \ldots, n\} \subset \mathbb{N}, m \bmod n := m - n \bmod n$ if $m > n$, $m \bmod n := m$, otherwise, for $n, m \in \mathbb{N}$. Naturally, our results can also be applied to the standard modulo operator when taking arrays of the form $[0..n-1]$, instead of $[1..n]$.

We call the array ψ_T with $\psi_T[i] := \mathsf{ISA}_T[\mathsf{SA}_T[i] - 1 \bmod |T|]$ for every $1 \leq i \leq |T|$ the **last-to-front** mapping of T.

A permutation S is called a **rotation** of R iff there exists exactly one $k \in \{1, \ldots, |R|\}$ such that $R[i] = S[(k + i) \bmod |R|]$. A permutation S is called a **reversed rotation** of R iff there exists one $k \in \{1, \ldots, |R|\}$ such that $R[i] = S[(k - i) \bmod |R|]$. In both cases, we call k the **shift** of S. If S is a rotation of R and there exists $1 \leq i \leq n$ such that $R[i] = S[i]$, then $S = R$.

Let $\Sigma_2 = \{a, b\}$ be a binary alphabet with $a < b$. The **complementation** $\overline{\cdot} : \Sigma_2^* \to \Sigma_2^*$ complements a string, i.e., $\bar{T}[i] = a$ if $T[i] = b$, $T[i] = b$ if $T[i] = a$.

Definition 1. *The n-th **Fibonacci word** $F_n \in \Sigma_2^*$ ($n \in \mathbb{N}$) is defined by $F_n = b$ if $n = 1$, a if $n = 2$, $F_n = F_{n-1}F_{n-2}$ otherwise. The sequence of lengths $f_n := |F_n|$ form the **Fibonacci numbers**. The sequence $\{\bar{F}_n\}_{n \in \mathbb{N}}$ is called **rabbit sequence** [7] and sometimes confused with the Fibonacci words (e.g., see [11]). The ending of the n-th Fibonacci word ($n \geq 3$) is given by $\delta_n := F_n[f_n - 1..f_n]$ such that $\delta_n = ba$ if n is even, $\delta_n = ab$ if n is odd.*

A **factorization** partitions T into z substrings $T = w_1 \cdots w_z$. These substrings are called **factors**. In particular, we have:

Definition 2 ([23]). *A factorization $w_1 \cdots w_z = T$ is called the **LZ77 factorization** of T iff w_x is the shortest prefix of $w_x \cdots w_z$ that occurs exactly once in $w_1 \cdots w_x$.*

Table 1. For each string T of a given sequence, we show the relationship between SA_T and ISA_T, as well as the shape of BWT_T. β is a variable character with $\beta \geq b$, and c is a character with $c > b$.

Sequence	$n \geq 4$	SA $\leftrightarrow$ ISA	shift	BWT
F_n (Definition 1)	even	rotation	$f_{n-2} + 1$	$b^{f_{n-2}} a^{f_{n-1}}$
Z_n (Definition 3)	even	rotation	$f_{n-2} + 1$	$b^{f_{n-2}} a^{f_{n-1}-1} b$
$B_n := \beta F_n$	even	equal	0	$b^{f_{n-2}} \beta a^{f_{n-1}}$
$D_n := \bar{F}_n c$	even	equal	0	$b^{f_{n-1}-1} c a^{f_{n-2}} b$
$C_n := F_n c$	odd	reversed rotation	f_n	$b^{f_{n-2}-1} c a^{f_{n-1}} b$

Definition 3 ([1,12,22]). *The n-th **singular** word is defined as $Z_n := F_n[f_n]F_n[1..f_n - 1]$. Alternatively, Z_n can be written as $Z_n = a$ if $n = 1$, $Z_n = b$ if $n = 2$, $Z_n = aa$ if $n = 3$, $Z_n = Z_{n-2}Z_{n-3}Z_{n-2}$ otherwise.*

Lemma 1. *For any $n \geq 3$, $F_n = Z_1 \cdots Z_{n-2}\gamma$ is the LZ77 factorization of F_n, where $\gamma := \delta_n[1]$, i.e., $\gamma = b$ if n is odd, a if n is even.*

Proof. – [**basis**] $F_3 = Z_1 b = ab$, $F_4 = Z_1 Z_2 a = aba$.
 – [**hypothesis**] Assume the claim holds for $n-1$ and $n-2$.
 – [**induction proof**] Let $\gamma := F_{n-2}[f_{n-2}] = F_n[f_n]$. Then $Z_{n-2} = \bar{\gamma} F_{n-2}[1..f_{n-2}-1] = \bar{\gamma} Z_1 \cdots Z_{n-4}$, and $F_n = Z_1 \cdots Z_{n-3} \bar{\gamma} Z_1 \cdots Z_{n-4} \gamma = Z_1 \cdots Z_{n-2} \gamma$.

$\square$

Table 1 gives a summary of the properties shown in this paper.

Table 2. Instances of the string sequences considered in Table 1. We additionally examine F_7 that does not show any of the attractive properties we study. Neither F_7 nor $F_6 c$ possesses any interesting properties we focus on.

i	1	2	3	4	5	6	7	8
$F_6[i]$	a	b	a	a	b	a	b	a
$SA_{F_6}[i]$	8	3	6	1	4	7	2	5
$ISA_{F_6}[i]$	4	7	2	5	8	3	6	1
Rotation Shift: 4								

i	1	2	3	4	5	6	7	8
$Z_6[i]$	b	a	b	a	a	b	a	b
$SA_{Z_6}[i]$	4	7	2	5	8	3	6	1
$ISA_{Z_6}[i]$	8	3	6	1	4	7	2	5
Rotation Shift: 4								

i	1	2	3	4	5	6	7	8	9
$(\bar{F_6}c)[i]$	b	a	b	b	a	b	a	b	c
$SA_{\bar{F_6}c}[i]$	5	2	7	4	1	6	3	8	9
$ISA_{\bar{F_6}c}[i]$	5	2	7	4	1	6	3	8	9

i	1	2	3	4	5	6	7	8	9	10	11	12	13	14
$(F_7 c)[i]$	a	b	a	a	b	a	b	a	a	b	a	a	b	c
$SA_{F_7 c}[i]$	8	3	11	6	1	9	4	12	7	2	10	5	13	14
$ISA_{F_7 c}[i]$	5	10	2	7	12	4	9	1	6	11	3	8	13	14
Reverse Rotation Shift: 13														

i	1	2	3	4	5	6	7	8	9
$(\beta F_6)[i]$	β	a	b	a	a	b	a	b	a
$SA_{\beta F_6}[i]$	9	4	7	2	5	8	3	6	1
$ISA_{\beta F_6}[i]$	9	4	7	2	5	8	3	6	1

i	1	2	3	4	5	6	7	8	9	10	11	12	13
F_7	a	b	a	a	b	a	b	a	a	b	a	a	b
SA_{F_7}	11	8	3	12	9	6	1	4	13	10	7	2	5
ISA_{F_7}	7	12	3	8	13	6	11	2	5	10	1	4	9
BWT_{F_7}	b	b	b	a	a	b	b	a	a	a	a	a	a

i	1	2	3	4	5	6	7	8	9
$F_6 c$	a	b	a	a	b	a	b	a	c
$SA_{F_6 c}$	3	1	4	6	8	2	5	7	9
$ISA_{F_6 c}$	2	6	1	3	7	4	8	5	9
$BWT_{F_6 c}$	b	c	a	b	b	a	a	a	a

Remark 1. The arithmetic progression that characterizes SA and ISA is not restricted to the family of Fibonacci-like strings. For example, the sequences $S_1 = abaa$, $S_n = aS_{n-1}a$ and $E_0 = bb, E_1 = bbab, E_n = b^{n+1}a^n b$ have a suffix array that is a reverse rotation of its inverse. Instances of both sequences are depicted in Table 3.

4   The Suffix Array and Its Inverse

We examine the suffix array structure of each sequence considered in Table 1. Besides this, we are interested in revealing some relationship between SA and ISA. Some examples are depicted in Table 2. Lemma 3 gives us some rules that determine whether a specific array is a rotation or reversed rotation of its inverse. For our sequences, Lemma 5 shows that the suffix array has the form of the array dealt in Definition 4 for some certain n.

Table 3. Instances of the string sequences given in Remark 1. Both instances have an SA that is reverse rotated to its inverse.

i	1 2 3 4 5 6 7 8		i	1 2 3 4 5 6 7 8
$S_2[i]$	$a\ a\ a\ b\ a\ a\ a\ a$		$E_3[i]$	$b\ b\ b\ b\ a\ a\ a\ b$
$\mathrm{SA}_{S_2}[i]$	8 7 6 5 1 2 3 4		$\mathrm{SA}_{E_3}[i]$	5 6 7 8 4 3 2 1
$\mathrm{ISA}_{S_2}[i]$	5 6 7 8 4 3 2 1		$\mathrm{ISA}_{E_3}[i]$	8 7 6 5 1 2 3 4
$\mathrm{BWT}_{S_2}[i]$	$a\ a\ a\ b\ a\ a\ a\ a$		$\mathrm{BWT}_{E_3}[i]$	$b\ a\ a\ a\ b\ b\ b\ b$
Rev. Rot. Shift: 5			Rev. Rot. Shift: 5	

Definition 4. *Let R be an array of integers with length $n \in \mathbb{N}$. We call R* **arithmetic progressed** *iff there exists $m < n$ and $q \in \{1,\dots,n\}$ such that $R[i] = q$ if $i = 1$, $(R[i] = R[i-1] + m) \bmod n$ if $i > 1$, for $1 \le i \le n$.*

Lemmas 2 to 4 consider an array R that is arithmetic progressed, and let n, m and q be defined as in Definition 4.

Lemma 2. *R is a permutation iff $\gcd(m, n) = 1$.*

Proof. By construction, R is an endomorphism. Let us take any $r \in \{1\dots,n-1\}$. By $R[(i+r) \bmod n] = (R[i] + rm) \bmod n$ we see that

$$\gcd(n, m) = 1 \Leftrightarrow rm \bmod n \ne n\ \forall 1 \le r \le n-1$$
$$\Leftrightarrow (R[i] + rm) \bmod n \ne R[i]\ \forall 1 \le i \le n\ \text{and}\ \forall 1 \le r \le n-1.$$

$\square$

Lemma 3. *Considering the inverse R^{-1} of R, the following properties hold:*

(a) The array R^{-1} is a rotation of R with shift $(q + (q-1)m - 1) \bmod n$ if and only if $m^2 \bmod n = 1$ and $\gcd(m, n) = 1$ holds.
(b) If R^{-1} is a rotation of R and $q \in \{1, m\}$, then $R^{-1} = R$.
(c) The array R^{-1} is a reversed rotation of R with shift $(q + (q-1)m + 1) \bmod n$ if and only if $m^2 \bmod n = n-1$ and $\gcd(m, n) = 1$ holds.

Proof. (a) Let $x := R[i]$ for an arbitrary, fixed $1 \le i \le n$. Then $R[(i + m) \bmod n] = (x + m^2) \bmod n$. We conclude that $R^{-1}[x] = i$ and $R^{-1}[(x + m^2) \bmod n] = (i + m) \bmod n$ holds. Now we yield the equivalence

$$R^{-1}[(x+1) \bmod n] = (i+m) \bmod n \Leftrightarrow m^2 \bmod n = 1.$$

Since $R[1] = q$, the shift is $R[q] - R^{-1}[q] \bmod n = (q + (q-1)m - 1) \bmod n$.
(b) If $q = 1$, then $R[1] = 1$; hence 1 is a fix point. If $q = m$, then $R[m+1] = (q + m^2) \bmod n = m + 1$; hence $m + 1$ is a fix point.

(c) Let x, i be defined as in proof of Item (a). Then we yield the equivalence

$$R^{-1}[(x-1) \bmod n] = (i+m) \bmod n \Leftrightarrow m^2 \bmod n = n - 1.$$

Since $R[1] = q$, the shift is $R[q] + R^{-1}[q] \bmod n = (q + (q-1)m + 1) \bmod n$.

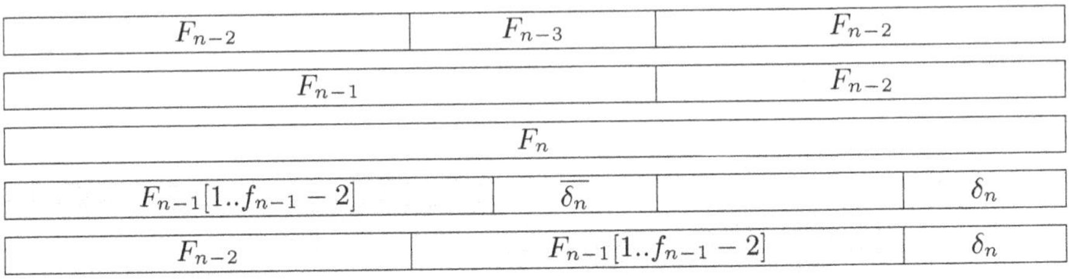

$\square$

F_{n-2}		F_{n-3}	F_{n-2}	
F_{n-1}			F_{n-2}	
F_n				
$F_{n-1}[1..f_{n-1}-2]$		$\overline{\delta_n}$		δ_n
F_{n-2}		$F_{n-1}[1..f_{n-1}-2]$		δ_n

Fig. 1. Overview over the different split-ups considered for F_n with $n \geq 4$. The lower part is shown in proof of Lemma 7 and used by Lemma 10.

Lemma 4. *The last-to-front mapping $\psi_R[i] := R^{-1}[R[i] - 1 \bmod n]$ shows the following characterizations:*

(a) If R^{-1} is a rotation of R, then $\psi_R[i] = (i - m) \bmod n$.
(b) If R^{-1} is a reversed rotation of R, then $\psi_R[i] = (i + m) \bmod n$.

Proof. We follow the observations in Lemma 3.

(a) Let k denote the shift of R^{-1}. Then $\psi_R[i] = R^{-1}[R[i] - 1 \bmod n] = R^{-1}[q + (i-1)m - 1 \bmod n] = R[q + (i-1)m - 1 - k \bmod n] = q + (q + (i-1)m - k - 2)m \bmod n = i - m \bmod n$.

(b) Let k denote the shift of R^{-1}. Then $\psi_R[i] = R^{-1}[R[i] - 1 \bmod n] = R^{-1}[q + (i-1)m - 1 \bmod n] = R[k - q - (i-1)m + 1 \bmod n] = q + km - qm - (i-1)m^2 \bmod n = q + qm + (q-1)m^2 + m - qm - (i-1)m^2 \bmod n = i + m \bmod n$.

$\square$

The following, well-known properties of the Fibonacci numbers allow us to apply Lemmas 2 to 4 to the suffix array and the last-to-front mapping of some instances of the string sequences under consideration:

Lemma 5 ([9,21]). *The following statements hold:*

– $\gcd(f_n, f_{n-1}) = 1$ *for $n \in \mathbb{N}$.*
– *For $n > 1$ even, $f_{n-1}^2 \bmod f_n = 1$ holds.*
– *For $n > 1$ odd, $f_{n-1}^2 \bmod f_n = n - 1$ holds.*
– *Since $(n - m)^2 \bmod n = m^2 \bmod n$ for any $m, n \in \mathbb{N}$, we can exchange f_{n-1} by f_{n-2} in the items above.*

Since $\mathsf{SA}_{F_1}, \mathsf{SA}_{F_2}$ and SA_{F_3} are the identity, we focus on the strings F_n with $n \geq 4$.

Lemma 6 (Christodoulakis et al. [3, Lemma 2.8]). *For $n > 3$, $F_n = F_{n-2}F_{n-3}\cdots F_2\delta_n$.*

Lemma 7. *For $n \geq 4$ and $1 \leq i < f_{n-1}$, we have $F_n[i..] \prec F_n[i + f_{n-2}..]$ if n is even, $F_n[i..] \succ F_n[i + f_{n-2}..]$ if n is odd.*

Proof. It follows from Lemma 6 that $F_n = F_{n-2}F_{n-3}\cdots F_2\delta_n$ and $F_n[1..f_{n-1}] = F_{n-1}$ and $F_{n-1} = F_{n-3}F_{n-4}\cdots F_2\overline{\delta_n}$. So $\mathsf{lcp}(F_n[1..], F_n[1+f_{n-2}..]) = f_{n-1}-|\delta_n|$. For any $1 \leq i < f_{n-1}$, the order $\prec$ of $F_n[i..]$ and $F_n[i + f_{n-2}..]$ is determined by comparing $F_n[f_{n-1} - 1] = \overline{\delta_n}[1]$ with $F_n[f_n - 1] = \delta_n[1]$.   □

Lemma 8. *For $n \geq 4$ even, $F_n[f_n..] < F_n[f_n + f_{n-2} \bmod f_n..] < F_n[f_n + 2f_{n-2} \bmod f_n..] < \cdots < F_n[f_n + (f_n - 1)f_{n-2} \bmod f_n..]$.*

Proof. The conclusion of the inequations is divided up into two intervals and a starting position:

- It follows by Lemma 7 that $F_n[i..] \prec F_n[i + f_{n-2}..]$ for all $1 \leq i < f_{n-1}$.
- Since $F_n = F_{n-1}F_{n-2} = F_{n-2}F_{n-3}F_{n-2}$, $F_n[i..]$ is a prefix of $F_n[i - f_{n-1}..] = F_n[i + f_{n-2} \bmod f_n..]$ for every $f_{n-1} < i \leq f_n$. Hence, $F_n[i..] < F_n[i + f_{n-2} \bmod f_n..]$.
- By Lemma 6, $F_n[f_n..]$ is the lexicographically smallest suffix of F_n. Since f_n and f_{n-2} are coprime, there is a lexicographically increasing chain starting at $F_n[f_n..]$ that visits every suffix of F_n by a step of f_{n-2} (taking modulo f_n). The lexicographically largest suffix is $F_n[f_n + (f_n - 1)f_{n-2} \bmod f_n..] = F_n[f_{n-1}..]$.   □

Theorem 1. *For $n \in \mathbb{N}$ even, ISA_{F_n} is a rotation of SA_{F_n} with a shift of $f_{n-2}+1$. SA_{F_n} is given by $\mathsf{SA}_{F_n}[i] = f_n$ if $i = 1$, $\mathsf{SA}_{F_n}[i] = (\mathsf{SA}_{F_n}[i - 1] + f_{n-2}) \bmod f_n$ otherwise.*

Proof. The arithmetic characterization of SA_{F_n} follows directly from Lemma 8. By Lemma 3(a), ISA_{F_n} is a rotation of SA_{F_n}.   □

Theorem 2. *Let $B_n := \beta F_n$ with a character $\beta \geq b$. For $n \in \mathbb{N}$ even, ISA_{B_n} is equal to SA_{B_n}, which is given by $\mathsf{SA}_{B_n}[i] = f_n + 1$ if $i = 1$, $\mathsf{SA}_{B_n}[i] = (\mathsf{SA}_{B_n}[i - 1] + f_{n-2}) \bmod f_n$ if $2 < i < f_n + 1$,*

Proof. By Lemma 8 we know the lexicographical order of the suffixes $B_n[i..]$ for $i > 1$. It remains to pigeonhole $B_n[1..]$. Theorem 1 tells us that $F_n[f_{n-1}..]$ is the largest suffix of f_n. By transitivity it suffices to show that $F_n[f_{n-1}..] < B_n$: this is clear if $\beta > b$. Otherwise ($\beta = b$), $B_n[1..] = bF_n[1..]$ and $F_n[f_{n-1}..] = bF_n[1 + f_{n-1}..] = bF_{n-2}$. But we saw in proof of Lemma 8 that $F_{n-2} < F_n$, hence $F_n[f_{n-1}..] < B_n$. Together we get that the step is f_{n-2}.

We complete the arithmetic characterization of SA_{B_n} by showing a fix point:

$$\mathsf{SA}_{B_n}[f_{n-2}+2] = (f_{n-2}^2 + f_{n-2} + 1) \bmod f_n = f_{n-2} + 2,$$

where we used that $f_{n-2}^2 \bmod f_n = 1$ by Lemma 5, and $|F_{m-2}| + 2 < |F_m|$ for every $m > 4$. Hence, by Lemma 3(b), $\mathsf{ISA}_{B_n}[2..f_n-1]$ is equal to $\mathsf{SA}_{B_n}[2..f_n-1]$. □

For the sequences C_n and D_n we have results similar to Lemma 8:

Corollary 1. *(a) Let $C_n := F_n c$ with a character $c > b$. For $n \geq 5$ odd,*
$$C_n[f_{n-1}..] < C_n[2f_{n-1} \bmod f_n..] < \ldots < C_n[f_n f_{n-1} \bmod f_n..] = C_n[f_n..].$$
(b) Let $D_m := \bar{F}_m c$ with a character $c > b$. For $m \geq 4$ even, $D_m[f_{m-1}..] < D_m[2f_{m-1} \bmod f_m..] < \ldots < D_m[f_m f_{m-1} \bmod f_m..] = D_m[f_m..].$

Proof. We follow the steps of the proof of Lemma 8: It follows from Lemma 7 that $C_n[i..] > C_n[i+f_{n-2}..]$ for all $1 \leq i < f_{n-1}$. Hence $C_n[i..] < C_n[i+f_{n-1} \bmod f_n..]$ for all $f_{n-2} < i \leq f_n$. By the same argument, $\bar{F}_m[i..] > \bar{F}_m[i+f_{m-2}..]$ for all $1 \leq i < f_{m-1}$, thus $D_m[i..] < D_m[i+f_{m-1} \bmod f_m..]$ for all $f_{m-2} < i \leq f_m$. $F_n[i+f_{n-1}..]$ is a prefix of $F_n[i..]$ for every $1 \leq i \leq f_{n-2}$. So the order $C_n[i+f_{n-1}..] > C_n[i..]$ is determined by comparing c with $F_{n-3}[1]$. Since $D_m = \bar{F}_{m-2}\bar{F}_{m-3}\bar{F}_{m-2}c$, $\bar{F}_m[i+f_{n-1}..]$ is a prefix of $\bar{F}_m[i..]$ for every $1 \leq i \leq f_{m-2}$. So the order $D_m[i+f_{n-1}..] > D_m[i..]$ is determined by comparing c with $\bar{F}_{m-3}[1]$. So far, we have $C_n[i..] < C_n[i+f_{n-1} \bmod f_n..]$ for every $1 \leq i < f_n$, and $D_m[i..] < D_m[i+f_{m-1} \bmod f_m..]$ for every $1 \leq i < f_m$. Since f_n and f_{n-1} are coprime, there is a lexicographically increasing chain starting at $C_n[f_{n-1}..]$ that visits every suffix of C_n, except $C_n[f_n+1..]$, by a step of f_{n-1} (taking modulo f_n); the same holds for D_m. □

Theorem 3. *Let $C_n := F_n c$ with a character $c > b$. For $n \in \mathbb{N}$ odd, $\mathsf{ISA}_{C_n}[1..f_n]$ is a reversed rotation of $\mathsf{SA}_{C_n}[1..f_n]$ with a shift of f_n. SA_{C_n} is given by $\mathsf{SA}_{C_n}[i] = f_n+1$ if $i = f_n+1$, $\mathsf{SA}_{C_n}[i] = f_n$ if $i = f_n$, $\mathsf{SA}_{C_n}[i] = (\mathsf{SA}_{C_n}[i+1]+f_{n-1}) \bmod f_n$ if $1 \leq i < f_n$.*

Proof. Since $C_n[f_n+1..] = c$ is the largest suffix of C_n, the arithmetic characterization of SA_{C_n} follows from Corollary 1(a). By Lemma 3(c), $\mathsf{ISA}_{C_n}[1..f_n]$ is a reversed rotation of $\mathsf{SA}_{C_n}[1..f_n]$. □

Theorem 4. *Let $D_n := \bar{F}_n c$ with a character $c > b$. For $n \in \mathbb{N}$ even, ISA_{D_n} is equal to SA_{D_n}; both are given by $\mathsf{SA}_{D_n}[i] = f_n+1$ if $i = f_n+1$, $\mathsf{SA}_{D_n}[i] = f_n$ if $i = f_n$, $\mathsf{SA}_{D_n}[i] = (\mathsf{SA}_{D_n}[i+1] - f_{n-2}) \bmod f_n$ if $1 \leq i \leq f_n$.*

Proof. Since $D_n[f_n+1..] = c$ is the largest suffix of D_n, the arithmetic characterization of SA_{D_n} follows from Corollary 1(b). By Lemma 3(b), ISA_{D_n} is equal to SA_{D_n}. □

Lemma 9. *For $n \geq 4$ even, $Z_n[1+f_{n-2}..] < Z_n[1+2f_{n-2} \bmod f_n..] < \ldots < Z_n[1+f_n f_{n-2} \bmod f_n..] = Z_n[1..]$, where Z_n is the n-th singular word, defined in Definition 3.*

Table 4. We divide the suffixes of the considered strings in Sect. 5 into blocks of ba-,aa- and ba-type (to some extent). These types are arranged (mostly) like for the text T in this table.

$T[\mathsf{SA}_T[i]]$	$a \ldots a$	$a \ldots a$	$b \ldots b$
$T[\mathsf{SA}_T[i] - 1]$	$b \ldots b$	$a \ldots a$	$a \ldots a$
Blocks	ba-type	aa-type	ab-type

Proof. Let $G := F_n[1..f_n - 1]$. For n even, $Z_n = bF_n[1..f_n - 1]$. Following the proof of Lemma 8 for G (Lemma 6 is still applicable for G since it depends on the $\delta_n[1]$-value, not on $\delta_n[2]$) yields $G[i..] \prec G[i + f_{n-2} \bmod f_n..]$ for every $1 \leq i < f_{n-1} - 1$. Thus, $Z_n[i..] \prec Z_n[i + f_{n-2} \bmod f_n..]$ for every $1 < i < f_{n-1}$. For the other i-values we consider that $Z_n = Z_{n-2}Z_{n-3}Z_{n-2}$; hence $Z_n[i+f_{n-1}..]$ is a prefix of $Z_n[i..]$ for every $1 \leq i \leq f_{n-2}$. So $Z_n[i..] < Z_n[i + f_{n-2} \bmod f_n..]$ for every $f_{n-1} < i \leq f_n$. To sum up, there is a lexicographically increasing chain starting at $Z_n[1 + f_{n-1}..]$ that visits every suffix of Z_n by a step of f_{n-2} (taking modulo f_n). □

Theorem 5. *For $n \in \mathbb{N}$ even, ISA_{Z_n} is a rotation of SA_{Z_n} with a shift of $f_{n-2}+1$. SA_{Z_n} is given by $\mathsf{SA}_{Z_n}[i] = f_{n-2}+1$ if $i = 1$, $\mathsf{SA}_{Z_n}[i] = (\mathsf{SA}_{Z_n}[i-1]+f_{n-2}) \bmod f_n$ otherwise.*

Proof. The arithmetic characterization of SA_{Z_n} follows from Lemma 9. ISA_{Z_n} is a rotation of SA_{Z_n}, due to Lemma 3(a). □

5   Burrows-Wheeler Transform

In this section, we give a characterization of the BWT for the string sequences displayed by Table 1. We generate BWT_T of any string T by taking the preceding character of the suffix $T[\mathsf{SA}_T[i]..]$ while successively incrementing $1 \leq i \leq |T|$. Fortunately, the acquired results for SA_T can be directly applied for constructing BWT_T: Consider any $T \in \Sigma_2^*$ whose SA_T is a rotation (or reversed rotation) of its inverse. Further, consider that SA_T is arithmetic progressed; then the previous/next entry in SA_T is determined by a step of m or $n - m$ (we introduce m, n like in Definition 4). If the b-entries in T are distributed in such a way that we find a b at position $i + m$ or $i + n - m$ for each $1 \leq i \leq |T|$ with $T[i] = b$, then the suffixes $T[i..]$ that succeed a b ($= T[i - 1]$) are aligned successively in SA_T, which means that the BWT_T generates a homogenous block of b's, like in Table 4. This stepping-characteristic is caught by the following

Lemma 10 (Rytter [17]). *For any $n \geq 4$, we have*

- *$F_n[i] = F_n[i + f_{n-2}]$ for any $i \in \{1, \ldots, f_{n-1} - 2\}$, and*
- *$F_n[i] = F_n[i + f_{n-1}]$ for any $1 \leq i \leq f_{n-2}$.*

Proof. By Lemma 6, $F_n = F_{n-2}F_{n-3}\cdots F_1\delta_n$. Also, $F_n[1..f_{n-1}] = F_{n-3}$
$F_{n-4}\cdots F_1\overline{\delta_n}$. The second claim follows by splitting $F_{n-2}F_{n-3}F_{n-2}$. Figure 1
illustrates the proven properties. □

Theorem 6. *For n even, we get $\psi_{F_n}[i] = (i + f_{n-1}) \bmod f_n$ and $\mathsf{BWT}_{F_n} = b^{f_{n-2}}a^{f_{n-1}}$.*

Proof. Because F_n does not contain the string bb, the only substrings of length
two are aa, ab and ba. We will focus on the suffixes that start with an a that
are preceded by a b. We call these of type ba; they are depicted in Table 4. If
we show that these suffixes have successive numbers in the beginning of SA_{F_n},
we yield the claimed structure of BWT_{F_n}. We find these suffixes by tracking the
chain given in the proof of Lemma 8. The chain starts at the smallest suffix
$F_n[f_n..]$ and takes steps of length f_{n-2} (modulo f_n). Fortunately, $F_n[f_n..]$ is
exactly a ba-type suffix with $F_n[f_n - 1..] = \delta_n = ba$. By Lemma 10, the chain
will visit iteratively the next ba-type suffix, until it accesses the ba-type suffix
$\mathsf{SA}_{F_n}[f_{n-2}] = f_{n-1} + 1$. This is the last ba-type suffix: $\mathsf{SA}_{F_n}[f_{n-2} + 1] = 1$,
and by definition of the BWT, the preceding character of the suffix $\mathsf{SA}_{F_n}[1..]$
is $F_n[f_n] = a$. Because F_n contains exactly f_{n-2} many b's, we will never again
meet a ba-type suffix while continuing traversing the chain. The structure of ψ_{F_n}
follows by Lemma 4(a). □

Theorem 7. *For $n \in \mathbb{N}$ even and $\beta \geq b$, let $B_n := \beta F_n$. $\mathsf{BWT}_{B_n} = b^{f_{n-2}}\beta a^{f_{n-1}}$
and $\psi_{B_n}[i] = f_n + 1$ if $i = f_{n-2} + 1$, $\psi_{B_n}[i] = 1$ if $i = f_n + 1$, $\psi_{B_n}[i] = (i + f_{n-1}) \bmod f_n$ otherwise.*

Proof. The proof is conducted analogously to proof of Theorem 6: By Theorem 2,
the smallest suffix is $B_n[f_n + 1..]$ and the step of SA_{B_n} is f_{n-2}; $B_n[f_n + 1..]$ is a ba-
type suffix. By proof of Theorem 2, the largest suffix is $B_n[1..]$. By following the
chain of the proof of Theorem 6, we traverse successively ba-type suffixes until we
visit the ba-type suffix $\mathsf{SA}_{B_n}[f_{n-2}] = f_{n-1} + 2$. If $\beta \neq b$, we have already visited
all suffixes of ba-type. Again, by a step of f_{n-2}, we find $\mathsf{SA}_{B_n}[f_{n-2} + 1] = 2$. The
suffix $B_n[2..]$ is preceded by a β. □

Theorem 8. *For $n \geq 5$ odd, let $C_n := F_n c$. $\mathsf{BWT}_{C_n} = b^{f_{n-2}-1}ca^{f_{n-1}}b$ and
$\psi_{C_n}[i] = f_n + 1$ if $i = f_{n-2}$, $\psi_{C_n}[i] = f_n$ if $i = f_n + 1$, $\psi_{C_n}[i] = (i + f_{n-1}) \bmod f_n$
otherwise.*

Proof. The proof is conducted analogously to proof of Theorem 6: By
Corollary 1(a), the smallest suffix is $C_n[f_{n-1}..]$ and the step of SA_{C_n} is f_{n-1};
$C_n[f_{n-1}..]$ is a ba-type suffix. By Theorem 3, the largest suffix is $C_n[f_n + 1..]$;
it is preceded by a b character. Since $C_n[f_n..]$ is the second largest suffix,
$\mathsf{BWT}_{C_n}[f_n + 1] = b$. By following the chain of the proof of Theorem 6, we traverse
successively ba-type suffixes until we visit the ba-type suffix $\mathsf{SA}_{C_n}[f_{n-2} - 1] = f_{n-2} + 1$. Again, by a step of f_{n-1}, we find $\mathsf{SA}_{C_n}[f_{n-2}] = 1$. The suffix $C_n[1..]$ is
preceded by a c. With Lemma 4(b) we yield the structure of $\psi_{F_n c}$. □

Theorem 9. *For* $n \geq 4$ *even, let* $D_n := \bar{F}_n c$. $\mathsf{BWT}_{D_n} = b^{f_{n-1}-1} ca f^{f_{n-2}} b$ *and* $\psi_{D_n}[i] = f_n + 1$ *if* $i = f_{n-1}$, $\psi_{D_n}[i] = f_n$ *if* $i = f_n + 1$, $\psi_{D_n}[i] = (i + f_{n-2}) \bmod f_n$ *otherwise.*

Proof. The proof is conducted by complementing the results of Theorem 6: Substrings of length two in D_n are of the form bc, bb, ba and ab; D_n does not contain the string aa. By Corollary 1(b), the smallest suffix is $D_n[f_{n-1}..]$ and the step of SA_{D_n} is f_{n-1}; $D_n[f_{n-1}..]$ is a ba-type suffix.

Because the suffixes starting with an a are always preceded by a b (suffixes of ba-type), it suffices to show that the suffixes starting with a b that are preceded by a b (suffixes of bb-type) are consecutively aligned after the block of ba-type suffixes. By Theorem 4, the largest suffix is $D_n[f_n + 1..]$; it is preceded by a b character. Since $D_n[f_n..]$ is the second largest suffix, $\mathsf{BWT}_{D_n}[f_n + 1] = b$. Moreover, the largest ba-type suffix is $D_n[f_n - 2..]$. By following a chain similar in the proof of Theorem 6, we traverse successively ba-type suffixes until we visit the ba-type suffix $\mathsf{SA}_{D_n}[f_{n-2}] = f_n - 1$. We next visit the bb-type suffixes starting from $\mathsf{SA}_{D_n}[f_{n-2} + 1] = f_{n-1} - 1$ to $\mathsf{SA}_{D_n}[f_{n-1} - 1] = f_{n-2} + 1$. By a step of f_{n-1}, we find $\mathsf{SA}_{D_n}[f_{n-1}] = 1$. The suffix $D_n[1..]$ is preceded by a c. $\square$

Theorem 10. *For* $n \geq 4$ *even,* $\mathsf{BWT}_{Z_n} = b^{f_{n-2}} a f^{f_{n-1}-1} b$ *and* $\psi_{Z_n}[i] = (i + f_{n-1}) \bmod f_n$, *where* Z_n *is the* n-*th singular word, defined in Definition 3.*

Proof. Like F_n, the string Z_n does not contain bb as a substring. The proof is conducted analogously to proof of Theorem 6: By Lemma 9, the largest suffix is $Z_n[1..]$. It is preceded by $Z_n[|Z_n|] = b$; hence $\mathsf{BWT}_{Z_n}[f_n] = b$. Moreover, the largest ba-type suffix is $Z_n[2..]$. By Theorem 5, the smallest suffix is $Z_n[f_{n-2} + 1..]$ and the step of SA_{Z_n} is f_{n-2}; $Z_n[f_{n-2}..]$ is a ba-type suffix since $Z_n[f_{n-2} - 1..f_{n-2}] = \delta_{n-2}$. By following the chain of the proof of Theorem 6, we traverse successively ba-type suffixes until we visit the last ba-type suffix $\mathsf{SA}_{Z_n}[f_{n-2} - 1] = 2$. $\square$

6   Outlook

We presented a family of string sequences based on the intriguing Fibonacci words, and highlighted some interesting combinatorial properties of the suffix array, its inverse, and the BWT of those sequences. It is an open question whether we can specify a general class of strings having the studied properties, like the BWT based on rotations [14]. One problem is that the conditions are not symmetric. Although we considered appending c or prepending β to a Fibonacci string, sequences like $\beta \bar{F}_n$, $\bar{F}_n \alpha$ and $F_n \alpha$ with $\alpha \leq a$ do not show any nice properties. Even giving a characterization of the set of strings whose SA and ISA are identical seems hard for us. The here presented techniques could be useful for further research.

Acknowledgement. We are grateful to Gabriele Fici for helpful discussion, and to our student Sven Schrinner who discovered one rotation property while solving an exercise.

References

1. Berstel, J., Savelli, A.: Crochemore factorization of sturmian and other infinite words. In: Královič, R., Urzyczyn, P. (eds.) MFCS 2006. LNCS, vol. 4162, pp. 157–166. Springer, Heidelberg (2006)
2. Burrows, M., Wheeler, D.J., Burrows, M., Wheeler, D.J.: A block-sorting lossless data compression algorithm. Digital Equipment Corporation, Technical report (1994)
3. Christodoulakis, M., Iliopoulos, C.S., Ardila, Y.J.P.: Simple algorithm for sorting the Fibonacci string rotations. In: Wiedermann, J., Tel, G., Pokorný, J., Bieliková, M., Štuller, J. (eds.) SOFSEM 2006. LNCS, vol. 3831, pp. 218–225. Springer, Heidelberg (2006)
4. Droubay, X.: Palindromes in the Fibonacci word. Inf. Process. Lett. **55**(4), 217–221 (1995)
5. Ferragina, P., Manzini, G.: Indexing compressed text. J. ACM **52**(4), 552–581 (2005)
6. Gagie, T., Gawrychowski, P., Kärkkäinen, J., Nekrich, Y., Puglisi, S.J.: LZ77-based self-indexing with faster pattern matching. In: Pardo, A., Viola, A. (eds.) LATIN 2014. LNCS, vol. 8392, pp. 731–742. Springer, Heidelberg (2014)
7. Gramss, T.: Entropy of the symbolic sequence for critical circle maps. Phys. Rev. E **50**, 2616–2620 (1994)
8. Iliopoulos, C.S., Moore, D., Smyth, W.F.: A characterization of the squares in a Fibonacci string. Theor. Comput. Sci. **172**(1–2), 281–291 (1997)
9. Hoggatt Jr., V.E., Bicknell-Johnson, M.: Composites and primes among powers of Fibonacci numbers increased or decreased by one. Fibonacci Q. **15**, 2 (1977)
10. Kärkkäinen, J.: Fast bwt in small space by blockwise suffix sorting. Theor. Comput. Sci. **387**(3), 249–257 (2007)
11. de Luca, A.: A combinatorial property of the Fibonacci words. Inf. Process. Lett. **12**(4), 193–195 (1981)
12. de Luca, A., de Luca, A.: Combinatorial properties of sturmian palindromes. Int. J. Found. Comput. Sci. **17**(03), 557–573 (2006)
13. Manber, U., Myers, G.: Suffix arrays: A new method for on-line string searches. In: Proceedings of the First Annual ACM-SIAM Symposium on Discrete Algorithms. SODA 1990, pp. 319–327, Philadelphia, PA, USA (1990)
14. Mantaci, S., Restivo, A., Sciortino, M.: Burrows-Wheeler transform and Sturmian words. Inf. Process. Lett. **86**(5), 241–246 (2003)
15. Monnerot-Dumaine, A.: The Fibonacci Word fractal, 24 pages, 25 figures, February 2009
16. Pirillo, G.: Fibonacci numbers and words. Discrete Math. **173**(1–3), 197–207 (1997)
17. Rytter, W.: The structure of subword graphs and suffix trees of Fibonacci words. Theor. Comput. Sci. **363**(2), 211–223 (2006)
18. Saari, K.: Periods of factors of the Fibonacci word. In: Proceedings of WORDS 2007. Institut de Mathematiques de Luminy (2007)
19. Sadakane, K.: Compressed suffix trees with full functionality. Theory Comput. Syst. **41**(4), 589–607 (2007)
20. Simpson, J., Puglisi, S.J.: Words with simple burrows-wheeler transforms. Electr. J. Comb. 15(1) (2008)
21. Wells, D.: Prime Numbers: The Most Mysterious Figures in Math. Wiley, Hoboken (2011)
22. Wen, Z.X., Wen, Z.Y.: Some properties of the singular words of the Fibonacci word. Eur. J. Comb. **15**(6), 587–598 (1994)
23. Ziv, J., Lempel, A.: A universal algorithm for sequential data compression. IEEE Trans. Inf. Theory **23**(3), 337–343 (1977)

Prefix-Suffix Square Completion

Marius Dumitran[1] and Florin Manea[2]([✉])

[1] Faculty of Mathematics and Computer Science, University of Bucharest,
Str. Academiei 14, 010014 Bucharest, Romania
marius.dumitran@fmi.unibuc.ro

[2] Department of Computer Science, Christian-Albrechts University of Kiel,
Christian-Albrechts-Platz 4, 24118 Kiel, Germany
flm@informatik.uni-kiel.de

Abstract. We consider a family of new formal operation on words: the prefix square completion, the suffix square completion, and the prefix-suffix square completion. By suffix square completion (respectively, prefix square completion), one can derive from a word w any word wx (respectively, xw) if w has a suffix (respectively, prefix) yxy; by prefix-suffix square completion we derive from a word w any word w' that is obtained either by prefix square completion or by suffix square completion from w. We discuss two main aspects of these operations. On the one hand, we study the derivation of infinite words by iterated prefix-suffix square completion, and show that, although any word generated by square completion operations contains squares, we can generate infinite words that do not contain any repetition of exponent greater than 2. On the other hand, focusing on finite words, we give a linear time procedure that, given two words, decides whether the longer can be generated by iterated prefix-suffix square completion from the shorter.

1 Introduction

The *prefix duplication, suffix duplication, and prefix-suffix duplication* are formal operations on words that were introduced in [1] following a biological motivation. Essentially, by suffix duplication, from a word w one can derive any word wx where x is a suffix of w; that is, suffixes of w are duplicated. Prefix duplication is the symmetric operation: by this operation one can duplicate prefixes of w. Finally, the prefix-suffix duplication is just a combination of the previous two operations: by it, one can duplicate either a suffix or a prefix of the initial words. As it is the case in the study of many word operations, one is mainly interested in the words that can be generated by iteratively applying these operations to an initial word. The results of [1] show that the set of words that can be generated in arbitrarily many prefix-suffix duplication steps from some initial word is not context-free, unless the initial word is unary. However, the language of the words that can be generated by iterated prefix-suffix duplication from a word is semi-linear and belongs to **NL**. Finally, an algorithm deciding in $\mathcal{O}(n^2 \log n)$

The work of Florin Manea was supported by the DFG grant 596676.

F. Manea and D. Nowotka (Eds.): WORDS 2015, LNCS 9304, pp. 147–159, 2015.
DOI: 10.1007/978-3-319-23660-5_13

time, for two given words, whether the longer word (whose length is n) can be generated from the shorter word by prefix-suffix duplication was given.

Restrictions of the above operations where considered in [2]: the *bounded suffix, prefix and prefix-suffix duplication*. For these operations, the length of the duplicated prefix or suffix was bounded by a predefined constant. Again, the study of these operations addressed two main directions. On the one hand, language theoretic properties of the set of words generated by iterated bounded prefix-suffix duplication from a certain initial set were obtained. On the other hand, efficient algorithms solving the membership problem for such languages were devised. In particular, one can decide in $\mathcal{O}(nk \log k)$ whether some word (of length n) can be generated from a shorter word by bounded prefix-suffix duplication where the length of the duplicated factors is at most k.

We introduce here a new family of word operations, closely related to the prefix-suffix duplication. The initial motivation of studying the prefix-suffix duplication were some biological processes that essentially create repetitions at the ends of the genetic sequences (see [1] for a longer discussion on this motivation); however, the formal operations defined in [1] assumed that such repetitions are created by replicating their root. Here we assume a different point of view: we consider the possibility of creating squares (the simplest type of repetition) at one of the ends of the word by completing a prefix or suffix of the considered sequence to a square. More precisely, by *suffix square completion*, we derive from a word w a word wx if w has a suffix yxy; note that the suffix we complete to a square must contain the root of the square (here, yx), and that suffix duplication is obtained by restricting square suffix to the case when y is the empty word. The *prefix square completion* is symmetrical: from a word w we derive xw if w has a prefix yxy. The *prefix-suffix square completion* just combines the previous two operations: from a word w we derive w' if w' can be obtained either by prefix square completion or by suffix square completion from w.

Most of the language theoretic results obtained for the prefix, suffix, and prefix-suffix duplication operations seem to hold also for the newly defined square completion operations. Therefore, our investigation of the prefix-suffix square completion operations is aimed in the following two directions.

Firstly (and quite differently from what was done so far) we investigate how square completion operations can be used to generate infinite words. We say that a right-infinite word $\mathbf{w}$ can be generated by one of the square completion operations if there is an infinite sequence of finite prefixes of $\mathbf{w}$, namely $w_0, w_1, \ldots$, such that w_i is obtained by the respective operation from w_{i-1}. For instance, the infinite Fibonacci word or the Period-doubling word can be generated by suffix square completion. Furthermore, we show that the infinite Thue-Morse word can also be generated by suffix square completion. This exhibits a property that seems interesting to us: every (infinite) word generated by suffix completion contains squares, but there are (infinite) words generated by this operation (which basically creates squares) that avoid any repetition of (rational) exponent higher than 2. In comparison, we show that the Thue-Morse infinite word cannot be generated by prefix-suffix duplication. However, we show that one can generate an infinite cube-free word by suffix-duplication. This is a weaker version of the result obtained for square completion: every (infinite) word generated by suffix duplication contains squares, but there are (infinite) words generated by this operation that avoid any repetition of integer exponent higher than 2.

Secondly, we shift our attention to finite words, and, in this case, try to see whether a word can be generated by prefix-suffix square completion from one of its factors. We give an algorithm that identifies in linear time, for a given word w, all its prefixes from which w can be generated by iterated suffix square completion. This algorithm is an essential step in showing that we can identify in linear time (and produce a compact representation of) all the factors of a word from which it can be generated by iterated prefix-suffix square completion. This leads immediately to a linear-time algorithm deciding, for two given words, whether the longer word can be generated from the shorter word by prefix-suffix square completion. This algorithm is much faster than the corresponding one from the case of prefix-suffix duplication.

2   Definitions

§ **Basic Facts.** We define $\Sigma_k = \{0, \ldots, k-1\}$ to be an alphabet with k letters. For an alphabet Σ we denote by Σ^* (respectively, Σ^+) the set of finite (respectively, non-empty finite) words over Σ, and by Σ^ω the set of (right-)infinite words over Σ. For a word w and some $1 \leq i \leq |w|$ we denote the i-th letter of w by $w[i]$. We also denote the factor that starts with the i-th letter and ends with the j-th letter in w by $w[i..j]$.

The powers of a word w are defined recursively by $w^0 = \lambda$ and $w^n = ww^{n-1}$ for $n \geq 1$. If w cannot be expressed as a nontrivial power (i.e., w is not a repetition) of another word, then w is *primitive*. A *period* of a word w over V is a positive integer p such that $w[i] = w[j]$ for all i and j with $i \equiv j(\bmod\ p)$; if p is a period of w, then w is called p-periodic. Let $per(w)$ be the smallest period of w. The notion of repetition can be extended to the case of rational exponents: a word of length n and period p is called a repetition with exponent $\frac{n}{p}$. A word w with $per(w) \leq \frac{|w|}{2}$ is called run; a run $w[i..j]$ (so, $p = per(w[i..j]) < \frac{i-i+1}{2}$) is maximal if and only if it cannot be extended to the left or right to get a word with period p, i.e., $i = 1$ or $w[i-1] \neq w[i+p-1]$, and, $j = n$ or $w[j+1] \neq w[j-p+1]$.

The infinite Thue-Morse word **t** is defined as $\mathbf{t} = \lim_{n \to \infty} \phi_t^n(0)$, for the morphism $\phi_t : \Sigma_2^* \to \Sigma_2^*$ where $\phi_t(0) = 01$ and $\phi_t(1) = 10$ (see [3,4]). It is well-known (see, for instance, [5]) that **t** does not contain any factor of the form $xyxyx$ (overlaps). Consequently, the infinite word **t** does not contain any repetition of (rational) exponent greater than 2.

The infinite Fibonacci word **f** is defined as $\mathbf{f} = \lim_{n \to \infty} f_n$, where $f_0 = 0$, $f_1 = 01$, and $f_k = f_{k-1}f_{k-2}$ for $k \geq 2$ (see, e.g., [5,6]). The Fibonacci word contains cubes, but it does not contain repetitions of exponent 4; it is also a Sturmian word. Finally, $\mathbf{f} = \lim_{n \to \infty} \phi_f^n(0)$, where $\phi_f(0) = 01$ and $\phi_f(1) = 0$.

The Period doubling word **d** is defined as $\mathbf{d} = \lim_{n \to \infty} \phi_d^n(0)$, for the morphism $\phi_d : \Sigma_2^* \to \Sigma_2^*$ where $\phi_t(0) = 01$ and $\phi_t(1) = 00$. This infinite word was studied both in combinatorics on words and, just like the Fibonacci word, in quasicrystal spectral theory (see, [6–9] and the references therein). To generate this sequence, note that to obtain $\phi_d^n(0)$ we can concatenate $\phi^{n-1}(0)$ to itself and then change the last digit (1 becomes 0 and vice versa). Interestingly, this sequence is obtained by replacing the 2 letters by 0 in the square-free Hall sequence $\mathbf{h} = \lim_{n\to\infty} \phi_h(2)$, with $\phi_h(2) = 210, \phi_h(1) = 20, \phi_h(0) = 1$ (see [10]). This sequence has label $A096268$ in The Online Dictionary of Integer Sequences.

Stewart's choral sequence $\mathbf{s}$ is defined as $\mathbf{s} = \lim_{n\to\infty} s_n$, where $s_0 = 0$ and $s_{k+1} = s_k s_k s_k^*$ for $k \geq 0$, with s_k^* being a copy of s_k with the middle letter changed from 0 to 1 or vice versa. It is known that $\mathbf{s}$ does not contain cubes. For more details, see The Online Dictionary of Integer Sequences, where this sequence has the label $A116178$.

See [5] for further details on the concepts discussed here.

§**Square Completion Operations**. The prefix, suffix, and prefix-suffix duplication operations were defined in [1]. Given a word $w \in \Sigma^*$, we have:

- *Prefix Duplication*: $PD(w) = \{xw \mid w = xw' \text{ for some } x \in \Sigma^+\}$.
- *Suffix Duplication*: $SD(w) = \{wx \mid w = w'x \text{ for some } x \in \Sigma^+\}$.
- *Prefix-suffix Duplication*: $PSD(w) = PD(w) \cup SD(w)$.

Further, we define the prefix, suffix, and prefix-suffix square completion operations. Given a word $w \in \Sigma^*$, we define:

- *Prefix Square Completion*: $PSC(w) = \{xw \mid w = yxyw', \text{ with } x \in \Sigma^+, y \in \Sigma^*\}$.
- *Suffix Square Completion*: $SSC(w) = \{wx \mid w = w'yxy, \text{ with } x \in \Sigma^+, y \in \Sigma^*\}$.
- *Prefix-suffix Square Completion*: $PSSC(w) = PSC(w) \cup SSC(w)$.

We further define, for $\Theta \in \{PD, SD, PSD, PSC, SSC, PSSC\}$, its iteration: $\Theta_k^0(x) = \{x\}, \Theta_k^{n+1}(x) = \Theta_k^n(x) \cup \Theta_k(\Theta_k^n(x))$, for $n \geq 0$, $\Theta_k^*(x) = \bigcup_{n\geq 0} \Theta_k^n(x)$.

Finally, we say that a right-infinite word $\mathbf{w}$ is generated by the operation Θ and we write $\mathbf{w} \in \Theta^\omega$, where $\Theta \in \{PD, SD, PSD, PSC, SSC, PSSC\}$, if there exists a sequence of finite words $(w_n)_{n\in\mathbb{N}}$ such that: w_i is a prefix of $\mathbf{w}$ and $w_{i+1} \in \Theta(w_i)$ for all $i \in \mathbb{N}$.

Example 1. Consider the words $w_0 = ab$, $w_{2i+1} = w_i w_i$ and $w_{2i+2} = w_{2i+1}b$ for $i \geq 0$. Let $\mathbf{w} = \lim_{n\to\infty} w_n$. Clearly, $w_{2i+1} \in PD(w_{2i})$, and, as all the words w_i end with b, $w_{2i+2} \in SD(w_{2i+1})$. Therefore, $\mathbf{w} \in PSD^\omega$.

§**Algorithmic Prerequisites.** The computational model we use to design and analyse our algorithms is the standard unit-cost RAM (Random Access Machine) with logarithmic word size, which is generally used in the analysis of algorithms. In the upcoming algorithmic problems, we assume that the words we process are sequences of integers (called letters, for simplicity). In general, if the input word has length n then we assume its letters are in $\{1, \ldots, n\}$, so each letter fits in a single memory-word. This is a common assumption in stringology (see, e.g., the discussion in [11]).

First, we recall the *range minimum query* data structure (see [12]). Given an array of n integers $T[\cdot]$, we can produce in $\mathcal{O}(n)$ time several data-structures for this array, allowing us to answer in constant time to queries $RMQ(i, j)$ and $posRMQ(i, j)$, asking for the minimum value and, respectively, its position among $T[i], T[i + 1], \ldots, T[j]$.

For a word u, $|u| = n$, over $V \subseteq \{1, \ldots, n\}$ we build in $\mathcal{O}(n)$ time its suffix tree and suffix array, as well as LCP-data structures, allowing us to retrieve in constant time the length of the longest common prefix of any two suffixes $u[i..n]$ and $u[j..n]$ of u, denoted $LCP_u(i, j)$ (the subscript u is omitted when there is no danger of confusion). See, e.g., [11,13], and the references therein.

We will also use in our algorithms the fact that the number of maximal runs of a word is linear and their list (with a run $w[i..j]$ represented as the triple $(i, j, per(w[i..j]))$) can be computed in linear time in the RAM with logarithmic word size model (see [14,15]). The exponent of a maximal run $w[i..j]$ occurring in w is defined as $\frac{j-i+1}{per(w[i..j])}$; in [15] it is shown that the sum of the exponents of maximal runs in a word of length n is upper bounded by n.

3   Generating Infinite Words

Lemma 1. *Let $x, y, z \in \Sigma^*$ such that x is a prefix of y and y a prefix of z. If $z \in SSC^*(x)$ then $z \in SSC^*(y)$.*

Proof. If $x = z$ the conclusion follows immediately. Otherwise, there exists a sequence of words $x_0, x_1, \ldots, x_n$, with $n \geq 1$, such that $x_0 = x, x_n = z$, and $x_i \in SSC(x_{i-1})$ for $1 \leq i \leq n$. As x_0 is a prefix of y and y is a prefix of x_n, there exists $1 \leq k \leq n$ such x_{k-1} is a prefix of y and y is a prefix of x_k. So, let $y = x_{k-1}u$. As $x_k \in SSC(x_{k-1})$ we get that $x_{k-1} = wvv'v$ and $x_k = wvv'vv'$; it follows that u is a prefix of v' and $v' = uv''$. So, $x_k = wvuv''vuv'' = yv''$. Therefore, x_k can be obtained by suffix square completion from y by appending v'' to y. It follows that $x_j \in SSC^*(y)$ for all $j \geq k$, so $z = x_n \in SSC^*(y)$.   □

The next three propositions show how the suffix square completion can be used to generate three important infinite words.

Proposition 1. *The Fibonacci word **f** is in SSC^ω.*

Proof. By Lemma 1, it is enough to show that for all $n \geq 4$, there exists f'_n such that $f'_n \in SSC^*(f_n)$, f_{n+1} is a prefix of f'_n, and f''_n is a prefix of f_{n+2}. Indeed, it follows that we can derive f'_{n+1} from f'_n (because f'_{n+1} can be derived from the prefix f_{n+1} of f'_n), and so on. For short, we would be able to derive, starting with f_4, an infinite sequence of prefixes of **f**. Now, $f_n = f_{n-1}f_{n-2}$; we can derive from f_n the word $f_{n-1}f_{n-2}f_{n-2}$, and then $f_{n-1}f_{n-2}f_{n-2}f_{n-2} = f_{n-1}f_{n-2}f_{n-2}f_{n-3}f_{n-4}$; but, $f_{n-1}f_{n-2}f_{n-2}f_{n-3}f_{n-4} = f_nf_{n-1}f_{n-4} = f_{n+1}f_{n-4}$. So, taking $f'_n = f_{n+1}f_{n-4}$ we get exactly what we wanted: $f'_n \in SSC^*(f_n)$, f_{n+1} is a prefix of f'_n, and f''_n is a prefix of f_{n+2}.   □

Proposition 2. *The Period-doubling word **d** is in SSC^ω.*

Proof. Let $d_n = \phi_d^n(0)$, for $n \geq 0$. It is not hard to note that if $d_{n+1} \in SSC^*(d_n)$, then $d_{n+2} \in SSC^*(d_{n+1})$; indeed, if in the ith step of the derivation of d_{n+1} from d_n we derived the word $wxyxy$ from the word $wxyx$, then in the ith step of the derivation of d_{n+2} from d_{n+1} we derive $\phi_d(wxyxy)$ from the word $\phi_d(wxyx)$. Therefore, it is enough to show that d_4 can be derived from $d_3 = 01000101$. Now, in the first step we derive $01000\underline{101}(01)$ (we place the root of the square between parentheses, and the suffix which we completed is underlined). Then, from 0100010101 we derive $0100010(10)(10)$. From 01000101010 we derive in two steps 0100010101000 (by duplicating twice the 0 letter occurring at the end of the word). Now, from 0100010101000 we derive $010001010100(0100) = d_4$. This concludes our proof.   □

Proposition 3. *The Thue-Morse word* **t** *is in* SSC^ω.

Proof. First we fix some notations. Let $\bar{t}_n$ be t_n in which we change all the 1 letters into 0 and all the 0 letters into 1. It is well known that $t_{n+1} = t_n\bar{t}_n$. Also, let t'_n be the word obtained from t_n by deleting its last letter. Let us first note that t_6 ends with 10110 and $\bar{t}_6$ starts with 1; so, from t'_6, which has the suffix 1011, we can derive in one SSC-step the word t_61, which is a prefix of t_7. Moreover, for all $n \geq 7$ we have that t_n ends with $\phi_t^{n-6}(101)\phi_t^{n-6}(10)$, so from t'_n we can generate $t_n\phi_t^{n-6}(1)$, which is a prefix of t_{n+1}. We now note that $t_{2n} = t_{2n-1}\bar{t}_{2n-2}t_{2n-3}\bar{t}_{2n-4}\ldots t_1\bar{t}_0 t_0$. As t_{2i-1} ends with $\bar{t}_{2i-2}$ for all i, it is immediate that $t'_{2n} \in SSC^*(t_{2n-1})$. Similarly, $t_{2n+1} = t_{2n}\bar{t}_{2n-1}t_{2n-2}\bar{t}_{2n-3}\ldots \bar{t}_1 t_0\bar{t}_0$. Now, t_{2i} ends with $\bar{t}_{2i-1}$ for all i, so $t'_{2n+1} \in SSC^*(t_{2n})$.

Now we have all the ingredients to show that $\mathbf{t} \in SSC^\omega$. We start with t_5 and derive from it, in multiple steps, t'_6. From t'_6 we derive in one step t_61. From t_61 we can derive t'_7 by Lemma 1, because we can derive t'_7 from the prefix t_6 of t_61. Then we continue this process. Generally, at some point of our derivation we obtained t'_n. From this we derive $t_n\phi_t^{n-6}(1)$, and further we derive in multiple steps t'_{n+1} (again, we can do this according to Lemma 1, because we can derive t'_{n+1} from the prefix t_n of $t_n\phi_t^{n-6}(1)$). This concludes our proof.   □

The previous results shows the existence of infinite words which can be generated by iterated suffix square completion, which avoid any power of rational exponent strictly greater than 2. However, **t** cannot be generated by the prefix-suffix duplication, so we cannot get in the same way a similar result for this operation.

Proposition 4. *The Thue-Morse word* **t** *is not in* PSD^ω.

Proof. To begin with, let $\bar{t}_n$ be t_n in which we change all the 1 letters into 0 and all the 0 letters into 1, and let $\bar{\mathbf{t}} = \lim_{n\to\infty} \bar{t}_n$.

Our proof is based on a series of claims regarding the occurrences of squares inside the Thue-Morse word.

   Claim 1. For all $n \geq 0$, t_n does not start with a square. Same holds for $\bar{t}_n$.

   Claim 2. For all $n \geq 0$, t_n does not end with a square. Same holds for $\bar{t}_n$.

   Claim 3. For all $n \geq 0$, if $t_n[i..j]$ is a square such that $i \leq 2^{n-1}$ and $j > 2^{n-1}$ (i.e., this square goes over the centre of t_n) then $i > 2^{n-2}$ and $j \leq 2^{n-1} + 2^{n-2}$

(i.e., the square is completely contained in $t_n[2^{n-2}+1..2^{n-1}+2^{n-2}] = \bar{t}_{n-1}\bar{t}_{n-1}$). Same holds for $\bar{t}_n$.

Claim 4. For all $n \geq 0$, if $t_n[i..j]$ is a square such that $i \leq 2^{n-1}$ and $j > 2^{n-1}$ (i.e., this square goes over the centre of t_n) and either $i \neq 2^{n-2}+1$ or $j \neq 2^{n-1}+2^{n-2}$, then $i > 2^{n-2}+2^{n-3}$ and $j \leq 2^{n-1}+2^{n-3}$ (i.e., the square is completely contained in $t_n[2^{n-2}+2^{n-3}+1..2^{n-1}+2^{n-3}]$). Same holds for $\bar{t}_n$.

We now move on to the main proof.

According to Claim 1, it is enough to show that $\mathbf{t} \notin SD^\omega$. Indeed, if $\mathbf{t} \in PSD^\omega \setminus SD^\omega$, then $\mathbf{t}$ has a prefix x obtained by prefix duplication. Clearly, x has a square prefix, so $\mathbf{t}$ has a square prefix, as well; this is a contradiction.

We now show by induction that it is impossible to derive from a prefix of t_n (respectively, a prefix of $\bar{t}_n$) a prefix of $\mathbf{t}$ (respectively, of $\bar{\mathbf{t}}$) longer than $|t_{n+1}| + \frac{|t_n|}{2} = 2^{n+1} + 2^{n-1}$. This property can be manually checked for $n \leq 6$.

Let us assume the above property holds for all t_k and $\bar{t}_k$ with $k \leq n-1$ and show that it holds for t_n. Clearly, we have that:

$$t_{n+2} = t_{n-2}\bar{t}_{n-2}\bar{t}_{n-2}t_{n-2}\bar{t}_{n-2}t_{n-2}t_{n-2}\bar{t}_{n-2}\bar{t}_{n-2}t_{n-2}t_{n-2}\bar{t}_{n-2}t_{n-2}\bar{t}_{n-2}\bar{t}_{n-2}t_{n-2}.$$

Let $\tau_i = t_{n+2}[(i-1)2^{n-2}+1]$, for $1 \leq i \leq 16$. Basically, the factors τ_i are the factors t_{n-2} or $\bar{t}_{n-2}$ emphasised in the decomposition of t_{n+2} from above. That is: $\tau_1 = t_{n-2}$, $\tau_2 = \bar{t}_{n-2}$, and so on. Clearly, $\tau_1 \cdots \tau_4 = t_n$, $\tau_5 \cdots \tau_8 = \bar{t}_n$, $\tau_7 \cdots \tau_{10} = t_n$, and $\tau_9 \cdots \tau_{16} = \bar{t}_{n+1}$.

We assume, for the sake of a contradiction, that we can derive from a prefix of $\tau_1 \cdots \tau_4$ a word longer than $\tau_1 \ldots \tau_{10}$. Such a derivation starts with the initial prefix of $\mathbf{t}$, and reaches in several steps a word y that is shorter than $\tau_1 \ldots \tau_{10}$ but in the next step we produce a word yz, which is longer than $\tau_1 \ldots \tau_{10}$. So zz has the centre somewhere in $\tau_1 \ldots \tau_{10}$. We want to see where it may begin and where it centre might be.

If yz ends inside t_{n+2}, we note that, according to Claim 3, applied to $t_{n+2} = \tau_1 \cdots \tau_{16}$, the factor zz cannot begin inside $\tau_1 \ldots \tau_4$; similarly if yz ends inside t_{n+3}, then we apply Claim 3 to t_{n+3}. It also cannot start at the same position as τ_5: in that case, yz must end inside t_{n+2}, so zz could only be the square $\tau_5 \cdots \tau_{12} = \bar{t}_n t_n$; this would mean that $y = \tau_1 \cdots \tau_8 = t_{n+1}$ and this word can be obtained by suffix duplication. Moreover, zz cannot start somewhere else inside $\tau_5\tau_6$: such a square (centred in $\tau_1 \cdots \tau_{10}$) would end in $\tau_1 \cdots \tau_{16} = t_{n+2}$, and this leads to a contradiction with Claim 4 (zz would not be contained in $t_{n+2}[2^n+2^{n-1}+1..2^{n-1}+2^{n-3}]$, as it should). So, the first z factor of the square zz is completely contained in $\tau_7 \ldots \tau_{10}$. By Claim 4 (applied to t_{n+2}), there is no square that starts in $\tau_7\tau_8$ and ends strictly after the ending position of τ_{10}. In conclusion, both the starting position and the centre of the square zz occur in $\tau_9\tau_{10}$. Moreover, as zz ends after the ending position of τ_{10}, its centre cannot occur in the first half of τ_9.

So, y ends inside $\tau_9\tau_{10}$, but not in the first half of τ_9. We repeat the reasoning above for y. This word was obtained from some y' by appending v to it. That is, $y = y'v$ and y' ends with v. Just like before, vv cannot start in $\tau_1 \cdots \tau_6$, so it must start in $\tau_7\tau_8$; also, vv cannot be $\tau_8\tau_9$ (again, this would mean that t_{n+1} ends with a square). Moreover, vv cannot start in τ_7, by Claim 3 applied to

$\tau_7\tau_8\tau_9\tau_{10} = t_{n-1}$. If y' ends in $\tau_7\tau_8$ it is fine; if not, we repeat the procedure and consider the string from which y' was obtained in the role of y'. Generally, we repeat this reduction until we reach an intermediate word obtained in the derivation of y from x which ends inside $\tau_7\tau_8$. So, suppose y' ended inside $\tau_7\tau_8$.

Assume first that y' does not end inside τ_7 (so it ends inside τ_8). Say that y' was derived from some y'' by duplicating the factor u (i.e., $y' = y''u$ and u is a suffix of y''). Then, by applying Claim 3 to $\bar{t}_n = \tau_5 \cdots \tau_8$, we get that the square uu cannot start in $\tau_5\tau_6$. This means y'' ends inside $\tau_7\tau_8$, just like y'. We repeat the process with y'' in the role of y' until we reach a word that ends inside τ_7.

So, let us assume that y' ends inside τ_7. Take w_1 be the suffix of y' contained in τ_7 and w_2 the suffix of y contained in $\tau_7 \cdots \tau_{10}$. It is immediate that the duplication steps that were used to produce y from the current y' can be used to produce w_2 from w_1. But w_1 is a prefix of $\bar{t}_{n-2}$ and w_1 is a prefix of $\bar{t}$ of strictly longer than $|t_{n-1}| + \frac{|t_{n-2}|}{2}$ (as y after the first half of τ_9). This is a contradiction with our induction hypothesis.

Clearly our claim also holds for $\bar{t}_n$, so our induction proof is complete.

Hence, it is impossible to derive from a prefix of t_n (respectively, a prefix of $\bar{t}_n$) a prefix of $\mathbf{t}$ (respectively, of $\mathbf{t}$) longer than $|t_{n+1}| + \frac{|t_n|}{2} = 2^{n+1} + 2^{n-1}$, for all n. This means that we cannot construct by suffix duplication an infinite sequence of finite prefixes of $\mathbf{t}$ whose limit is $\mathbf{t}$.   □

Despite the negative result of the previous Proposition, we show that, in fact, the duplication operation can still be used to generate infinite words that do not contain repetitions of large exponent. More precisely, while suffix square completion was enough to generate words that avoid rational powers greater than 2, the suffix duplication is enough to generate words that avoid integer powers greater than 2 (so, automatically, rational powers greater or equal to 3).

Proposition 5. *Stewart's choral sequence* $\mathbf{s}$ *is in* SD^ω.

Proof. It is enough to show that for all $n \geq 2$ we have $s_{n+1} \in SD^*(s_n)$. To show this, note first that, because $s_{n+1} = s_n s_n s_n^*$, we have $s_{n+1} = s_n s_n^* s_n^*$. Now, from s_n we derive $s_n s_n = (s_{n-1}s_{n-1}s_{n-1}^*)(s_{n-1}s_{n-1}s_{n-1}^*)$, in one duplication step. Then we duplicate the suffix $s_{n-1}s_{n-1}^*$ of s_n^2, and get $s_n s_n s_{n-1} s_{n-1}^*$. We now duplicate the suffix s_{n-1}^* of the word we obtained, and derive $s_n s_n s_{n-1} s_{n-1}^* s_{n-1}^* = s_n s_n s_n^* = s_{n+1}$. This concludes our proof: we derived, starting with s_w, an infinite sequence of prefixes of $\mathbf{s}$.   □

4   Finite Words: Algorithms

In this section, we need the following immediate extension of Lemma 1.

Lemma 2. *Let* $w \in \Sigma^*$ *be a word, and consider two factors of this word* $w[i_1..j_1]$ *and* $w[i_2..j_2]$, *such that* $i_1 \leq i_2 \leq j_2 \leq j_1$. *If* $w \in SSC^*(w[i_2..j_2])$ *then* $w \in SSC^*(w[i_1..j_1])$.

The following lemma states that we can construct efficiently a data structure providing insight in the structure of the squares occurring inside a word.

Lemma 3. *Given a word w of length n we can compute in $\mathcal{O}(n)$ time the values*

$$MinRightEnd[i] = \min\left\{j \mid \exists\, w[\ell..j] \text{ a square, such that } \ell \leq i < \ell + \frac{j - \ell + 1}{2}\right\}.$$

Note that $MinRightEnd[i]$ denotes the minimum position $j > i$ such that there exists a square ending on position j, whose first half (that is, the square's left root) contains position i.

Alternatively, we can compute in linear time an array

$$MaxLeftEnd[i] = \max\{\ell \mid \exists\, w[\ell..j] \text{ a square, such that } \ell + \frac{j - \ell + 1}{2} \leq i \leq j\}.$$

$MaxLeftEnd[i]$ is the maximum position $\ell < i$ such that there exists a square starting on position ℓ, whose second half (i.e., right root) contains position i.

These two arrays will be used in the following way. The array $MinRightEnd$ (respectively, $MaxLeftEnd$) tells us that from a factor $w[i + 1..j]$ with $j \geq MinRightEnd[i]$ (respectively, a factor $w[j..i - 1]$ with $j \leq MaxLeftEnd[i]$) we can generate $w[i'..j]$ (respectively, $w[j..i']$) for some $i' \leq i$ (respectively, $i' \geq i$); moreover, from a factor $w[i+1..j']$ with $j' < MinRightEnd[i]$ we cannot generate any factor $w[i'..j']$ (respectively, from $w[j'..i - 1]$ with $j' > MaxLeftEnd[i]$ we cannot generate $w[j'..i']$) with $i' \leq i$ (respectively, $i' \geq i$).

We can now move on to the main results of this section.

Theorem 1. *Given a word w of length n we can identify the minimum $i \leq n$ such that $w \in SSC^*(w[1..i])$.*

Proof. For each $1 \leq i \leq n$, let $L[i]$ be 1 if $w \in SSC^*(w[1..i])$ and 0 otherwise. We show how to compute the values of the array $L[\cdot]$.

We first compute all the runs of the input word in linear time. We sort the runs with respect to their ending position; as these ending positions are between 1 and n, and we have $\mathcal{O}(n)$ runs, we can clearly do this sorting in linear time, using, e.g., count sort.

Now, we observe that if there is a square $w[i..i+2\ell-1]$ then, if $L[i+2\ell-1] = 1$, we have $L[j] = 1$, for all j such that $i + \ell - 1 \leq j \leq i + 2\ell - 2$, as well. Indeed, from $w[1..j]$ we can obtain in one suffix square completion step $w[1..i + 2\ell - 1]$, if $i + \ell - 1 \leq j \leq i + 2\ell - 2$, and then from $w[1..i + 2\ell - 1]$ we can obtain w. Generally, if $w[i..j]$ is a maximal run of period ℓ, and $L[j] = 1$ then, for all $i + \ell - 1 \leq k \leq j - 1$, we have $L[j] = 1$. On the other hand, if $L[j] = 1$, then j must necessarily fall inside a run of w.

Based on the previous observations, and on Lemma 1, we get that the sequence of 1 values in the array L forms, in fact, a contiguous non-empty suffix of this array. It is, thus, enough to know the starting point of this suffix.

We are now ready to present the general approach we use in our algorithm. As a first step in our algorithm, we set $L[n] = 1$ as $w \in SSC^*(w)$ clearly holds.

Also, during the computation, we maintain the value ℓ of the leftmost position we discovered so far such that $L[\ell] = 1$; initially $\ell = n$. We now go through all positions of w from n to 1 in decreasing order. When we reach some position j of the word w, for every maximal run $r = w[i..j]$ of period p we test whether $i + p - 1 < \ell$; if so, we update ℓ and set $\ell = i + p - 1$.

After this traversal of the word, we make $L[i] = 1$ for all $i \geq \ell$. It is clear that our algorithm runs in linear time. □

There are words (for example a^n) which can be derived by prefix-suffix completion from all their factors. However, by Lemma 2, given a word w, for each position i of w where a factor generating w starts, there is a minimum $j_i \geq i$ such that $w \in PSC^*(w[i..j_i])$; then, for all $j \geq j_i$ we have that $w \in PSC^*(w[i..j])$. Therefore, we are interested in computing for each position of a word, the shortest factors starting there, from which w can be derived.

Theorem 2. *Given a word w of length n, we can identify in $\mathcal{O}(n)$ time the shortest factor $w[i..j_i]$ such that $w \in SSC^*(w[i..j])$ for all $i \leq n$.*

Proof. We start by computing the arrays *MinRightEnd* and *MaxLeftEnd* and preprocess them so that we can answer range maximum and, respectively, minimum queries on their ranges in constant time. For the reminder of the demonstration we will write $RMQ(i, k)$ for the minimum value among $MaxLeftEnd[i..k] = \{MaxLeftEnd[i], MaxLeftEnd[i + 1], \ldots, MaxLeftEnd[k]\}$ and $posRMQ(i, k)$ for respective position of that minimum value (if there are more positions with the same value we will consider the rightmost).

Our approach will be to find for each i the value of j_i defined as above. By Lemma 2 it follows that $j_1 \leq j_2 \leq \cdots \leq j_{n-1} \leq j_n$.

We start with computing j_1, using the algorithm in Theorem 1. We then explain how to compute j_2. Note that a sufficient condition for $w \in w[2..j_1]$ is that $MinRightEnd[1] \leq j_1$. Indeed, if there is a square that starts on position 1 and finishes before or on j_1, then we can derive $w[1..j_1]$ from $w[2..j_1]$ in one prefix square completion step, and, thus, we can derive w from $w[2..j_1]$. This means that we can set, in that case, $j_2 = j_1$. The condition is not necessary, as there could be several suffix square completion operations doable starting from $w[2 \ldots j_1]$ that would expand this factor to $w[2 \ldots p]$ with $p \geq MinRightEnd[1]$; so we could start by applying these completions first, and then derive $w[1..p]$, and then do the completions that allow us to construct w. However, we can easily test whether this is the case. More precisely, if $RMQ(j_1, MinRightEnd[1]) \geq 2$ then each position j with $j_1 \leq j \leq MinRightEnd[1]$ is contained in a second half of a square that starts at least on position 2. Thus, by successive suffixes completions we can obtain, in order, longer and longer factors $w[2..m]$ that contain position j, for each $j_1 \leq j \leq MinRightEnd[1]$. In the end, we get a factor $w[2..m]$ that contains $MinRightEnd[1]$; we then derive $w[1..m]$ and from this one we can derive the entire w. So, also in this case, we can simply set $j_2 = j_1$. If this is not the case, then we set $j_2 = posRMQ(j_1, MinRightEnd[1])$. Indeed, we could not derive a factor that contains 1 from any factor that does not contain $MinRightEnd[1]$, and, in order to produce $MinRightEnd[1]$ from a factor that starts on 2, we must be able to derive factors containing all the positions j between j_1 and $MinRightEnd[1]$ with $MinRightEnd[j] = 1$ (otherwise, we could not derive any factor covering these positions from smaller factors of our word).

Further, we consider the case where we know $j_1, \ldots j_{i-1}$ and want to compute j_i. We first verify if $MinRightEnd[i-1] \leq j_{i-1}$. Just like above, this means that we could derive $w[j..j_{i-1}]$ with $j \leq i-1$ from $w[i..j_{i-1}]$; then we could continue and derive the whole w. In this case we simply set $j_i = j_{i-1}$ and continue with the computation of j_{i+1}.

Secondly, we test whether $RMQ(j_{i-1}, MinRightEnd[i-1]) \geq i$. If this is the case, we set $j_i = j_{i-1}$, just like in the case when $i = 2$. Indeed, we can first obtain $w[i..p]$ for some $p \geq MinRightEnd[i-1]$ from $w[i..j_{i-1}]$; then we obtain $w[i-1..p]$ in one step, and from this factor we can derive w in multiple steps.

On the other hand, if $RMQ(j_{i-1}, MinRightEnd[i]) < i$ then we cannot proceed as above. Let $\ell = posRMQ(j_{i-1}, MinRightEnd[i])$. We keep updating $\ell = posRMQ(\ell, MinRightEnd[i])$ while $RMQ(\ell, MinRightEnd[i]) < i$. At the end of this while cycle, ℓ will give us the rightmost position j between j_{i-1} and $MinRightEnd[i]]$ such that any square that contains j in the second half starts on a position before i. So, basically, it is impossible to derive a factor that contains ℓ from a factor starting on i: to obtain ℓ we need a factor that contains (at least) $i-1$, while to obtain $i-1$ we need a factor that ends after ℓ. Thus, we set $j_i = \ell$. It is immediate that w can be derived from $w[i..j_i]$.

Repeating this process for all i, we compute correctly all the values j_i as defined in the statement.

To compute the complexity of this algorithm, it is enough to evaluate the time needed to compute j_i. Clearly, when computing j_i we execute $\mathcal{O}(j_i - j_{i-1})$ RMQ and $posRMQ$ queries and a constant number of other comparisons. Hence, our algorithm runs in linear time.   □

A direct consequence of the previous theorem is that we can decide in linear time whether a shorter word generates a longer word by iterated prefix-suffix square completion.

Theorem 3. *Given two words w and x, with $|x| < |w| = n$, we can decide in $\mathcal{O}(n)$ time whether $w \in PSC^*(x)$.*

Proof. We first run a linear time pattern matching algorithm (e.g., the Knuth-Morris-Pratt algorithm [16]) to locate all the occurrences of x in w. Then we run the algorithm from the proof of Theorem 2 to find, for each position i of w, the shortest factor $w[i..j_i]$ such that $w \in PSC^*(w[i..j_i])$. Finally, we just have to check whether one of these factors is contained in an occurrence of x starting at the same position.   □

5   Future Work

There are several directions in which the results presented in this paper could be extended. One direction that seems very interesting to us is to see which is the minimum exponent of a repetition avoidable by an infinite word constructed by iterated (prefix-)suffix duplication. We saw that we can construct words that avoid cubes, and every such word contains squares. Is it the same case as for prefix-suffix square completion, and we can avoid any power greater than two?

Most of the language theoretic results obtained for prefix-suffix duplication can be immediately replicated for prefix-suffix square completion. However, it is not immediate whether the language of finite words obtained by iterated prefix-suffix square completion from a single word remains semi-linear. It would be interesting to settle this fact. Also, a thorough study of the languages of finite words obtained by iterated prefix-suffix square completion from special sets of initial words (singleton sets, finite sets, regular sets, etc.) would be interesting.

Finally, it would be interesting to check whether our algorithms can be easily extended to measure the prefix-suffix square completion distance between two words: what is the minimum number of steps of square completion needed to obtain a word from one of its factors.

References

1. García-López, J., Manea, F., Mitrana, V.: Prefix-suffix duplication. J. Comput. Syst. Sci. **80**(7), 1254–1265 (2014)
2. Dumitran, M., Gil, J., Manea, F., Mitrana, V.: Bounded prefix-suffix duplication. In: Holzer, M., Kutrib, M. (eds.) CIAA 2014. LNCS, vol. 8587, pp. 176–187. Springer, Heidelberg (2014)
3. Thue, A.: Über unendliche Zeichenreihen. Norske Vid. Skrifter I. Mat.-Nat. Kl., Christiania **7**, 1–22 (1906)
4. Thue, A.: Über die gegenseitige Lage gleicher Teile gewisser Zeichenreihen. Norske Vid. Skrifter I. Mat.-Nat. Kl., Christiania **1**, 1–67 (1912)
5. Lothaire, M.: Comb. Words. Cambridge University Press, Cambridge (1997)
6. Allouche, J., Shallit, J.O.: Automatic Sequences - Theory, Applications, Generalizations. Cambridge University Press, Cambridge (2003)
7. Damanik, D.: Local symmetries in the period-doubling sequence. Discrete Appl. Math. **100**(1–2), 115–121 (2000)
8. Currie, J.D., Rampersad, N., Saari, K., Zamboni, L.Q.: Extremal words in morphic subshifts. Discrete Math. **322**, 53–60 (2014)
9. Endrullis, J., Hendriks, D., Klop, J.W.: Degrees of streams. Integers Electron. J. Comb. Number Theor. **11B**(A6), 1–40 (2011)
10. Hall, M.: Generators and Relations in Groups - The Burnside Problem. Lectures on Modern Mathematics, vol. 2. Wiley, New York (1964). 42–92
11. Kärkkäinen, J., Sanders, P., Burkhardt, S.: Linear work suffix array construction. J. ACM **53**, 918–936 (2006)
12. Bender, M.A., Farach-Colton, M.: The LCA problem revisited. In: Gonnet, G.H., Viola, A. (eds.) LATIN 2000. LNCS, vol. 1776, pp. 88–94. Springer, Heidelberg (2000)
13. Gusfield, D.: Algorithms on Strings, Trees, and Sequences: Computer Science and Computational Biology. Cambridge University Press, New York (1997)
14. Kolpakov, R., Kucherov, G.: Finding maximal repetitions in a word in linear time. In: Proceedings of FOCS, pp. 596–604 (1999)

15. Bannai, H., I, T., Inenaga, S., Nakashima, Y., Takeda, M., Tsuruta, K.: A new characterization of maximal repetitions by Lyndon trees. In: Proceedings of SODA, pp. 562–571 (2015)
16. Knuth Jr., D.E., Morris, J.H., Pratt, V.R.: Fast pattern matching in strings. SIAM J. Comput. **6**(2), 323–350 (1977)

Square-Density Increasing Mappings

Florin Manea[1] and Shinnosuke Seki[4,3,2(✉)]

[1] Department of Computer Science, Christian-Albrechts-Universität zu Kiel,
Christian-Albrechts-Platz 4, 24118 Kiel, Germany
flm@informatik.uni-kiel.de
[2] Department of Computer Science, Aalto University, P. O. Box 15400,
00076 Aalto, Finland
[3] Helsinki Institute for Information Technology (HIIT), Espoo, Finland
[4] Department of Communication Engineering and Informatics,
University of Electro-Communications,
1-5-1, Chofugaoka, Chofu, Tokyo 1828585, Japan
s.seki@uec.ac.jp

Abstract. The square conjecture claims that the number of distinct squares, factors of the form xx, in a word is at most the length of the word. Being associated with it, it is also conjectured that binary words have the largest square density. That is, it is sufficient to solve the square conjecture for words over binary alphabet. We solve this subsidiary conjecture affirmatively, or more strongly, we prove the irrelevance of the alphabet size in solving the square conjecture, as long as the alphabet is not unary. The tools we employ are homomorphisms with which one can convert an arbitrary word into a word with strictly larger square density over an intended alphabet.

1 Introduction

Let Σ be an alphabet and Σ^* be the set of all words over Σ. Let $w \in \Sigma^*$. By $|w|$, we denote its length. Let

$$\mathrm{Sq}(w) = \{uu \mid w = xuuy \text{ for some } x, y \in \Sigma^* \text{ with } w \neq xy\}$$

be the set of all squares occurring in w. Its size, denoted by $\#\mathrm{Sq}(w)$, has been conjectured to be upper-bounded by the length of w [1].

Conjecture 1 (Square Conjecture). For any nonnegative integer $n \geq 0$, a word of length n contains at most n distinct squares.

This conjecture was considered in a series of papers so far, and the upper bound on $\#\mathrm{Sq}(w)$ was improved over the years. Already in [1] it was shown that for a word w of length n we have $\#\mathrm{Sq}(w) \leq 2n$. A simpler proof of this result

---

F. Manea—His work is in part supported by the DFG grant 596676.
S. Seki—His work is in part supported by the Academy of Finland, Postdoctoral Researcher Grant 13266670/T30606.

F. Manea and D. Nowotka (Eds.): WORDS 2015, LNCS 9304, pp. 160–169, 2015.
DOI: 10.1007/978-3-319-23660-5_14

was given in [2]; this upper-bound was further improved by Ilie to $\#\mathrm{Sq}(w) \leq 2n - \log n$ [3]. The best upper bound known so far is shown in [4], where it is shown that $\#\mathrm{Sq}(w) \leq \frac{11n}{6}$.

A slightly stronger conjecture is that for a word w of length n, over an alphabet Σ with at least two letters, we have $\#\mathrm{Sq}(w) \leq n - |\Sigma|$ (see [5]). Essentially, this conjecture is supported by the fact that the known examples of words with a large number of distinct squares are always binary. For instance, an infinite word, over the binary alphabet $\Sigma_2 = \{a, b\}$, whose finite factors have a relatively large number of distinct squares compared to their length was given by Fraenkel and Simpson [1]:

$$w_{\mathrm{fs}} = a^1 b a^2 b a^3 b a^2 b a^3 b a^4 b a^3 b a^4 b a^5 b a^4 b a^5 b a^6 b \cdots . \tag{1}$$

None of its factors of length n with k letters b contain more than $\frac{2k-1}{2k+2}n$ distinct squares. A structurally simpler infinite word that has the same property was given in [6]:

$$w_{\mathrm{jms}} = a^1 b a^2 b a^3 b a^4 b a^5 b a^6 b \cdots \tag{2}$$

Inspired by the properties of the above two infinite words, in [6] an even stronger conjecture was proposed for binary words. Let $k \geq 2$. For any binary word $w \in \Sigma_2^+$ of length n with k b's where $k \leq \lfloor \frac{n}{2} \rfloor$, we have $\#\mathrm{Sq}(w) \leq \frac{2k-1}{2k+2}n$. As explained in [6], this conjecture does not consider words with at most one b because they are, just like unary words, square sparse. This conjecture was shown to hold for some classes of binary words (for instance, for words $a^{i_0} b a^{i_1} b \cdots a^{i_{k-1}} b a^{i_k}$ where the exponents $i_1, \ldots, i_k$ are pairwise distinct, or for words with at most 9 b's).

Among all the words of length n for an integer n, the one(s) with largest number of distinct squares might not be binary (see an example in [5]). It was however widely believed that the words having the largest ratio of the number of distinct squares on it to its length (square density) are binary. This brings us to the following conjecture.

Conjecture 2. For a word over an arbitrary alphabet, there exists a binary word with strictly larger square density.

Conjecture 2 will be solved affirmatively as Theorem 2. In fact, we show a much stronger result: for a word over an arbitrary alphabet and for another alphabet Σ with at least two letters, there exists a word over Σ with strictly larger square density, in which every letter in Σ occurs.

There are several important consequences of this result. First of all, it shows that if one can prove that Conjecture 1 holds for words over some fixed non-unary alphabet Σ (e.g., for binary words), then it holds in general. Then, it shows that no upper bound on the square density is sharp. So, once the existence of a constant c with $\frac{\#\mathrm{Sq}(w)}{|w|} \leq c$ for any word w is proved, this inequality immediately turns out to be strict. For instance, the upper-bound given in [4] hence implies that there does not exist any word whose square density reaches $\frac{11}{6}$.

Moreover, our result also puts in a new light on the conjecture of [6]. If one would be able to show that for any binary word $w \in \Sigma_2^+$ of length n with k b's where $k \leq \lfloor \frac{n}{2} \rfloor$, we have $\#\mathrm{Sq}(w) \leq \frac{2k-1}{2k+2}n$, then Conjecture 1 would hold not only for binary alphabets, but in general. Thus, the techniques of [6] might be useful in an attack on the Square Conjecture of Fraenkel and Simpson.

The techniques we use to show Conjecture 2 are quite different from those that were used so far in the works related to Conjecture 1. We first show that given a binary word we can effectively construct a word over an alphabet of size 2^i, with $i \geq 2$, that has exactly the same square density. Essentially, this new word is just the image of the initial word under a carefully constructed mapping. Using the same basic idea, but with much more complicated technical details, we show how one can design a new mapping from an alphabet Σ to an alphabet of size $|\Sigma| - 1$ such that the image of any word under this mapping has a strictly larger square density than the initial word. These constructions show that Conjecture 2 holds: starting with some arbitrary word, we first produce using the first mapping a word with the same square density over an alphabet with at least 4 letters, and then we apply iteratively the second mapping to decrease the size of the alphabet one by one, till we reach a binary word, while strictly increasing, at each step, the square density of the obtained words.

This paper is organized as follows. Section 2 is for preliminaries. In Sect. 3, we propose mappings to expand the alphabet that preserve or increase fractional-power density in general. Using these mappings, in Sect. 4, we engineer strict square-density increasing mappings that convert a given word into another word over an alphabet of intended size. Section 5 concludes this paper with discussion on future research directions.

2   Preliminaries

An *alphabet* is a finite set of letters. For an integer $\ell \geq 1$, let Σ_ℓ be an alphabet of ℓ letters. The binary alphabet Σ_2 plays a significant role in this paper; we always let its letters be a and b. When the number of letters does not matter, we omit the subscript and simply write Σ.

By Σ^*, we denote the set of all words over Σ, and let $\Sigma^+ = \Sigma^* \setminus \{\lambda\}$, where λ is the empty word. Let $w \in \Sigma^*$ be a word. The *length* of w is denoted by $|w|$. For a letter $a \in \Sigma$, $|w|_a$ denotes the number of occurrences of a in w. Let $\mathrm{alph}(w) = \{a \in \Sigma \mid |w|_a \geq 1\}$, that is, the set of letters occurring in w. If $|\mathrm{alph}(w)| = m$, then w is said to be m-ary. A *factor* of w is a word $v \in \Sigma^*$ such that $w = xvy$ for some $x, y \in \Sigma^*$; if $y = \lambda$ (resp. $x = \lambda$), then the factor v is especially called a *prefix* (resp. *suffix*) of w.

For coprime positive integers n, d, the $\frac{n}{d}$-*power* of a word x is defined as $x^{\lfloor n/d \rfloor} x_p$ for some prefix x_p of x such that $\lfloor \frac{n}{d} \rfloor + \frac{|x_p|}{|x|} = \frac{n}{d}$ (if such x_p does not exist, then the power is undefined). The power is denoted by $x^{n/d}$. A word is a $\frac{n}{d}$-*power* if it is the $\frac{n}{d}$-*power* of some word. For a positive rational number $k \geq 1$ and a word w, by $\#\mathrm{Pow}_k(w)$, we denote the number of distinct k-powers in w, that is,

$$\#\mathrm{Pow}_k(w) = \left| \left\{ x^k \in \Sigma^+ \mid x^k \text{ is a factor of } w \right\} \right|.$$

The ratio of this number to the length of w is called the k-*power density* of w, and denoted by $\rho_{\mathrm{pow},k}(w)$, that is, $\rho_{\mathrm{pow},k}(w) = \frac{\#\mathrm{Pow}_k(w)}{|w|}$. We collectively call the k-power densities the *fractional-power density*, or more simply, fpd.

The 2-power of a word is usually called a *square*. As an alias of the 2-power density, we employ the *square density*, or more simply, sqd. Let us use $\rho_{\mathrm{sq}}(w)$ as a simpler notation of $\rho_{\mathrm{pow},2}(w)$. As already mentioned in the Introduction, the currently best upper bound on the square density is $\frac{11}{6}$ (see [4]).

For a mapping $f : \Sigma^* \to \Sigma^*$, we say that f is k-*power density increasing* if $\rho_{\mathrm{pow},k}(f(w)) \geq \rho_{\mathrm{pow},k}(w)$ holds for all words $w \in \Sigma^*$. It is *strictly* k-*power density increasing* if the inequality holds strictly.

3 Fractional-Power Density Increasing Mappings to Expand the Alphabet

3.1 Fractional-Power Density Preserving Quaternarizer

Let $\Sigma_4 = \Sigma_2 \cup \{c, d\}$. Given a binary word $x \in \Sigma_2^*$, let us define a homomorphism $h_{1,x} : \Sigma_2^* \to \Sigma_4^*$, which is parameterized by x, as:

$$h_{1,x}(a) = xc$$
$$h_{1,x}(b) = xd. \tag{3}$$

Since x is free from c or d, the next code property holds. A language $L \subseteq \Sigma^*$ is a *comma-free code* if $LL \cap \Sigma^+ L \Sigma^+ = \emptyset$. A language $L \subseteq \Sigma^*$ is a *block code* if all words in L are of the same length. For definitions and basic facts regarding codes in general, see, e.g.,[7].

Lemma 1. *For any binary word $x \in \Sigma_2^*$, the set $\{h_{1,x}(a), h_{1,x}(b)\}$ is a comma-free block code.*

Based on $h_{1,x}$, we define a mapping $\delta_{1,x} : \Sigma_2^* \to \Sigma_4^*$ as: for a binary word $w \in \Sigma_2^*$,

$$\delta_{1,x}(w) = h_{1,x}(w)x.$$

The subscript 1 of this mapping plays a role in explicitly stating the (logarithm) of the domain size; later in Sect. 3.2 we will generalize this mapping into a mapping from the 2^i-letters alphabet to the 2^{i+1}-letters alphabet and refer to it by a notation with subscript i. A particular interest lies in applying to w the self-parameterized mapping $\delta_{1,w}$. To this end, let us define the mapping $\delta_1 : \Sigma_2^* \to \Sigma_4^*$ as: for $w \in \Sigma_2^*$,

$$\delta_1(w) = \delta_{1,w}(w).$$

Lemma 2. *For any rational number $k \geq 1$ and binary word $w \in \Sigma_2^*$, the quaternary word $\delta_1(w)$ satisfies $\rho_{\mathrm{pow},k}(\delta_1(w)) \geq \rho_{\mathrm{pow},k}(w)$. If $k \geq 2$, then $\rho_{\mathrm{pow},k}(\delta_1(w)) = \rho_{\mathrm{pow},k}(w)$.*

Proof. For any k-power x^k in w, $\delta_1(w)$ contains the k-power $h_{1,w}(x)^k$ as well as its first $|w|$ cyclic shifts, and these $|w| + 1$ k-powers are pairwise distinct due to Lemma 1. In this way, we get that the $\#\mathrm{Pow}_k(w)$ k-powers in w account for $(|w| + 1)\#\mathrm{Pow}_k(w)$ k-powers in $\delta_1(w)$, and all of them are pairwise distinct due to Lemma 1. These k-powers are of length at least $|w| + 1$. All the k-powers in w also occur in $\delta_1(w)$; clearly, they cannot be longer than w. Now we have

$$
\begin{aligned}
\rho_{\mathrm{pow},k}(\delta_1(w)) &= \frac{\#\mathrm{Pow}_k(\delta_1(w))}{|\delta_1(w)|} \\
&\geq \frac{(|w| + 1)\#\mathrm{Pow}_k(w) + \#\mathrm{Pow}_k(w)}{(|w| + 2)|w|} = \rho_{\mathrm{pow},k}(w). \qquad (4)
\end{aligned}
$$

Next, we prove that if $k \geq 2$, then k-powers in $\delta_1(w)$ are one of these types. Let y^k be a k-power in $\delta_1(w)$. If y is free from c or d, then y^k is a factor of w, and hence, a k-power of w. Otherwise, we can let $y = w_s cv$ or $y = w_s dv$ for some suffix w_s of w and $v \in \Sigma_4^*$. It should suffice to analyze the case when $y = w_s cv$. For $k \geq 2$, $y^2 = w_s cvw_s cv$ is a factor of y^k, and hence, a factor of $\delta_1(w)$. This implies that $vw_s \in \{wc, wd\}^* w$ because $\delta_1(w) \in \{wc, wd\}^* w$. Hence, $v \in \{wc, wd\}^* w_p = \{h_{1,w}(a), h_{1,w}(b)\}^* w_p$ for the prefix w_p of w that satisfies $w = w_p w_s$. Thus, y^k is a cyclic shift of the $h_{1,w}$-image of some k-power in w. In conclusion, in this case, the inequality (4) turns into an equation. □

Remark 1. Since d occurs in the h_x-image of b but not in that of a for any x, if a word w is unary, then $\delta_1(w)$ is binary. Lemma 2 hence implies that for any rational number $k \geq 1$, k-powers are sparser on unary words than on non-unary words, though this is trivial at least intuitively.

In Sect. 3.2, we focus on 2-powers, that is, squares. Hence, we provide the special case of Lemma 2 with $k = 2$ as an independent lemma.

Lemma 3. *For any binary word $w \in \Sigma_2^*$, the quaternary word $\delta_1(w)$ satisfies $\rho_{\mathrm{sq}}(\delta_1(w)) = \rho_{\mathrm{sq}}(w)$.*

Remark 2. For reference, it is worth introducing one more fpd-preserving quaternarizer ξ. Given a binary word $w \in \{a, b\}^*$, ξ renames it using a homomorphism f which maps a to c and b to d, and appends $f(w)$ to w, that is, $\xi(w) = wf(w)$. For any rational number k and a binary word $w \in \{a, b\}^*$, we have $\rho_{\mathrm{pow},k}(\xi(w)) = \rho_{\mathrm{pow},k}(w)$. In order to achieve the goal of this paper, we can employ both of the quaternarizers. They differ in the frequency of the four letters a, b, c, d in the resulting word. For ξ, we have $|\xi(w)|_c = |\xi(w)|_a$ and $|\xi(w)|_d = |\xi(w)|_b$. In contrast, on $\delta_1(w)$, c and d occur quadratically less frequently than a and b as:

$$
\begin{aligned}
|\delta_1(w)|_c &= |w|_a \\
|\delta_1(w)|_d &= |w|_b \\
|\delta_1(w)|_a &= |w|_a(|w| + 1) \\
|\delta_1(w)|_b &= |w|_b(|w| + 1).
\end{aligned}
$$

As detailed in Remark 4 later, biased frequency of letters improves on the performance of the strictly square-density increasing mapping we will give in Sect. 4.

3.2 Fractional-Power Density Preserving Mappings to Double the Alphabet Size

Let us generalize the quaternarizer just proposed to a k-power density preserving mapping from an alphabet $\Sigma_{2^i} = \{a_0, a_1, a_2, \ldots, a_{2^i-2}, a_{2^i-1}\}$ of size 2^i to an alphabet $\Sigma_{2^{i+1}} = \Sigma_{2^i} \cup \{a_{2^i}, a_{2^i+1}, a_{2^i+2}, \ldots, a_{2^{i+1}-2}, a_{2^{i+1}-1}\}$ of double size for arbitrary $i \geq 1$.

Given a 2^i-ary word $x \in \Sigma_{2^i}^*$, we define a homomorphism $h_{i,x} : \Sigma_{2^i}^* \to \Sigma_{2^{i+1}}^*$ as: for $0 \leq k \leq 2^i - 1$,

$$h_{i,x}(a_k) = xa_{2^i+k}.$$

This is just a generalization of (3). Lemma 1 is generalized accordingly.

Lemma 4. *For any 2^i-ary word $x \in \Sigma_{2^i}^*$, the set $\{h_{i,x}(a) \mid a \in \Sigma_{2^i}\}$ is a comma-free block code.*

Based on $h_{i,x}$, we define a mapping $\delta_{i,x} : \Sigma_{2^i}^* \to \Sigma_{2^{i+1}}^*$ as: for a 2^i-ary word $w \in \Sigma_{2^i}^*$,

$$\delta_{i,x}(w) = h_{i,x}(w)x.$$

For a 2^i-ary word $w_i \in \Sigma_{2^i}^*$, let $w_{i+1} = \delta_{i,w_i}(w_i)$.

As done in the proof of Lemma 2, we can prove that $\rho_{\mathrm{pow},k}(w_{i+1}) \geq \rho_{\mathrm{pow},k}(w_i)$ for $k \geq 1$, or $\rho_{\mathrm{pow},k}(w_{i+1}) = \rho_{\mathrm{pow},k}(w_i)$ if $k \geq 2$.

Starting from a binary word $w_1 = w \in \Sigma_2^*$, we can create an infinite sequence of words with the same square density as:

$$w = w_1 \xrightarrow{\delta_{1,w_1}} w_2 \xrightarrow{\delta_{2,w_2}} w_3 \longrightarrow \cdots \longrightarrow w_i \xrightarrow{\delta_{i,w_i}} w_{i+1} \longrightarrow \cdots.$$

4 Strictly Square-Density Increasing Mappings

The aim of this section is to design a mapping from an arbitrary word to a binary word that strictly increases square density, which amounts to an affirmative solution to Conjecture 2.

The mapping is a composition of the alphabet expanding mapping proposed in Sect. 3.2 and a strictly sqd-increasing mapping that shrinks the alphabet by one letter, which we propose now in Sect. 4.1.

4.1 Strictly Square-Density Increasing Alphabet Downsizer

For $m \geq 3$, assume that the m-ary alphabet Σ_m contains three letters $a, b,$ and c. Let us propose a strictly sqd-increasing mapping $g_i : \Sigma_m^* \to (\Sigma_m \setminus \{c\})^*$ such that, for all $w \in \Sigma_m^*$ in which a, b, and c occur, the inequality

$\rho_{sq}(w) < \rho_{sq}(g_i(w))$ holds. We define g_i based on a homomorphism $f_i : \Sigma_m^* \to (\Sigma_m \setminus \{c\})^*$ parameterized by an integer $i \geq 1$, which is defined as

$$f_i(a) = (aab)^i a$$
$$f_i(b) = (aab)^i b$$
$$f_i(c) = (aab)^i ab$$
$$f_i(d) = (aab)^i d \quad \text{for any } d \in \Sigma_m \setminus \{a, b, c\}.$$

If $m = 3$, then Σ_m contains exactly three letters a, b, and c, so the function f_i is only defined for these values. Clearly, in that case, $f_i(w)$ is a binary word, for every word $w \in \Sigma_3$.

The next lemma follows immediately.

Lemma 5. *For any integer $i \geq 1$, $\{h_i(e) \mid e \in \Sigma_m\}$ is suffix-free, and hence, it is a code.*

Now we define a mapping $g_i : \Sigma_m^* \to \Sigma_{m-1}^*$ as: for a word $w \in \Sigma_m^*$,

$$g_i(w) = f_i(w)(aab)^i.$$

For an m-ary word $w \in \Sigma_m^*$ in which a, b, and c occur, let us count distinct squares in the resulting $(m-1)$-ary word $g_i(w)$. For a square x^2 in w, its f_i-image $f_i(x)^2$ is a square in $g_i(w)$, and its first $3i$ cyclic shifts are also squares in $g_i(w)$. In this way, we can find $(3i + 1)\#\mathrm{Sq}(w)$ squares in $g_i(w)$, but some of them might be identical. We show that this is not the case.

Claim. These $(3i + 1)\#\mathrm{Sq}(w)$ squares are pairwise distinct.

Proof. Consider two squares $(a_1 \cdots a_n)^2, (b_1 \cdots b_m)^2$ in w for some $n, m \geq 1$ and letters $a_1, \ldots, a_n, b_1, \ldots, b_m \in \Sigma_m$. We will show that if some cyclic shifts of their f_i-images are equal, then these squares are identical, that is, $n = m$, and $a_j = b_j$ for all $1 \leq j \leq n$. Hence, assume

$$\beta f_i(a_2) \cdots f_i(a_n) f_i(a_1) \cdots f_i(a_n)\alpha = y f_i(b_2) \cdots f_i(b_m) f_i(b_1) \cdots f_i(b_m)x \quad (5)$$

for some $\alpha, x \leq_p (aab)^i$ and $\beta, y \in \Sigma_m^+$ such that $f_i(a_1) = \alpha\beta$ and $f_i(b_1) = xy$. Without loss of generality, we may assume $|\beta| \geq |y|$, and if their lengths are the same, then $|\alpha| \geq |x|$. Being of length at most $3i$, x is a suffix of $f_i(a_n)\alpha$. Then (5) implies that $(f_i(b_1) \cdots f_i(b_m))^2$ is a factor of $f_i(a_n)(f_i(a_1) \cdots f_i(a_n))^2$.

If $|\beta| = |y|$, i.e., $|\alpha| \geq |x|$, (5) implies that x is a suffix of α, and hence, $f_i(b_1)$ is a suffix of $f_i(a_1)$. Lemma 5 gives $f_i(a_1) = f_i(b_1)$, and this derives $f_i(a_2) \cdots f_i(a_n) f_i(a_1) \cdots f_i(a_n) = f_i(b_2) \cdots f_i(b_m) f_i(b_1) \cdots f_i(b_m)$ from (5). The code property stated in Lemma 5 decodes this equation as $n = m$ and $f_i(a_j) = f_i(b_j)$ for all $1 \leq j \leq n$. Since f_i is injective, we have $a_j = b_j$ for all $1 \leq j \leq n$.

Otherwise, i.e., if $|\beta| > |y|$, α is a suffix of x since the length of the image of a letter is either $3i + 1$ or $3i + 2$. Note that x is a proper suffix of $f_i(a_n)\alpha$. Thus, there exists a nonempty prefix w_p of $f_i(a_n)$ such that $f_i(a_n)\alpha = w_p x$. Let $f_i(a_n) = w_p w_s$ for some w_s. Then (5) implies

$$(w_s f_i(a_1) \cdots f_i(a_{n-1}) w_p)^2 = (f_i(b_1) \cdots f_i(b_m))^2. \quad (6)$$

This equation means that w_s is a prefix of $f_i(b_1)$ because $|w_s| \le 3i + 1$ due to the nonemptiness of w_p. No suffix of $f_i(d)$ for any $d \ne a$ can be a prefix of f_i-images, and hence, $a_n = a$, and more strongly, $w_s \in (aab)^*a$. Let $w_s = (aab)^k a$ for some $0 \le k < i$. Now we have $f_i(b_1) = xy \le_p (aab)^k a f_i(a_1)$, that is, $(aab)^i \le_p (aab)^k a(aab)^i$. Deleting the prefix $(aab)^k$ from the both sides of this equation, however, results in the contradictory relation $(aab)^{i-k} \le_p a(aab)^i$. Therefore, w_s is empty. Then (6) turns into the equation $(f_i(a_1) \cdots f_i(a_n))^2 = (f_i(b_1) \cdots f_i(b_m))^2$. The claim has been thus proved.

Note that the squares found so far are of length at least $6i + 2$. Let us enumerate shorter squares in $g_i(w)$ next. For simplicity, we assume that i is even. Since a occurs in w, $f_i(a) = (aab)^i a$ is a factor of $g_i(w)$. We claim that $f_i(b)(aab)^i$ is also a factor of $g_i(w)$. Indeed, b occurs in w and $(aab)^i$ is a prefix of the f_i-image of any letter and it is also a suffix of $g_i(w)$. These factors contain the following $3i/2$ squares:

$$
\begin{aligned}
&(aab)^2, ((aab)^2)^2, ((aab)^3)^2, \ldots, ((aab)^{i/2})^2, \\
&(aba)^2, ((aba)^2)^2, ((aba)^3)^2, \ldots, ((aba)^{i/2})^2, \\
&(baa)^2, ((baa)^2)^2, ((baa)^3)^2, \ldots, ((baa)^{i/2})^2.
\end{aligned} \tag{7}
$$

More short squares can be found in the factors $f_i(a)(aab)^i$, $f_i(b)(aab)^i$, and $f_i(c)(aab)^i$ of $g_i(w)$. The factor $f_i(a)(aab)^i = (aab)^i a(aab)^i$ contains the following $2i - 1$ squares:

$$
\begin{aligned}
&a^2, (aaba)^2, ((aab)^2a)^2, \ldots, ((aab)^{i-1}a)^2, \\
&(abaa)^2, ((aba)^2a)^2, \ldots, ((aba)^{i-1}a)^2.
\end{aligned} \tag{8}
$$

The second factor $f_i(b)(aab)^i = (aab)^i b(aab)^i$ contains the following i squares:

$$
b^2, (baab)^2, ((baa)^2 b)^2, \ldots, ((baa)^{i-1}b)^2. \tag{9}
$$

The third factor $f_i(c)(aab)^i = (aab)^i ab(aab)^i$ contains the following $2i$ squares:

$$
\begin{aligned}
&(ab)^2, (abaab)^2, (ab(aab)^2)^2, \ldots, (ab(aab)^{i-1})^2, \\
&(ba)^2, (baaba)^2, (b(aab)^2a)^2, \ldots, (b(aab)^{i-1}a)^2.
\end{aligned} \tag{10}
$$

These $13i/2 - 1$ short squares listed in (7)–(10) are obviously pairwise distinct. Moreover, their length is at most $6i - 2$, and hence, they are strictly shorter than the $(3i + 1)\#\mathrm{Sq}(w)$ squares listed at the beginning.

Having found the $(3i + 1)\#\mathrm{Sq}(w) + 13i/2 - 1$ squares in $g_i(w)$, we obtain

$$
\begin{aligned}
\rho_{\mathrm{sq}}(g_i(w)) - \rho_{\mathrm{sq}}(w) &= \frac{\#\mathrm{Sq}(g_i(w))}{|g_i(w)|} - \frac{\#\mathrm{Sq}(w)}{|w|} \\
&\ge \frac{(3i + 1)\#\mathrm{Sq}(w) + 13i/2 - 1}{(3i + 2)|w| + 3i} - \frac{\#\mathrm{Sq}(w)}{|w|} \\
&= \frac{(13|w|/2 - 3\#\mathrm{Sq}(w))i - |w|(\#\mathrm{Sq}(w) + 1)}{|w|((3i + 2)|w| + 3i)}.
\end{aligned} \tag{11}
$$

Since $\rho_{\mathrm{sq}}(w) \le 2$ (according to [1]), we immediately get that $13|w|/2 - 3\#\mathrm{Sq}(w)$ is positive. Thus, for any integer $i > \frac{|w|(\#\mathrm{Sq}(w)+1)}{13|w|/2 - 3\#\mathrm{Sq}(w)}$, the strict inequality $\rho_{\mathrm{sq}}(g_i(w)) > \rho_{\mathrm{sq}}(w)$ holds. □

Remark 3. Needless to say, this strictly sqd-increasing mapping does not contradict the constant upper-bound on the sqd. The limit of the lowerbound given in (11) as i approaches infinity is $\frac{13/6 - \rho_{\mathrm{sq}}(w)}{|w| + 1}$. Since the mapping g_i lengthens a word quadratically, if we apply it iteratively as $w, g_i(w), g_i(g_i(w)), \ldots$, then the improvement of square density is very likely to deteriorate rapidly and the square density of the words obtained in this manner remains smaller than a constant upper-bound.

Remark 4. In the first inequality of (11), we bounded $|g_i(w)|$ from above very roughly by $(3i + 2)|w| + 3i$, considering the imaginable "worst" case when all letters in w are c (among all the letters in Σ_m, only c is mapped to a word of length $3i + 2$, and all the others are mapped to words of length $3i + 1$). Since a and b are assumed to occur in w, $|g_i(w)|$ can be bounded from above by $(3i + 2)|w| + 3i - 2$. It can be bounded more sharply by $\left(3i + \frac{4}{3}\right)|w| + 3i$ because, modulo some renaming, we can assume that c is the least frequent letter in w. However, this would not improve in a significant way the final result of our evaluation.

The above argument works for any $m \geq 3$. Repeating it as many times as it is necessary, it eventually produces, starting from w and going through $g_i(w), g_i(g_i(w)), \ldots$, a binary word with strictly larger square density.

Thus, the next theorem holds.

Theorem 1. *For any word w over a ternary or larger alphabet, there exists a binary word $u \in \Sigma_2^*$ with $\rho_{\mathrm{sq}}(u) > \rho_{\mathrm{sq}}(w)$.*

Combining the sqd-preserving quaternarizer (Lemma 3, but only when we start with a binary word) with the mapping used in the proof of Theorem 1, we conclude that for any word we can effectively construct a binary word with strictly larger square density.

Needless to say, the number of squares on a unary word is exactly (the floor of) the half of its length, and a construction of a binary word with larger square density is known (see, e.g., the word constructed in [6], given in the Introduction). Conjecture 2 has been thus solved affirmatively as follows.

Theorem 2. *For any word w over an arbitrary alphabet Σ, there exists a binary word $u \in \Sigma_2^*$ such that $\rho_{\mathrm{sq}}(u) > \rho_{\mathrm{sq}}(w)$.*

Beyond this affirmative solution to the conjecture, the mappings we proposed so far allow us to choose the size of (non-unary) alphabet for our convenience in order to solve the square conjecture.

Theorem 3. *For any word $w \in \Sigma^*$ and an integer $m \geq 2$, there exists a word $u \in \Sigma^*$ with $\rho_{\mathrm{sq}}(u) > \rho_{\mathrm{sq}}(w)$ in which m distinct letters occur.*

As a side comment, note that in this homomorphism-based algebraic approach to the square conjecture, larger alphabets may be more useful because they enable homomorphisms to embed more "micro" squares, as just exemplified by the construction of the homomorphism f_i.

Finally, note the following corollary.

Corollary 1. *No upper bound on the square density is sharp.*

Clearly, if we obtain a word with $\rho_{sq}(w) = c$, by Theorem 2 we can construct a binary word w' with $\rho_{sq}(w') > c$.

5   Future Work

There are several directions in which this work can be continued.

On one hand, in the algebraic framework that we introduced here, solving the following task is enough to solve the Square Conjecture.

Task 1 (Construction of a Square-Density Amplifier). *Can we find a mapping $f : \Sigma^* \to \Sigma^*$ for which there exists a constant $c > 1$, such that for all $w \in \Sigma^*$, if $\rho_{sq}(w) \geq 1$ then $\rho_{sq}(f(w)) \geq c\rho_{sq}(w)$?*

On another hand, the results of Sect. 3 show that the part of our approach where the size of the alphabet is increased can be used also when one is interested in counting distinct k-powers inside a word. However, it would be interesting to show an equivalent of Theorems 2 and 3 for k-powers, also when $k \neq 2$.

Finally, it seems interesting to investigate deeper whether the fact that we only have to show Conjecture 1 for binary words has some immediate consequence. For instance, can we get better results using the techniques of [4] for binary words? Can we develop the approach of [6] and see whether their conjecture holds (even in some weaker form), thus solving the Squares Conjecture?

Acknowledgements. We appreciate fruitful discussions with James Currie at Christian-Albrechts-University of Kiel, with Hideo Bannai and Simon Puglisi at the University of Helsinki, and with Nataša Jonoska when she visited Aalto University.

References

1. Fraenkel, A.S., Simpson, J.: How many squares can a string contain? J. Comb. Theory Ser. A **82**, 112–120 (1998)
2. Ilie, L.: A simple proof that a word of length n has at most $2n$ distinct squares. J. Comb. Theory Ser. A **112**(1), 163–164 (2005)
3. Ilie, L.: A note on the number of squares in a word. Theoret. Comput. Sci. **380**, 373–376 (2007)
4. Deza, A., Franek, F., Thierry, A.: How many double squares can a string contain? Discrete Appl. Math. **180**, 52–69 (2015)
5. Deza, A., Franek, F., Jiang, M.: A d-step approach for distinct squares in strings. In: Giancarlo, R., Manzini, G. (eds.) CPM 2011. LNCS, vol. 6661, pp. 77–89. Springer, Heidelberg (2011)
6. Jonoska, N., Manea, F., Seki, S.: A stronger square conjecture on binary words. In: Geffert, V., Preneel, B., Rovan, B., Štuller, J., Tjoa, A.M. (eds.) SOFSEM 2014. LNCS, vol. 8327, pp. 339–350. Springer, Heidelberg (2014)
7. Jürgensen, H., Konstantinidis, S.: Codes. In: Rozenberg, G., Salomaa, A. (eds.) Handbook of Formal Languages, vol. 1, pp. 511–607. Springer, Heidelberg (1997)

Mechanical Proofs of Properties
of the Tribonacci Word

Hamoon Mousavi and Jeffrey Shallit[(✉)]

School of Computer Science, University of Waterloo, Waterloo, ON N2L 3G1, Canada
hamoon.mousavi@gmail.com, shallit@uwaterloo.ca

Abstract. We implement a decision procedure for answering questions
about a class of infinite words that might be called (for lack of a better name) "Tribonacci-automatic". This class includes, for example, the
famous Tribonacci word $\mathbf{T} = 0102010010201\cdots$, the fixed point of the
morphism $0 \rightarrow 01$, $1 \rightarrow 02$, $2 \rightarrow 0$. We use our decision procedure to
reprove some old results about the Tribonacci word from the literature,
such as assertions about the occurrences in $\mathbf{T}$ of squares, cubes, palindromes, and so forth. We also obtain some new results, including on
enumeration.

1 Introduction

In several previous papers (e.g., [1,21–24,32] we have explored the ramifications
of a decision procedure for the logical theory $\mathrm{Th}(\mathbb{N}, +, a(n))$, where $(a(n))_{n\geq 0}$
is an infinite sequence specified by a finite-state machine M. Furthermore, in
many cases we can explicitly enumerate various aspects of such sequences, such
as subword complexity [8]. Roughly speaking, given a predicate P of one or more
variables in the logical theory, the method transforms M to a new automaton
M' that accepts the representations of those variables making the predicate true.
The ideas are based on extensions of the logical theory $\mathrm{Th}(\mathbb{N}, +)$, sometimes
called *Presburger arithmetic* [28,29]. See, for example, [6].

A critical point is what sort of representations are allowed. According to the
results in [5], it suffices that the representations be based on a Pisot number.
The critical point is that there must be an automaton that can perform addition
of two numbers in the appropriate representation [15,16].

In the papers mentioned above, we applied our method to the so-called k-
automatic sequences, which correspond to automata that work with the ordinary
base-k expansions of numbers. More recently, we also proved a number of new
results using Fibonacci (or "Zeckendorf") representation [13], which is based on
writing integers as a sum of Fibonacci numbers.

It is our contention that the power of this approach has not been widely
appreciated, and that many results, previously proved using long and involved
ad hoc techniques, can be proved with much less effort by phrasing them as logical predicates and employing a decision procedure. Furthermore, many enumeration questions can be solved with a similar approach. Although the worst-case

© Springer International Publishing Switzerland 2015
F. Manea and D. Nowotka (Eds.): WORDS 2015, LNCS 9304, pp. 170–190, 2015.
DOI: 10.1007/978-3-319-23660-5_15

running time of the procedure is enormously large (of the form $2^{2^{\cdot^{\cdot^{\cdot 2^{p(n)}}}}}$ where the number of 2's is the number of quantifier alternations), n is the size of the polynomial, and p is a polynomial), in practice the procedure often terminates in a reasonable time.

In this paper we discuss our implementation of an analogous algorithm for Tribonacci representation. We use it to reprove some old results from the literature purely mechanically, as well as obtain some new results. The implementation of the decision procedure was created by the first author. It is called Walnut, and is available for free download at https://www.cs.uwaterloo.ca/~shallit/papers.html.

We have not rigorously proved the correctness of this implementation, but it has been tested in a large number of different ways (including some results verified with independently-written programs). In this, we are well in the tradition of many other results in combinatorics on words that have been verified with machine computations — despite a lack of formal verification of the code. Even if the code were formally verified, one could reasonably ask for a proof of the correctness of the verification code! We believe that publication of our code, allowing checking by any interested reader, serves as an adequate check.

We view our work as part of a modern trend in mathematics. For other works on using computerized formal methods to prove theorems see, for example, [25, 27].

2 Tribonacci Representation

Let the Tribonacci numbers be defined, as usual, by the linear recurrence $T_n = T_{n-1} + T_{n-2} + T_{n-3}$ for $n \geq 3$ with initial values $T_0 = 0$, $T_1 = 1$, $T_2 = 1$. (We caution the reader that some authors use a different indexing for these numbers.) Here are the first few values of this sequence.

n	0	1	2	3	4	5	6	7	8	9	10	11	12	13	14	15	16
T_n	0	1	1	2	4	7	13	24	44	81	149	274	504	927	1705	3136	5768

From the theory of linear recurrences we know that

$$T_n = c_1 \alpha^n + c_2 \beta^n + c_3 \gamma^n$$

where α, β, γ are the zeros of the polynomial $x^3 - x^2 - x - 1$. The only real zero is $\alpha \doteq 1.83928675521416113$; the other two zeros are complex and are of magnitude $< 3/4$. Solving for the constants, we find that $c_1 \doteq 0.336228116994941094225$, the real zero of the polynomial $44x^3 - 2x - 1 = 0$. It follows that $T_n = c_1 \alpha^n + O(.75^n)$. In particular $T_n/T_{n-1} = \alpha + O(.41^n)$.

It is well-known that every non-negative integer can be represented, in an essentially unique way, as a sum of Tribonacci numbers $(T_i)_{i \geq 2}$, subject to the constraint that no three consecutive Tribonacci numbers are used [7]. For example, $43 = T_7 + T_6 + T_4 + T_3$.

Such a representation can be written as a binary word $a_1 a_2 \cdots a_n$ representing the integer $\sum_{1 \le i \le n} a_i T_{n+2-i}$. For example, the binary word 110110 is the Tribonacci representation of 43.

Let $\Sigma_2 = \{0, 1\}$. For $w = a_1 a_2 \cdots a_n \in \Sigma_2^*$, we define $[a_1 a_2 \cdots a_n]_T :=$ $\sum_{1 \le i \le n} a_i T_{n+2-i}$, even if $a_1 a_2 \cdots a_n$ has leading zeros or occurrences of the word 111.

By $(n)_T$ we mean the *canonical* Tribonacci representation for the integer n, having no leading zeros or occurrences of 111. Note that $(0)_T = \epsilon$, the empty word. The language of all canonical representations of elements of $\mathbb{N}$ is $\epsilon + (1 + 11)(0 + 01 + 011)^*$.

Just as Tribonacci representation is an analogue of base-k representation, we can define the notion of *Tribonacci-automatic sequence* as the analogue of the more familiar notation of k-automatic sequence [2,11]. We say that an infinite word $\mathbf{a} = (a_n)_{n \ge 0}$ is Tribonacci-automatic if there exists an automaton with output $M = (Q, \Sigma_2, q_0, \delta, \kappa, \Delta)$ for a coding κ such that $a_n = \kappa(\delta(q_0, (n)_T))$ for all $n \ge 0$. An example of a Tribonacci-automatic sequence is the infinite Tribonacci word,

$$\mathbf{T} = T_0 T_1 T_2 \cdots = 0102010010201 \cdots$$

which is generated by the following 3-state automaton (Fig. 1):

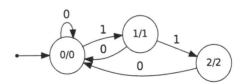

Fig. 1. Automaton generating the Tribonacci sequence

To compute T_i, we express i in canonical Tribonacci representation, and feed it into the automaton. Then T_i is the output associated with the last state reached (denoted by the symbol after the slash).

A basic fact about Tribonacci representation is that addition can be performed by a finite automaton. To make this precise, we need to generalize our notion of Tribonacci representation to r-tuples of integers for $r \ge 1$. A representation for $(x_1, x_2, \ldots, x_r)$ consists of a string of symbols z over the alphabet Σ_2^r, such that the projection $\pi_i(z)$ over the i'th coordinate gives a Tribonacci representation of x_i. Notice that since the canonical Tribonacci representations of the individual x_i may have different lengths, padding with leading zeros will often be necessary. A representation for $(x_1, x_2, \ldots, x_r)$ is called canonical if it has no leading $[0, 0, \ldots 0]$ symbols and the projections into individual coordinates have no occurrences of 111. We write the canonical representation as

$(x_1, x_2, \ldots, x_r)_T$. Thus, for example, the canonical representation for $(9, 16)$ is $[0, 1][1, 0][0, 0][1, 1][0, 1]$.

Thus, our claim about addition in Tribonacci representation is that there exists a deterministic finite automaton (DFA) M_{add} that takes input words of the form $[0, 0, 0]^*(x, y, z)_T$, and accepts if and only if $x+y = z$. Thus, for example, M_{add} accepts $[1, 0, 1][0, 1, 1][0, 0, 0]$ since the three words obtained by projection are 100, 010, and 110, which represent, respectively, 4, 2, and 6 in Tribonacci representation.

Since this automaton does not appear to have been given explicitly in the literature and it is essential to our implementation, we give it below in Table 1. This automaton actually works even for non-canonical expansions having three consecutive 1's. The initial state is state 1. The state 0 is a "dead state" that can safely be ignored.

We briefly sketch a proof of the correctness of this automaton. States can be identified with certain sequences, as follows: if x, y, z are the identical-length words arising from projection of a word that takes M_{add} from the initial state 1 to the state t, then t is identified with the integer sequence $([x0^n]_T + [y0^n]_T - [z0^n]_T)_{n \geq 0}$. State 0 corresponds to sequences that can never lead to 0, as they are too positive or too negative.

When we intersect this automaton with the appropriate regular language (ruling out input triples containing 111 in any coordinate), we get an automaton with 149 states accepting $0^*(x, y, z)_T$ such that $x + y = z$.

Another basic fact about Tribonacci representation is that, for canonical representations containing no three consecutive 1's or leading zeros, the radix order on representations is the same as the ordinary ordering on $\mathbb{N}$. It follows that a very simple automaton can, on input $(x, y)_T$, decide whether $x < y$.

Putting this all together, we get the following decision procedure:

Procedure 1 (Decision Procedure for Tribonacci-Automatic Words)
Input:

- $m, n \in \mathbb{N}$;
- m DFA's generating the Tribonacci-automatic words $\mathbf{w}_1, \mathbf{w}_2, \ldots, \mathbf{w}_m$;
- a first-order proposition with n free variables $\varphi(v_1, v_2, \ldots, v_n)$ using constants and relations definable in $Th(\mathbb{N}, 0, 1, +)$ and indexing into $\mathbf{w}_1, \mathbf{w}_2, \ldots, \mathbf{w}_m$.

Output: DFA with input alphabet Σ_2^n accepting
$\{(k_1, k_2, \ldots, k_n)_T : \varphi(k_1, k_2, \ldots, k_n) \text{ holds}\}$.

3   Mechanical Proofs of Properties of the Infinite Tribonacci Word

Recall that a word x, whether finite or infinite, is said to have period p if $x[i] = x[i + p]$ for all i for which this equality is meaningful. Thus, for example, the English word `alfalfa` has period 3. The *exponent* of a finite word x, written $\exp(x)$, is $|x|/P$, where P is the smallest period of x. Thus $\exp(\texttt{alfalfa}) = 7/3$.

Table 1. Transition table for M_{add} for Tribonacci addition

q	[0,0,0]	[0,0,1]	[0,1,0]	[0,1,1]	[1,0,0]	[1,0,1]	[1,1,0]	[1,1,1]	acc/rej
0	0	0	0	0	0	0	0	0	0
1	1	2	3	1	3	1	0	3	1
2	4	0	5	4	5	4	6	5	0
3	0	7	0	0	0	0	0	0	0
4	0	0	0	0	0	0	8	0	0
5	9	0	10	9	10	9	11	10	0
6	12	13	0	12	0	12	0	0	1
7	0	14	0	0	0	0	0	0	0
8	0	0	9	0	9	0	10	9	0
9	0	0	4	0	4	0	5	4	0
10	2	15	1	2	1	2	3	1	0
11	7	16	0	7	0	7	0	0	1
12	14	17	0	14	0	14	0	0	1
13	18	19	20	18	20	18	21	20	0
14	3	1	0	3	0	3	0	0	0
15	0	0	0	0	0	0	22	0	0
16	20	18	21	20	21	20	0	21	1
17	5	4	6	5	6	5	23	6	1
18	0	0	8	0	8	0	24	8	0
19	0	0	0	0	0	0	25	0	0
20	10	9	11	10	11	10	0	11	1
21	0	12	0	0	0	0	0	0	0
22	0	0	26	0	26	0	27	26	0
23	0	28	0	0	0	0	0	0	0
24	13	29	12	13	12	13	0	12	0
25	0	0	0	0	0	0	26	0	0
26	0	0	0	0	0	0	4	0	0
27	15	0	2	15	2	15	1	2	0
28	0	30	0	0	0	0	0	0	0
29	0	0	31	0	31	0	32	31	0
30	0	3	0	0	0	0	0	0	0
31	0	0	0	0	0	0	33	0	0
32	26	0	27	26	27	26	34	27	0
33	0	0	0	0	0	0	9	0	0
34	16	35	7	16	7	16	0	7	0
35	31	0	32	31	32	31	36	32	0
36	37	38	39	37	39	37	0	39	1
37	17	40	14	17	14	17	0	14	0
38	19	0	18	19	18	19	20	18	0
39	0	41	0	0	0	0	0	0	1
40	0	0	22	0	22	0	42	22	0
41	21	20	0	21	0	21	0	0	0
42	38	43	37	38	37	38	39	37	0
43	0	0	0	0	0	0	31	0	0

If **x** is an infinite word with a finite period, we say it is *ultimately periodic*. An infinite word **x** is ultimately periodic if and only if there are finite words u, v such that $x = uv^\omega$, where $v^\omega = vvv \cdots$.

A nonempty word of the form xx is called a *square*, and a nonempty word of the form xxx is called a *cube*. More generally, a nonempty word of the form x^n is called an n'th power. By the *order* of a square xx, cube xxx, or n'th power x^n, we mean the length $|x|$.

The infinite Tribonacci word $\mathbf{T} = 0102010 \cdots = T_0 T_1 T_2 \cdots$ can be described in many different ways. In addition to our definition in terms of automata, it is also the fixed point of the morphism $\varphi(0) = 01$, $\varphi(1) = 02$, and $\varphi(1) = 0$. This word has been studied extensively in the literature; see, for example, [3,9,14,17,30,31,33,34].

It can also be described as the limit of the finite Tribonacci words $(Y_n)_{n\geq0}$, defined as follows:

$$Y_0 = \epsilon \qquad\qquad Y_1 = 2 \qquad\qquad\qquad Y_2 = 0$$
$$Y_3 = 01 \qquad\qquad Y_n = Y_{n-1}Y_{n-2}Y_{n-3} \text{ for } n \geq 4.$$

Note that Y_n, for $n \geq 2$, is the prefix of length T_n of **T**.

In the next subsection, we use our implementation to prove a variety of results about repetitions in **T**.

3.1   Repetitions

It is known that all strict epistandard words (or Arnoux-Rauzy words), are not ultimately periodic (see, for example, [19]). Since **T** is in this class, we have the following known result which we can reprove using our method.

Theorem 2. *The word* **T** *is not ultimately periodic.*

Proof. We construct a predicate asserting that the integer $p \geq 1$ is a period of some suffix of **T**:

$$(p \geq 1) \;\wedge\; \exists n \; \forall i \geq n \; \mathbf{T}[i] = \mathbf{T}[i + p].$$

(Note: unless otherwise indicated, whenever we refer to a variable in a predicate, the range of the variable is assumed to be $\mathbb{N} = \{0, 1, 2, \ldots\}$.) From this predicate, using our program, we constructed an automaton accepting the language

$$L = 0^* \{(p)_T \;:\; (p \geq 1) \;\wedge\; \exists n \; \forall i \geq n \; \mathbf{T}[i] = \mathbf{T}[i + p]\}.$$

This automaton accepts the empty language, and so it follows that **T** is not ultimately periodic.

Here is the log of our program:

```
p >= 1 with 5 states, in 426ms
i >= n with 13 states, in 3ms
i + p with 150 states, in 31ms
```

```
TR[i] = TR[i + p] with 102 states, in 225ms
i >= n => TR[i] = TR[i + p] with 518 states, in 121ms
Ai i >= n => TR[i] = TR[i + p] with 4 states, in 1098ms
En Ai i >= n => TR[i] = TR[i + p] with 2 states, in 0ms
p >= 1 & En Ai i >= n => TR[i] = TR[i + p] with 2 states, in 1ms
overall time: 1905ms
```

The largest intermediate automaton during the computation had 5999 states.

A few words of explanation are in order: here "T" refers to the sequence **T**, and "E" is our abbreviation for $\exists$ and "A" is our abbreviation for $\forall$. The symbol "=>" is logical implication, and "&" is logical and.   □

From now on, whenever we discuss the language accepted by an automaton, we will omit the 0* at the beginning.

We now turn to repetitions. As a particular case of [17, Theorem. 6.31 and Example 7.6, p. 130] and [18, Example 6.21] we have the following result, which we can reprove using our method.

Theorem 3. T *contains no fourth powers.*

Proof. A natural predicate for the orders of all fourth powers occurring in **T**:

$$(n > 0) \ \wedge \ \exists i \ \forall t < 3n \ \mathbf{T}[i + t] = \mathbf{T}[i + n + t].$$

However, this predicate could not be run on our prover. It runs out of space while trying to determinize an NFA with 24904 states.

Instead, we make the substitution $j = i + t$, obtaining the new predicate

$$(n > 0) \ \wedge \ \exists i \ \forall j \ ((j \geq i) \wedge (j < i + 3n)) \implies \mathbf{T}[j] = \mathbf{T}[j + n].$$

The resulting automaton accepts nothing, so there are no fourth powers. Here is the log.

```
n > 0 with 5 states, in 59ms
i <= j with 13 states, in 15ms
3 * n with 147 states, in 423ms
i + 3 * n with 799 states, in 4397ms
j < i + 3 * n with 1103 states, in 4003ms
i <= j & j < i + 3 * n with 1115 states, in 111ms
j + n with 150 states, in 18ms
TR[j] = TR[j + n] with 102 states, in 76ms
i <= j & j < i + 3 * n => TR[j] = TR[j + n] with 6550 states, in 1742ms
Aj i <= j & j < i + 3 * n => TR[j] = TR[j + n] with 4 states, in 69057ms
Ei Aj i <= j & j < i + 3 * n => TR[j] = TR[j + n] with 2 states, in 0ms
n > 0 & Ei Aj i <= j & j < i + 3 * n => TR[j] = TR[j + n] with 2 states, in 0ms
overall time: 79901ms
```

The largest intermediate automaton in the computation had 86711 states.   □

Next, we move on to a description of the orders of squares occurring in **T**. We reprove a result of Glen [17, Sect. 6.3.5].

Theorem 4. *All squares in* **T** *are of order* T_n *or* $T_n + T_{n-1}$ *for some* $n \geq 2$. *Furthermore, for all* $n \geq 2$, *there exists a square of order* T_n *and* $T_n + T_{n-1}$ *in* **T**.

Proof. A natural predicate for the lengths of squares is

$$(n > 0) \wedge \exists i \; \forall t < n \; \mathbf{T}[i + t] = \mathbf{T}[i + n + t].$$

but when we run our solver on this predicate, we get an intermediate NFA of 4612 states that our solver could not determinize in the allotted space. The problem appears to arise from the three different variables indexing T. To get around this problem, we rephrase the predicate, introducing a new variable j that represents $i + t$. This gives the predicate

$$(n > 0) \wedge \exists i \; \forall j \; ((i \leq j) \wedge (j < i + n)) \implies \mathbf{T}[j] = \mathbf{T}[j + n].$$

and the following log

```
n > 0 with 5 states, in 59ms
i <= j with 13 states, in 15ms
3 * n with 147 states, in 423ms
i + 3 * n with 799 states, in 4397ms
j < i + 3 * n with 1103 states, in 4003ms
i <= j & j < i + 3 * n with 1115 states, in 111ms
j + n with 150 states, in 18ms
TR[j] = TR[j + n] with 102 states, in 76ms
i <= j & j < i + 3 * n => TR[j] = TR[j + n] with 6550 states, in 1742ms
Aj i <= j & j < i + 3 * n => TR[j] = TR[j + n] with 4 states, in 69057ms
Ei Aj i <= j & j < i + 3 * n => TR[j] = TR[j + n] with 2 states, in 0ms
n > 0 & Ei Aj i <= j & j < i + 3 * n => TR[j] = TR[j + n] with 2 states, in 0ms
overall time: 79901ms
```

The resulting automaton accepts exactly the language $10^* + 110^*$. The largest intermediate automaton had 26949 states. □

We can easily get more information about the square occurrences in **T**. By modifying our previous predicate, we get

$$(n > 0) \wedge \forall j \; ((i \leq j) \wedge (j < i + n)) \implies \mathbf{T}[j] = \mathbf{T}[j + n]$$

which encodes those (i, n) pairs such that there is a square of order n beginning at position i of **T**.

This automaton has only 10 states and efficiently encodes the orders and starting positions of each square in **T**. During the computation, the largest intermediate automaton had 26949 states. Thus we have proved the following new result:

Theorem 5. *The language*

$$\{(i, n)_T \; : \; there \; is \; a \; square \; of \; order \; n \; beginning \; at \; position \; i \; in \; \mathbf{T}\}$$

is accepted by the automaton in Fig. 2.

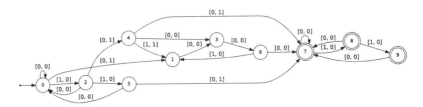

Fig. 2. Automaton accepting orders and positions of all squares in **T**

Next, we examine the cubes in **T**. Evidently Theorem 4 implies that any cube in **T** must be of order T_n or $T_n + T_{n-1}$ for some n. However, not every order occurs. We thus recover the following result of Glen [17, Sect. 6.3.7].

Theorem 6. *The cubes in* **T** *are of order T_n for $n \geq 5$, and a cube of each such order occurs.*

Proof. We use the predicate

$$(n > 0) \ \wedge \ \exists i \ \forall j \ ((i \leq j) \wedge (j < i + 2n)) \implies \mathbf{T}[j] = \mathbf{T}[j+n].$$

When we run our program, we obtain an automaton accepting exactly the language $(1000)0^*$, which corresponds to T_n for $n \geq 5$. The largest intermediate automaton had 60743 states. $\square$

4   Enumeration

Mimicking the base-k ideas in [8], we can also mechanically enumerate many aspects of Tribonacci-automatic sequences. We do this by encoding the factors having the property in terms of paths of an automaton. This gives the concept of *Tribonacci-regular sequence.* Roughly speaking, a sequence $(a(n))_{n \geq 0}$ taking values in $\mathbb{N}$ is Tribonacci-regular if the set of sequences

$$\{(a([xw]_T)_{w \in \Sigma_2^*} \ : \ x \in \Sigma_2^*\}$$

is finitely generated. Here we assume that $a([xw]_T)$ evaluates to 0 if xw contains the word 111. Every Tribonacci-regular sequence $(a(n))_{n \geq 0}$ has a *linear representation* of the form (u, μ, v) where u and v are row and column vectors, respectively, and $\mu : \Sigma_2 \to \mathbb{N}^{d \times d}$ is a matrix-valued morphism, where $\mu(0) = M_0$ and $\mu(1) = M_1$ are $d \times d$ matrices for some $d \geq 1$, such that

$$a(n) = u \cdot \mu(x) \cdot v$$

whenever $[x]_T = n$. The *rank* of the representation is the integer d.

Recall that if **x** is an infinite word, then the subword complexity function $\rho_{\mathbf{x}}(n)$ counts the number of distinct factors of length n. Then, in analogy with [8, Theorem. 27], we have

Theorem 7. *If* $\mathbf{x}$ *is Tribonacci-automatic, then the subword complexity function of* $\mathbf{x}$ *is Tribonacci-regular.*

Using our implementation, we can obtain a linear representation of the subword complexity function for $\mathbf{T}$. An obvious choice is to use the language

$$\{(n,i)_T \;:\; \forall j < i \; \mathbf{T}[i..i+n-1] \neq \mathbf{T}[j..j+n-1]\},$$

based on a predicate that expresses the assertion that the factor of length n beginning at position i has never appeared before. Then, for each n, the number of corresponding i gives $\rho_{\mathbf{T}}(n)$.

However, this does not run to completion in our implementation in the allotted time and space. Instead, let us substitute $u = j + t$ and $k = i - j$ to get the predicate

$$\forall k \; (((k > 0) \wedge (k \leq i)) \implies (\exists u \; ((u \geq j) \wedge (u < n + j) \wedge (\mathbf{T}[u] \neq \mathbf{T}[u+k])))).$$

This predicate is close to the upper limit of what we can compute using our program. The largest intermediate automaton had 1230379 states and the program took 12323.82 s, giving us a linear representation (u, μ, v) rank 22. When we minimize this using the algorithm in [4] we get the rank-12 linear representation

$$u = [1\ 0\ 0\ 0\ 0\ 0\ 0\ 0\ 0\ 0\ 0\ 0]$$

$$M_0 = \begin{bmatrix} 1 & 0 & 0 & 0 & 0 & 0 & 0 & 0 & 0 & 0 & 0 & 0 \\ 0 & 0 & 1 & 0 & 0 & 0 & 0 & 0 & 0 & 0 & 0 & 0 \\ 0 & 0 & 0 & 0 & 1 & 0 & 0 & 0 & 0 & 0 & 0 & 0 \\ -1 & 0 & 1 & 0 & 1 & 0 & 0 & 0 & 0 & 0 & 0 & 0 \\ 0 & 0 & 0 & 0 & 0 & 0 & 1 & 0 & 0 & 0 & 0 & 0 \\ -1 & 0 & 1 & 0 & 0 & 0 & 1 & 0 & 0 & 0 & 0 & 0 \\ -2 & 0 & 1 & 0 & 1 & 0 & 1 & 0 & 0 & 0 & 0 & 0 \\ -3 & 0 & 2 & 0 & 1 & 0 & 1 & 0 & 0 & 0 & 0 & 0 \\ -4 & 0 & 2 & 0 & 2 & 0 & 1 & 0 & 0 & 0 & 0 & 0 \\ -5 & 0 & 2 & 0 & 2 & 0 & 2 & 0 & 0 & 0 & 0 & 0 \\ -6 & 0 & 2 & 0 & 3 & 0 & 2 & 0 & 0 & 0 & 0 & 0 \\ -10 & 0 & 3 & 0 & 4 & 0 & 4 & 0 & 0 & 0 & 0 & 0 \end{bmatrix} \quad M_1 = \begin{bmatrix} 0 & 1 & 0 & 0 & 0 & 0 & 0 & 0 & 0 & 0 & 0 & 0 \\ 0 & 0 & 0 & 1 & 0 & 0 & 0 & 0 & 0 & 0 & 0 & 0 \\ 0 & 0 & 0 & 0 & 0 & 1 & 0 & 0 & 0 & 0 & 0 & 0 \\ 0 & 0 & 0 & 0 & 0 & 0 & 0 & 0 & 0 & 0 & 0 & 0 \\ 0 & 0 & 0 & 0 & 0 & 0 & 0 & 1 & 0 & 0 & 0 & 0 \\ 0 & 0 & 0 & 0 & 0 & 0 & 0 & 0 & 1 & 0 & 0 & 0 \\ 0 & 0 & 0 & 0 & 0 & 0 & 0 & 0 & 0 & 1 & 0 & 0 \\ 0 & 0 & 0 & 0 & 0 & 0 & 0 & 0 & 0 & 0 & 1 & 0 \\ 0 & 0 & 0 & 0 & 0 & 0 & 0 & 0 & 0 & 0 & 0 & 0 \\ 0 & 0 & 0 & 0 & 0 & 0 & 0 & 0 & 0 & 0 & 0 & 1 \\ 0 & 0 & 0 & 0 & 0 & 0 & 0 & 0 & 0 & 0 & 0 & 0 \\ 0 & 0 & 0 & 0 & 0 & 0 & 0 & 0 & 0 & 0 & 0 & 0 \end{bmatrix}$$

$$v' = [1\ 3\ 5\ 7\ 9\ 11\ 15\ 17\ 21\ 29\ 33\ 55]^R.$$

Comparing this to an independently-derived linear representation of the function $2n + 1$, we see they are the same. From this we get a well-known result (see, e.g., [12, Theorem 7]):

Theorem 8. *The subword complexity function of* $\mathbf{T}$ *is* $2n + 1$.

We now turn to computing the exact number of square occurrences in the finite Tribonacci words Y_n.

To solve this using our approach, we first generalize the problem to consider any length-n prefix of Y_n, and not simply the prefixes of length T_n.

The predicate represents the number of distinct squares in $\mathbf{T}[0..n-1]$:

$$L_{\mathrm{ds}} := \{(n,i,j)_T \; : \; (j \geq 1) \text{ and } (i+2j \leq n) \text{ and } \mathbf{T}[i..i+j-1] = \mathbf{T}[i+j..i+2j-1]$$
$$\text{and } \forall i' < i \; \mathbf{T}[i'..i'+2j-1] \neq \mathbf{T}[i..i+2j-1]\}.$$

This predicate asserts that $\mathbf{T}[i..i+2j-1]$ is a square occurring in $\mathbf{T}[0..n-1]$ and that furthermore it is the first occurrence of this particular word in $\mathbf{T}[0..n-1]$.

This represents the total number of occurrences of squares in $\mathbf{T}[0..n-1]$:

$$L_{\mathrm{dos}} := \{(n,i,j)_T \; : \; (j \geq 1) \text{ and } (i+2j \leq n) \text{ and } \mathbf{T}[i..i+j-1] = \mathbf{T}[i+j..i+2j-1]\}.$$

This predicate asserts that $\mathbf{T}[i..i+2j-1]$ is a square occurring in $\mathbf{T}[0..n-1]$.

Unfortunately, applying our enumeration method to this suffers from the same problem as before, so we rewrite it as

$$(j \geq 1) \wedge (i+2j \leq n) \wedge \forall u \, ((u \geq i) \wedge (u < i+j)) \implies \mathbf{T}[u] = \mathbf{T}[u+j]$$

When we compute the linear representation of the function counting the number of such i and j, we get a linear representation of rank 63. Now we compute the minimal polynomial of M_0 which is $(x-1)^2(x^2+x+1)^2(x^3-x^2-x-1)^2$. Solving a linear system in terms of the roots (or, more accurately, in terms of the sequences 1, n, T_n, T_{n-1}, T_{n-2}, nT_n, nT_{n-1}, nT_{n-2}) gives

Theorem 9. *The total number of occurrences of squares in the Tribonacci word Y_n is*

$$c(n) = \frac{n}{22}(9T_n - T_{n-1} - 5T_{n-2}) + \frac{1}{44}(-117T_n + 30T_{n-1} + 33T_{n-2}) + n - \frac{7}{4}$$

for $n \geq 5$.

In a similar way, we can count the occurrences of cubes in the finite Tribonacci word Y_n. Here we get a linear representation of rank 46. The minimal polynomial for M_0 is $x^4(x^3 - x^2 - x - 1)^2(x^2 + x + 1)^2(x-1)^2$. Using analysis exactly like the square case, we easily find

Theorem 10. *Let $C(n)$ denote the number of cube occurrences in the Tribonacci word Y_n. Then for $n \geq 3$ we have*

$$C(n) = \frac{1}{44}(T_n + 2T_{n-1} - 33T_{n-2}) + \frac{n}{22}(-6T_n + 8T_{n-1} + 7T_{n-2}) + \frac{n}{6}$$
$$- \frac{1}{4}[n \equiv 0 \ (\mathrm{mod} \ 3)] + \frac{1}{12}[n \equiv 1 \ (\mathrm{mod} \ 3)] - \frac{7}{12}[n \equiv 2 \ (\mathrm{mod} \ 3)].$$

Here $[P]$ is Iverson notation, and equals 1 if P holds and 0 otherwise.

5   Additional Results

Next, we encode the orders and positions of all cubes. We build a DFA accepting the language

$$\{(i, n)_T \; : \; (n > 0) \; \wedge \; \forall j \; ((i \le j) \wedge (j < i + 2n)) \implies \mathbf{T}[j] = \mathbf{T}[j + n]\}.$$

Theorem 11. *The language*

$$\{(n, i)_T \; : \; there \; is \; a \; cube \; of \; order \; n \; beginning \; at \; position \; i \; in \; \mathbf{T}\}$$

is accepted by the automaton in Fig. 3.

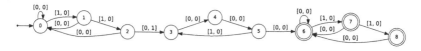

Fig. 3. Automaton accepting orders and positions of all cubes in **T**

We also computed an automaton accepting those pairs (p, n) such that there is a factor of **T** having length n and period p, and n is the largest such length corresponding to the period p. However, this automaton has 266 states, so we do not give it here.

5.1   Palindromes

We now turn to a characterization of the palindromes in **T**. Once again it turns out that the predicate we previously used in [13], namely,

$$\exists i \; \forall j < n \; \mathbf{T}[i + j] = \mathbf{T}[i + n - 1 - j],$$

resulted in an intermediate NFA of 5711 states that we could not successfully determinize.

Instead, we used two equivalent predicates. The first accepts n if there is an even-length palindrome, of length $2n$, centered at position i:

$$\exists i \ge n \; \forall j < n \; \mathbf{T}[i + j] = \mathbf{T}[i - j - 1].$$

The second accepts n if there is an odd-length palindrome, of length $2n + 1$, centered at position i:

$$\exists i \ge n \; \forall j \; (1 \le j \le n) \implies \mathbf{T}[i + j] = \mathbf{T}[i - j].$$

Theorem 12. *There exist palindromes of every length ≥ 0 in* **T**.

Proof. For the first predicate, our program outputs the automaton below. It clearly accepts the Tribonacci representations for all n (Fig. 4).

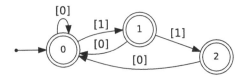

Fig. 4. Automaton accepting lengths of palindromes in **T**

The log of our program follows.

```
i >= n with 13 states, in 34ms
j < n with 13 states, in 8ms
i + j with 150 states, in 53ms
i - 1 with 7 states, in 155ms
i - 1 - j with 150 states, in 166ms
TR[i + j] = TR[i - 1 - j] with 664 states, in 723ms
j < n => TR[i + j] = TR[i - 1 - j] with 3312 states, in 669ms
Aj j < n => TR[i + j] = TR[i - 1 - j] with 24 states, in 5782274ms
i >= n & Aj j < n => TR[i + j] = TR[i - 1 - j] with 24 states, in 0ms
Ei i >= n & Aj j < n => TR[i + j] = TR[i - 1 - j] with 4 states, in 6ms
overall time: 5784088ms
```

The largest intermediate automaton had 918871 states. This was a fairly significant computation, taking about two hours' CPU time on a laptop.

We omit the details of the computation for the odd-length palindromes, which are quite similar.   □

Remark 1. A. Glen has pointed out to us that this follows from the fact that **T** is episturmian and hence rich, so a new palindrome is introduced at each new position in T.

We could also characterize the positions of all nonempty palindromes. To illustrate the idea, we generated an automaton accepting (i, n) such that $\mathbf{T}[i - n..i + n - 1]$ is an (even-length) palindrome (Fig. 5).

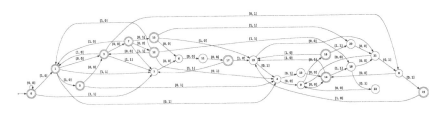

Fig. 5. Automaton accepting orders and positions of all nonempty even-length palindromes in **T**

The prefixes are factors of particular interest. Let us determine which prefixes are palindromes:

Theorem 13. *The prefix* $\mathbf{T}[0..n-1]$ *of length n is a palindrome if and only if* $n = 0$ *or* $(n)_T \in 1 + 11 + 10(010)^*(00 + 001 + 0011)$.

Proof. We use the predicate

$$\forall i < n\ \mathbf{T}[i] = \mathbf{T}[n-1-i].$$

The automaton generated is given below (Fig. 6). $\qquad\qquad\square$

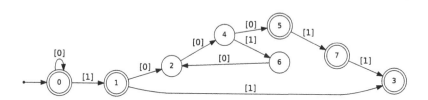

Fig. 6. Automaton accepting lengths of palindromes in $\mathbf{T}$

Remark 2. A. Glen points out to us that the palindromic prefixes of $\mathbf{T}$ are precisely those of the form Pal(w), where w is a finite prefix of the infinite word $(012)^\omega$ and Pal denotes the "iterated palindromic closure"; see, for example, [19, Example 2.6]. She also points out that these lengths are precisely the integers $(T_i + T_{i+2} - 3)/2$ for $i \geq 1$.

5.2 Quasiperiods

We now turn to quasiperiods. An infinite word $\mathbf{a}$ is said to be *quasiperiodic* if there is some finite nonempty word x such that $\mathbf{a}$ can be completely "covered" with translates of x. Here we study the stronger version of quasiperiodicity where the first copy of x used must be aligned with the left edge of $\mathbf{w}$ and is not allowed to "hang over"; these are called *aligned covers* in [10]. More precisely, for us $\mathbf{a} = a_0 a_1 a_2 \cdots$ is quasiperiodic if there exists x such that for all $i \geq 0$ there exists $j \geq 0$ with $i - n < j \leq i$ such that $a_j a_{j+1} \cdots a_{j+n-1} = x$, where $n = |x|$. Such an x is called a *quasiperiod*. Note that the condition $j \geq 0$ implies that, in this interpretation, any quasiperiod must actually be a prefix of $\mathbf{a}$.

Glen, Levé, and Richomme characterized the quasiperiods of a large class of words, including the Tribonacci word [20, Theorem 4.19]. However, their characterization did not explicitly give the lengths of the quasiperiods. We do that in the following new result.

Theorem 14. *A nonempty length-n prefix of $\mathbf{T}$ is a quasiperiod of $\mathbf{T}$ if and only if n is accepted by the following automaton (Fig. 7):*

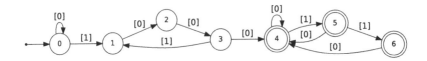

Fig. 7. Automaton accepting lengths of quasiperiods of the Tribonacci sequence

Proof. We write a predicate for the assertion that the length-n prefix is a quasiperiod:

$$\forall i \geq 0 \; \exists j \text{ with } i - n < j \leq i \text{ such that } \forall t < n \; \mathbf{T}[t] = \mathbf{T}[j+t].$$

When we do this, we get the automaton above. These numbers are those i for which $T_n \leq i \leq U_n$ for $n \geq 5$, where $U_2 = 0$, $U_3 = 1$, $U_4 = 3$, and $U_n = U_{n-1} + U_{n-2} + U_{n-3} + 3$ for $n \geq 5$.   □

5.3   Unbordered Factors

Next we look at unbordered factors. A word y is said to be a *border* of x if y is both a nonempty proper prefix and suffix of x. A word x is *bordered* if it has at least one border. It is easy to see that if a word y is bordered iff it has a border of length ℓ with $0 < \ell \leq |y|/2$.

Theorem 15. *There is an unbordered factor of length n of $\mathbf{T}$ if and only if $(n)_T$ is accepted by the automaton given below (Fig. 8).*

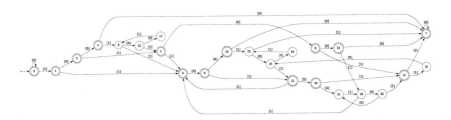

Fig. 8. Automaton accepting lengths of unbordered factors of the Tribonacci sequence

Proof. As in a previous paper [13] we can express the property of having an unbordered factor of length n as follows

$$\exists i \; \forall j, 1 \leq j \leq n/2, \; \exists t < j \; \mathbf{T}[i+t] \neq \mathbf{T}[i+n-j+t].$$

However, this does not run to completion within the available space on our prover. Instead, make the substitutions $t' = n - j$ and $u = i + t$. This gives the predicate

$$\exists i \; \forall t', \; n/2 \leq t' < n, \; \exists u, \; (i \leq u < i + n - t') \; \mathbf{T}[u] \neq \mathbf{T}[u + t'].$$

Here is the log:

```
2 * t with 61 states, in 276ms
n <= 2 * t with 79 states, in 216ms
t < n with 13 states, in 3ms
n <= 2 * t & t < n with 83 states, in 9ms
u >= i with 13 states, in 7ms
i + n with 150 states, in 27ms
i + n - t with 1088 states, in 7365ms
u < i + n - t with 1486 states, in 6041ms
u >= i & u < i + n - t with 1540 states, in 275ms
u + t with 150 states, in 5ms
TR[u] != TR[u + t] with 102 states, in 22ms
u >= i & u < i + n - t & TR[u] != TR[u + t] with 7489 states, in 3364ms
Eu u >= i & u < i + n - t & TR[u] != TR[u + t] with 552 states, in 5246873ms
n <= 2 * t & t < n => Eu u >= i & u < i + n - t & TR[u] != TR[u + t] with 944 states, in 38ms
At n <= 2 * t & t < n => Eu u >= i & u < i + n - t & TR[u] != TR[u + t] with 47 states, in 1184ms
Ei At n <= 2 * t & t < n => Eu u >= i & u < i + n - t & TR[u] != TR[u + t] with 25 states, in 2ms
overall time: 5265707ms                                                          □
```

5.4   Lyndon Words

Next, we turn to some results about Lyndon words. Recall that a nonempty word x is a *Lyndon word* if it is lexicographically less than all of its nonempty proper prefixes.[1]

Theorem 16. *There is a factor of length n of* **T** *that is Lyndon if and only if n is accepted by the automaton given below (Fig. 9).*

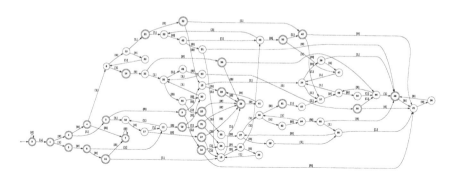

Fig. 9. Automaton accepting lengths of Lyndon factors of the Tribonacci sequence

Proof. Here is a predicate specifying that there is a factor of length n that is Lyndon:

$$\exists i\, \forall j, 1 \leq j < n,\ \exists t < n-j\ (\forall u < t\ \mathbf{T}[i+u] = \mathbf{T}[i+j+u]) \wedge \mathbf{T}[i+t] < \mathbf{T}[i+j+t].$$

Unfortunately this predicate did not run to completion, so we substituted $u' := i + u$ to get

$$\exists i\, \forall j, 1 \leq j < n,\ \exists t < n-j\ (\forall u', i \leq u' < i+t\ \mathbf{T}[u'] = \mathbf{T}[u'+j]) \wedge \mathbf{T}[i+t] < \mathbf{T}[i+j+t]. \ \square$$

---

[1] There is also a version where "prefixes" is replaced by "suffixes".

5.5   Critical Exponent

Recall from Sect. 3 that $\exp(w) = |w|/P$, where P is the smallest period of w. The *critical exponent* of an infinite word $\mathbf{x}$ is the supremum, over all factors w of $\mathbf{x}$, of $\exp(w)$.

Then Tan and Wen [33] proved that

Theorem 17. *The critical exponent of* $\mathbf{T}$ *is* $\rho \doteq 3.19148788395311874706$, *the real zero of the polynomial* $2x^3 - 12x^2 + 22x - 13$.

A. Glen points out that this result can also be deduced from [26, Theorem 5.2].

Proof. Let x be any factor of exponent ≥ 3 in $\mathbf{T}$. From Theorem 11 we know that such x exist. Let $n = |x|$ and p be the period, so that $n/p \geq 3$. Then by considering the first $3p$ symbols of x, which form a cube, we have by Theorem 11 that $p = T_n$. So it suffices to determine the largest n corresponding to every p of the form T_n. We did this using the predicate (Fig. 10).

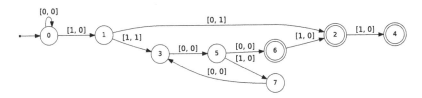

Fig. 10. Length n of longest factors having period $p = T_n$ of Tribonacci sequence

From inspection of the automaton, we see that the maximum length of a factor $n = U_j$ having period $p = T_j$, $j \geq 2$, is given by

$$U_j = \begin{cases} 2, & \text{if } j = 2; \\ 5, & \text{if } j = 3; \\ [110(100)^{i-1}0]_T, & \text{if } j = 3i+1 \geq 4; \\ [110(100)^{i-1}01]_T, & \text{if } j = 3i+2 \geq 5; \\ [110(100)^{i-1}011]_T, & \text{if } j = 3i+3 \geq 6. \end{cases}$$

A tedious induction shows that U_j satisfies the linear recurrence $U_j = U_{j-1} + U_{j-2} + U_{j-3} + 3$ for $j \geq 5$. Hence we can write U_j as a linear combination of Tribonacci sequences and the constant sequence 1, and solving for the constants we get

$$U_j = \frac{5}{2}T_j + T_{j-1} + \frac{1}{2}T_{j-2} - \frac{3}{2}$$

for $j \geq 2$.

The critical exponent of T is then $\sup_{j \geq 1} U_j/T_j$. Now

$$U_j/T_j = \frac{5}{2} + \frac{T_{j-1}}{T_j} + \frac{T_{j-2}}{2T_j} - \frac{3}{2T_j} = \frac{5}{2} + \alpha^{-1} + \frac{1}{2}\alpha^{-2} + O(1.8^{-j}).$$

Hence U_j/T_j tends to $5/2 + \alpha^{-1} + \frac{1}{2}\alpha^{-2} = \rho$. $\square$

We can also ask the same sort of questions about the *initial critical exponent* of a word **w**, which is the supremum over the exponents of all prefixes of **w**.

Theorem 18. *The initial critical exponent of* **T** *is* $\rho - 1$.

Proof. We create an automaton M_{ice} accepting the language

$$L = \{(n, p)_T \ : \ \mathbf{T}[0..n-1] \text{ has least period } p\}.$$

It is depicted in Fig. 11 below. An analysis similar to that we gave above for the critical exponent gives the result.   □

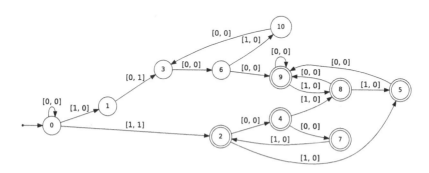

Fig. 11. Automaton accepting least periods of prefixes of length n

Theorem 19. *The only prefixes of the Tribonacci word that are powers are those of length* $2T_n$ *for* $n \geq 5$.

Proof. The predicate

$$\exists d < n \ (\forall j < n - d \ \mathbf{T}[j] = \mathbf{T}[d+j]) \ \wedge \ (\forall k < d \ \mathbf{T}[k] = \mathbf{T}[n-d+k])$$

asserts that the prefix $\mathbf{T}[0..n-1]$ is a power. When we run this through our program, the resulting automaton accepts 100010*, which corresponds to $F_{n+1} + F_{n-3} = 2T_n$ for $n \geq 5$.   □

6 Abelian Properties

We can derive some results about the abelian properties of the Tribonacci word **T** by proving the analogue of Theorem 63 of [13]:

Theorem 20. *Let* n *be a non-negative integer and let* $e_1 e_2 \cdots e_j$ *be a Tribonacci representation of* n*, possibly with leading zeros, with* $j \geq 3$*. Then*

(a) $|\mathbf{T}[0..n-1]|_0 = [e_1e_2\cdots e_{j-1}]_T + e_j.$
(b) $|\mathbf{T}[0..n-1]|_1 = [e_1e_2\cdots e_{j-2}]_T + e_{j-1}.$
(c) $|\mathbf{T}[0..n-1]|_2 = [e_1e_2\cdots e_{j-3}]_T + e_{j-2}.$

Proof. By induction, in analogy with the proof of [13, Theorem 63]. $\qquad\square$

Recall that the Parikh vector $\psi(x)$ of a word x over an ordered alphabet $\Sigma = \{a_1, a_2, \ldots, a_k\}$ is defined to be $(|x|_{a_1}, \ldots, |x|_{a_k})$, the number of occurrences of each letter in x. Recall that the abelian complexity function $\rho_{\mathbf{w}}^{\mathrm{ab}}(n)$ counts the number of distinct Parikh vectors of the length-n factors of an infinite word $\mathbf{w}$.

Using Theorem 20 we get another proof of a recent result of Turek [34].

Corollary 1. *The abelian complexity function of* $\mathbf{T}$ *is Tribonacci-regular.*

Proof. First, from Theorem 20 there exists an automaton TAB such that $(n, i, j, k)_T$ is accepted iff $(i, j, k) = \psi(\mathbf{T}[0..n-1])$. In fact, such an automaton has 32 states.

Using this automaton, we can create a predicate $P(n, i)$ such that the number of i for which $P(n, i)$ is true equals $\rho_{\mathbf{T}}^{\mathrm{ab}}(n)$. For this we assert that i is the least index at which we find an occurrence of the Parikh vector of $\mathbf{T}[i..i+n-1]$:

$\forall i' < i \ \exists a_0, a_1, a_2, b_0, b_1, b_2, c_0, c_1, c_2, d_0, d_1, d_2$

$\mathrm{TAB}(i+n, a_0, a_1, a_2) \wedge \mathrm{TAB}(i, b_0, b_1, b_2) \wedge \mathrm{TAB}(i'+n, c_0, c_1, c_2) \wedge \mathrm{TAB}(i', d_0, d_1, d_2) \wedge$
$((a_0-b_0 \neq c_0-d_0) \vee (a_1-b_1 \neq c_1-d_1) \vee (a_2-b_2 \neq c_2-d_2)). \quad \square$

Remark 3. Note that exactly the same proof would work for any word and numeration system where the Parikh vector of prefixes of length n is "synchronized" with n.

Remark 4. In principle we could mechanically compute the Tribonacci-regular representation of the abelian complexity function using this technique, but with our current implementation this is not computationally feasible.

Theorem 21. *Any morphic image of the Tribonacci word is Tribonacci-automatic.*

Proof. In analogy with Corollary 69 of [13]. $\qquad\square$

7 Things We Could Not Do Yet

There are a number of things we have not succeeded in computing with our prover because it ran out of space. These include

- Mirror invariance of $\mathbf{T}$ (that is, if x is a finite factor then so is x^R);
- Counting the number of special factors of length n (although it can be deduced from the subword complexity function);

- Statistics about, e.g., lengths of squares, cubes, etc., in the "flipped" Tribonacci sequence [31], the fixed point of $0 \to 01$, $1 \to 20$, $2 \to 0$;
- Recurrence properties of the Tribonacci word;
- Counting the number of distinct squares (not occurrences) in the finite Tribonacci word Y_n.
- Abelian complexity of the Tribonacci word.

In the future, an improved implementation may succeed in resolving these in a mechanical fashion.

Acknowledgments. We are very grateful to Amy Glen for her recommendations and advice. We thank Ondrej Turek and the referees for pointing out errors.

References

1. Allouche, J.P., Rampersad, N., Shallit, J.: Periodicity, repetitions, and orbits of an automatic sequence. Theoret. Comput. Sci. **410**, 2795–2803 (2009)
2. Allouche, J.P., Shallit, J.: Automatic Sequences: Theory, Applications, Generalizations. Cambridge University Press, Cambridge (2003)
3. Barcucci, E., Bélanger, L., Brlek, S.: On Tribonacci sequences. Fibonacci Quart. **42**, 314–319 (2004)
4. Berstel, J., Reutenauer, C.: Noncommutative Rational Series with Applications, Encyclopedia of Mathematics and Its Applications, vol. 137. Cambridge University Press, Cambridge (2011)
5. Bruyère, V., Hansel, G.: Bertrand numeration systems and recognizability. Theoret. Comput. Sci. **181**, 17–43 (1997)
6. Bruyère, V., Hansel, G., Michaux, C.,Villemaire, R.: Logic and p-recognizable sets of integers. Bull. Belgian Math. Soc. 1, 191–238 (1994), corrigendum. Bull. Belg. Math. Soc. **1**, 577 (1994)
7. Carlitz, L., Scoville, R., Hoggatt, Jr., V.E.: Fibonacci representations of higher order. Fibonacci Quart. **10**, 43–69, 94 (1972)
8. Charlier, E., Rampersad, N., Shallit, J.: Enumeration and decidable properties of automatic sequences. Int. J. Found. Comp. Sci. **23**, 1035–1066 (2012)
9. Chekhova, N., Hubert, P., Messaoudi, A.: Propriétés combinatoires, ergodiques et arithmétiques de la substitution de Tribonacci. J. Théorie Nombres Bordeaux **13**, 371–394 (2001)
10. Christou, M., Crochemore, M., Iliopoulos, C.S.: Quasiperiodicities in Fibonacci strings (2012), to appear in Ars Combinatoria. Preprint available at http://arxiv.org/abs/1201.6162
11. Cobham, A.: Uniform tag sequences. Math. Syst. Theory **6**, 164–192 (1972)
12. Droubay, X., Justin, J., Pirillo, G.: Episturmian words and some constructions of de Luca and Rauzy. Theoret. Comput. Sci. **255**, 539–553 (2001)
13. Du, C.F., Mousavi, H., Schaeffer, L., Shallit, J.: Decision algorithms for Fibonacci-automatic words, with applications to pattern avoidance (2014). http://arxiv.org/abs/1406.0670
14. Duchêne, E., Rigo, M.: A morphic approach to combinatorial games: theTribonacci case. RAIRO Inform. Théor. App. **42**, 375–393 (2008)
15. Frougny, C.: Representations of numbers and finite automata. Math. Systems Theory **25**, 37–60 (1992)

16. Frougny, C., Solomyak, B.: On representation of integers in linear numeration systems. In: Pollicott, M., Schmidt, K. (eds.) Ergodic Theory of $\mathbb{Z}^d$ Actions (Warwick, 1993–1994). London Mathematical Society Lecture Note Series, vol. 228, pp. 345–368. Cambridge University Press, Cambridge (1996)

17. Glen, A.: On sturmian and episturmian words, and related topics. Ph.D. thesis, University of Adelaide (2006)

18. Glen, A.: Powers in a class of a-strict episturmian words. Theoret. Comput. Sci. **380**, 330–354 (2007)

19. Glen, A., Justin, J.: Episturmian words: a survey. RAIRO Inform. Théor. App. **43**, 402–433 (2009)

20. Glen, A., Levé, F., Richomme, G.: Quasiperiodic and Lyndon episturmian words. Theoret. Comput. Sci. **409**, 578–600 (2008)

21. Goč, D., Henshall, D., Shallit, J.: Automatic theorem-proving in combinatorics on words. In: Moreira, N., Reis, R. (eds.) CIAA 2012. LNCS, vol. 7381, pp. 180–191. Springer, Heidelberg (2012)

22. Goč, D., Mousavi, H., Shallit, J.: On the number of unbordered factors. In: Dediu, A.-H., Martín-Vide, C., Truthe, B. (eds.) LATA 2013. LNCS, vol. 7810, pp. 299–310. Springer, Heidelberg (2013)

23. Goč, D., Saari, K., Shallit, J.: Primitive words and Lyndon words in automatic and linearly recurrent sequences. In: Dediu, A.-H., Martín-Vide, C., Truthe, B. (eds.) LATA 2013. LNCS, vol. 7810, pp. 311–322. Springer, Heidelberg (2013)

24. Goč, D., Schaeffer, L., Shallit, J.: Subword complexity and k-synchronization. In: Béal, M.-P., Carton, O. (eds.) DLT 2013. LNCS, vol. 7907, pp. 252–263. Springer, Heidelberg (2013)

25. Hales, T.C.: Formal proof. Notices Am. Math. Soc. **55**(11), 1370–1380 (2008)

26. Justin, J., Pirillo, G.: Episturmian words and episturmian morphisms. Theoret. Comput. Sci. **276**, 281–313 (2002)

27. Konev, B., Lisitsa, A.: A SAT attack on the Erdős discrepancy problem (2014). Preprint available at http://arxiv.org/abs/1402.2184

28. Presburger, M.: Über die Volständigkeit eines gewissen Systems derArithmetik ganzer Zahlen, in welchem die Addition als einzigeOperation hervortritt. In: Sprawozdanie z I Kongresu matematykówkrajów slowianskich, Warsaw, pp. 92–101, 395 (1929)

29. Presburger, M.: On the completeness of a certain system of arithmetic of whole numbers in which addition occurs as the only operation. Hist. Phil. Logic **12**, 225–233 (1991)

30. Richomme, G., Saari, K., Zamboni, L.Q.: Balance and Abelian complexity of the Tribonacci word. Adv. Appl. Math. **45**, 212–231 (2010)

31. Rosema, S.W., Tijdeman, R.: The Tribonacci substitution. INTEGERS 5 (3), Paper #A13 (2005). Available at http://www.integers-ejcnt.org/vol5-3.html

32. Shallit, J.: Decidability and enumeration for automatic sequences: a survey. In: Bulatov, A.A., Shur, A.M. (eds.) CSR 2013. LNCS, vol. 7913, pp. 49–63. Springer, Heidelberg (2013)

33. Tan, B., Wen, Z.Y.: Some properties of the Tribonacci sequence. Eur. J. Comb. **28**, 1703–1719 (2007)

34. Turek, O.: Abelian complexity function of the Tribonacci word. J. Integer Sequences **18**, Article 15.3.4 (2015). Available at https://cs.uwaterloo.ca/journals/JIS/VOL18/Turek/turek3.html

On Arithmetic Progressions in the Generalized Thue-Morse Word

Olga G. Parshina[⊠]

Novosibirsk State University, Pirogova Str. 2, 630090 Novosibirsk, Russia
parolja@gmail.com
http://www.nsu.ru/

Abstract. We consider the generalized Thue-Morse word on the alphabet $\Sigma = \{0, 1, 2\}$. The object of the research is arithmetic progressions in this word, which are sequences consisting of identical symbols. Let $A(d)$ be a function equal to the maximum length of an arithmetic progression with the difference d in the generalized Thue-Morse word. If n, d are positive integers and d is less than 3^n, then the upper bound for $A(d)$ is $3^n + 6$ and is attained with the difference $d = 3^n - 1$.

Keywords: Thue-Morse word · Arithmetic complexity · Arithmetic progression · Automatic word

1 Introduction

Arithmetic progressions in infinite words have been studied since the classical papers of Van der Waerden [1] and Szemerédi [2]. Results described in these papers say in particular, that we cannot constrain the length of a homogeneous arithmetic progression by a constant. For bounded differences and fixed word the question about the maximum possible length of an arithmetic progression depends on the structure of the word and the answer to it may be non-trivial. In some cases (e.g. Toeplitz words) the length of an arithmetic progression may be infinite. We are interested in ones having long progression with large differences only. Our research shows that such correlation between length of a progression and the value of its difference does not occur among automatic words.

The reported result refers to the Thue-Morse word, which was chosen because it is famous and the defining morphism is simple. More details on the Thue-Morse sequence and its applications can be found e.g. in [3]. It is known that the set of arithmetic progressions of the Thue-Morse word contains all binary words [4]. In [5] the results of [4] are generalized to symmetric D0L words including the generalized Thue-Morse word.

The original Thue-Morse word was studied earlier [6]. Here we considered its generalization on the alphabet $\Sigma_3 = \{0, 1, 2\}$, it was chosen to simplify the presentation, although the technique presented in the paper is easy to extend to Σ_q for an arbitrary prime q.

© Springer International Publishing Switzerland 2015
F. Manea and D. Nowotka (Eds.): WORDS 2015, LNCS 9304, pp. 191–196, 2015.
DOI: 10.1007/978-3-319-23660-5_16

We consider the maximal length of a homogeneous arithmetic progression of difference d. It is proven that the length grows quite rapidly *(see Theorem 1)* and reaches its maximums at points of specific kind. Since the Thue-Morse word over Σ_3 is easy to define by ternary representation of natural numbers, the proof of the theorem uses mostly this representation and arithmetic operations modulo 3. The proof has three stages: at first we reject the most part of the set of differences, which cannot provide the maximum of the length of a progression, then we present concrete values of the difference, of starting symbols and corresponding values of the length which are maxima of this function. The final part is exhaustion of the rest of initial symbols for this difference, there we give an upper bound of length for each type of starting number.

2   Preliminaries

Let $\Sigma_q = \{0, 1, ..., q-1\}$ be an alphabet. Consider a function $R_q : \mathbb{N} \rightarrow \Sigma_q$ which for every natural x gives its base-q expansion. The length of this word is denoted by $|R_q(x)|$. Also let $r_q(x)$ be the sum modulo q of the digits of x in its q-ary expansion. In other words $x = \sum_{i=0}^{n-1} x_i \cdot q^i$, $R_q(x) = x_{n-1} \cdots x_1 x_0$ and $r_q(x) = \sum_{i=0}^{n-1} x_i \bmod q$. The generalized Thue-Morse word is defined as: $w_{TM} = w_0 w_1 w_2 w_3 \cdots$, where $w_i = r_q(i) \in \Sigma$.

An arithmetic progression of length k with the starting number c and the difference d in arbitrary infinite word $v = v_0 v_1 v_2 v_3 \cdots$ is a sequence $v_c v_{c+d} v_{c+2d} \cdots v_{c+(k-1)d}$. We are interested in homogeneous progressions, i.e., in situations when $v_{c+id} = \alpha$ for each $i = 0, 1, ..., k-1$ and $\alpha \in \Sigma$.

Consider a function $A(c, d)$ which outputs the length of an arithmetic progression with starting symbol v_c and difference d for positive integers c and d. The function $A(d) = \max_c A(c, d)$ gives the length of the maximal arithmetic progression of difference d.

As we mentioned before, here we consider the alphabet Σ_3 and the generalized Thue-Morse word:

$$w_{TM} = 01212020112020101220101 2120 \cdots$$

Since the word is cube-free, $A(1) = 2$. We can see that $A(0, 5) = 2$, $A(2, 2) = A(20, 2) = 3$ and we state that $A(2) = 3$.

3   Main Result

Here we formulate the main theorem, the remainder of the paper is devoted to its proof.

Theorem 1. *For all numbers $n \geq 1$ the following holds:*

$$\max_{d < 3^n} A(d) = \begin{cases} 3^n + 6, & n \equiv 0 \bmod 3, \\ 3^n, & otherwise. \end{cases}$$

As we will see later, the function $A(d)$ reaches its maxima with differences of the form $3^n - 1$ for natural n. Let us prove that if $d \neq 3^n - 1$, the value of $A(d)$ will be not more than 3^n for fixed n.

3.1   Case of $d \neq 3^n - 1$

We need to prove that $A(d) \leq 3^n$.

At first we note that subsequences of the w_{TM} which are composed of letters with indexes having the same remainder of the division by three are equivalent to the word, so we do not need to consider differences which are divisible by three.

Every number may be represented in such a way: $c = y \cdot 3^n + x$, where x, y are arbitrary positive integers, $x < 3^n$. We will call x a suffix of c.

Consider the set $\mathbf{X} = \{0, 1, 2, ..., 3^n - 1\}$, $|\mathbf{X}| = 3^n$. Each difference d and suffix x belong to $\mathbf{X}$ and d is prime to $|\mathbf{X}|$. So the $\mathbf{X}$ is an additive cyclic group and d is a generator of $\mathbf{X}$, thus for every $x \in \mathbf{X}$ the set $\{x + i \cdot d\}_{i=0}^{3^n - 1}$ is precisely $\mathbf{X}$. We will prove the statement if for each $d \neq 3^n - 1$ we find an element $x \in \mathbf{X}$ with following properties:

(a) $x + d < 3^n$;
(b) $r_3(x + d) \neq r_3(x)$.

Indeed, consider the starting number of the form c. Because of (a), $c + d = y \cdot 3^n + (x + d)$. Hence, $r_3(c) = r_3(y) + r_3(x) \bmod 3$, $r_3(c + d) = r_3(y) + r_3(x + d) \bmod 3$. Because of (b), $r_3(c + d) \neq r_3(c)$, and the homogeneity of a progression will hold by at most 3^n steps.

If $r_3(d) \neq 0$, then a suitable x is a zero. In another case we use the inequation $d \neq 3^n - 1$ which means that $R_3(d) = d_{n-1} \cdots d_1 d_0$ has at least one letter $d_j, j \in \{0, 1, ..., n-1\}: d_j \neq 2$. There are two possibilities:

1. $\exists j : d_j \neq 2, d_{j-1} = 2$, in this case $x = 3^{j-1}$.
2. $\forall j \ (d_j \neq 2 \Rightarrow \forall k < j \ d_k \neq 2)$.

If $j > 1$, then, as soon as d is not divisible by three, $d_1 d_0$ may be equal to 01 or 11. In the first case we choose $x = 2$, in the second $x = 1$. But we can not find x satisfying (a) and (b) then $j = 1$, i.e., then $R_3(d) = \underbrace{2 \cdots 2}_{n-1} 1$. So we find three

suffixes x with properties (i) $x + d$, $x + 2d > 3^n$ and (ii) $r_3(x)$ is equal to the suffix of $x + d$ and $x + 2d$. Let us explain the necessity of these properties.

Consider a number c with such a suffix x, $r_3(c) = r_3(y) + r_3(x) \bmod 3$. The condition (i) gives to us equalities $r_3(c + d) = r_3(y + 1) + r_3(x + d) \bmod 3$ and $r_3(c + 2d) = r_3(y + 2) + r_3(x + 2d) \bmod 3$. With (ii) we need $r_3(y) = r_3(y + 1) = r_3(y + 2)$ for saving the homogeneity of a progression, but there is no number with such property. So we know that by at most three steps the value of r_3 will change. Suitable suffixes are $6, 15, 24$, and all described above guarantee that the arithmetic progression will be not longer than 3^n.

Of course, there are many suffixes satisfy (a),(b) or (i),(ii), but it is not necessary to consider all of them.

3.2   Case of $d = 3^n - 1$

Here we use the notation $R_3(x) = X = x_{n-1} \cdots x_1 x_0$, $R_3(y) = Y = y_{n-1} \cdots y_1 y_0$.

Lemma 1. *Let $d = 3^n - 1$, $c = z \cdot 3^{2n} + y \cdot 3^n + x$, where $x + y = 3^n - 1$, z is a non-negative integer, then*

$$\max_z A(c, d) = \begin{cases} 3^n + 3 - y, & n \equiv 0 \bmod 3, \\ 3^n - y, & otherwise. \end{cases}$$

Proof. Because of the value of d we can regard the action $c + d$ as two simultaneous actions: $x - 1$ and $y + 1$. Thus while $y \le 3^n - 1$ the sum of digits in YX (the concatenation of ternary notation of y and x) equals $2n$. This property provides us with the arithmetic progression of length $3^n - y$.

Let $y = 3^n - 1$, hence $x = 0$. If we add the difference d to such a number, the sum of digits in result's ternary notation will be equal to $4n$. To save the required property of members of the progression we need $2n \equiv 4n \bmod 3$, i.e., $n \equiv 0 \bmod 3$. After next addition of d, z increases to $z + 1$, y becomes 0 and $x = 3^n - 2$. We may define z arbitrary for holding the homogeneity of the progression (for example if $r_3(z) = 1$ we need $r_3(z + 1) = 2$, in this case z may be equal to 1). Now $r_3(y) + r_3(x) = 2n + 1 \bmod 3$. Let us add the difference once more: $Y = \underbrace{0 \cdots 0}_{n-1} 1$, $X = \underbrace{2 \cdots 2}_{n-1} 0$ and homogeneity holds. Next addition of d changes the value of $r_3(y) + r_3(x)$ because $Y = \underbrace{0 \cdots 0}_{n-1} 2$ and $X = \underbrace{2 \cdots 2}_{n-2} 12$.

So if $3 \mid n$, the length of an arithmetic progression equals $3^n + 3 - y$ and equals $3^n - y$ otherwise.

Lemma 2. *Let $n \equiv 0 \bmod 3$, $d = 3^n - 1$, $c = z \cdot 3^{2n} + y \cdot 3^n + x$, $y = 3^n - 2$, $x = 2$, z is arbitrary non-negative, then $\max_z A(c, d) = 3^n + 6$.*

Proof. Let us add to the c the difference d three times and look at the result:

1. $y = 3^n - 1$, $x = 1$;
2. $z \to z + 1$, $y = 0$, $x = 0$;
3. $y = 0$, $x = 3^n - 1$.

Sums of digits in ternary notation of described numbers are the same for a suitable z and coincide with the similar sum of initial c. After these steps we get into conditions of *Lemma 1* with $y = 0$ which provide us with an arithmetical progression of length $3^n + 3$. Now we subtract d from the initial c to make sure that $r_3(c - d) \ne r_3(c)$ and we cannot get an arithmetical progression longer. Indeed, $c - d = z \cdot 3^{2n} + (3^n - 3) \cdot 3^n + 3$ and the sum of digits in its ternary notation is $2n - 1$, while in c it is $2n + 1$.

So we find starting numbers for the difference $d = 3^n - 1$ which provide us with arithmetical progressions of the length mentioned in the Theorem.

Now let us prove that we can not get an arithmetical progression with difference $d = 3^n - 1$ longer than in the statement of the Theorem.

Here we represent a starting number of progression c like this: $c = y \cdot 3^n + x$, $x < 3^n$.

The case of initial number c with $x_j + y_j = 2$, $j = 0, 1, ..., n - 1$ is described in *Lemma* 1. In another case there is at least one index j: $x_j + y_j \neq 2$. We choose j which is the minimal. There are six possibilities of values (y_j, x_j): $(0,0), (0,1), (1,0), (1,2), (2,1), (2,2)$.

For c of each type we find numbers k and h: $r_3(c + k \cdot d) \neq r_3(c + h \cdot d)$.

We need two more parameters: l and m which are defined from these notations: $Y = y_{s-1} \cdots y_{j+l+1} 2 \cdots 2 y_j \cdots y_0$, $X = x_{n-1} \cdots x_{j+m+1} 0 \cdots 0 x_j \cdots x_0$. Of course m and l may be equal to zero, $y_{j+l+1} \neq 2$ and $x_{j+m+1} \neq 0$.

We will act such a way: we add $3^{j+1} \cdot d$ to the c, and the block of twos in Y transforms to the block of zeros, the block of zeros in X transforms to the block of twos. In cases $(y_j, x_j) \in \{(0,0), (1,0), (2,1), (2,2)\}$ after next $3^j \cdot d$ addition we get a number with the sum of digits different from the previous one. So suitable values of k and h are 3^{j+1} and $4 \cdot 3^j$. In cases $(y_j, x_j) \in \{(0,1), (1,2)\}$ to change the sum, we need to add $3^j \cdot d$ once more, and suitable k and h are $4 \cdot 3^j$ and $5 \cdot 3^j$. Let us consider an example for $(y_j, x_j) = (0,1)$.

Here $n = 7$, $m = l = 3$, $j = 1$, $d = 2186$, $R_3(d) = 2222222$.

number	R_3	r_3
$c = 2640685$	11222011100011	1
	+ 222222200	
$c + 3^{j+1} \cdot d$	12000011022211	1
	+ 22222220	
$c + 4 \cdot 3^j \cdot d$	12000111022201	1
	+ 22222220	
$c + 5 \cdot 3^j \cdot d$	12000211022121	0

But these values of k, h satisfy the Theorem if and only $j < n - 1$, the case $j = n - 1$ needs a special consideration.

We will act the following way: we add the $x \cdot d$ to the c and nullify x by that, then add d necessary number of times. So the worst case is then $X = x_{n-1} 2 \cdots 2$, $Y = y_{n-1} 0 \cdots 0$. Here is the table of values k, h.

(y_{n-1}, x_{n-1})	(0,0), (1,0)		(0,1), (2,1)		(2,2), (1,2)	
parameters	k	h	k	h	k	h
$3 \nmid n$	$3^{n-1} - 1$	3^{n-1}	$2 \cdot 3^{n-1} - 1$	$2 \cdot 3^{n-1}$	$3^n - 1$	3^n
$3 \mid n$	3^{n-1}	$3^{n-1} + 1$	$2 \cdot 3^{n-1}$	$2 \cdot 3^{n-1} + 1$	3^n	$3^n + 1$

One can see that these values satisfy the Theorem.

We have considered all the possible cases and thus completed the proof.

4 Conclusion

This result helps to better understand the structure of the well-known Thue-Morse word and its generalization. The result and the technique of the proof may be generalized on the Thue-Morse word over an arbitrary alphabet of prime cardinality q, the upper bound for the length of a progression in this case is $q^n + 2 \cdot q$.

References

1. Van der Waerden, B.L.: Beweis einer Baudetschen Vermutung. Nieuw Arch. Wisk. **15**, 212–216 (1927)
2. Szemerédi, E.: On sets of integers containing no k elements in arithmetic progression. Collection of articles in memory of Jurii Vladimirović Linnik. Acta Arith. **27**, 199–245 (1975)
3. Allouche, J.-P., Shallit, J.: The ubiquitous Prouhet-Thue-Morse sequence. In: Ding, C., Helleseth, T., Niederreiter, H. (eds.) Proceedings of SETA 1998, pp. 1–16. Springer, New York (1999)
4. Avgustinovich, S. V., Fon-der-Flaass, D. G., Frid, A. E.: Arithmetical complexity of infinite words. In: Proceedings of Words, Languages and Combinatorics III (2000), pp. 51–62 . World Scientific, Singapore (2003)
5. Frid, A.E.: Arithmetical complexity of symmetric D0L words. Theoret. Comput. Sci. **306**(1–3), 535–542 (2003)
6. Паршина, О.Г.: Однородные арифметические прогрессии в слове Туэ-Морса. Новосибирск: Материалы 53-й Международной научной студенческой конференции МНСК-2015, Математика, С. 218. (2015)

A Square Root Map on Sturmian Words

(Extended Abstract)

Jarkko Peltomäki[1,2]($\boxtimes$) and Markus Whiteland[2]

[1] Turku Centre for Computer Science TUCS, Turku, Finland
`jspelt@utu.fi`
[2] Department of Mathematics and Statistics, University of Turku, Turku, Finland
`mawhit@utu.fi`

Abstract. We introduce a square root map on Sturmian words and study its properties. Given a Sturmian word of slope α, there exists exactly six minimal squares in its language. A minimal square does not have a square as a proper prefix. A Sturmian word s of slope α can be written as a product of these six minimal squares: $s = X_1^2 X_2^2 X_3^2 \cdots$. The square root of s is defined to be the word $\sqrt{s} = X_1 X_2 X_3 \cdots$. We prove that $\sqrt{s}$ is also a Sturmian word of slope α. Moreover, we describe how to find the intercept of $\sqrt{s}$ and an occurrence of any prefix of $\sqrt{s}$ in s. Related to the square root map, we characterize the solutions of the word equation $X_1^2 X_2^2 \cdots X_m^2 = (X_1 X_2 \cdots X_m)^2$ in the language of Sturmian words of slope α where the words X_i^2 are minimal squares of slope α.

1 Introduction

Kalle Saari studies in [8] optimal squareful words and, more generally, α-repetitive words. Optimal squareful words are aperiodic words containing the least number of minimal squares (that is, squares with no proper square prefixes) such that every position starts a square. Saari proves that an optimal squareful word always contains exactly six minimal squares and characterizes these squares. From his considerations it follows that Sturmian words are a proper subclass of optimal squareful words.

In this paper we propose a square root map for Sturmian words. Let s be a Sturmian word of slope α, and write it as a product of the six minimal squares in its language: $s = X_1^2 X_2^2 X_3^2 \cdots$. The square root of s is defined to be the word $\sqrt{s} = X_1 X_2 X_3 \cdots$. The main result of this paper is that the word $\sqrt{s}$ is also a Sturmian word of slope α. In addition to proving that the square root map preserves the language $\mathcal{L}(\alpha)$ of a Sturmian word s of slope α, we show how to locate any prefix of $\sqrt{s}$ in s.

Section 3 contains a proof of the fact that the square root map preserves the language of a Sturmian word. For a given Sturmian word $s_{x,\alpha}$ of slope α and intercept x, we prove that $\sqrt{s_{x,\alpha}}$ equals the word $s_{\psi(x),\alpha}$ where $\psi(x) = \frac{1}{2}(x + 1 - \alpha)$. As a corollary, we obtain a description of those Sturmian words which are fixed points of the square root map.

© Springer International Publishing Switzerland 2015
F. Manea and D. Nowotka (Eds.): WORDS 2015, LNCS 9304, pp. 197–209, 2015.
DOI: 10.1007/978-3-319-23660-5_17

In Sect. 3 we observe that given a minimal square w^2 occurring in some Sturmian word, the interval $[w]$ on the circle corresponding to the factor w satisfies the square root condition $\psi([w^2]) \subseteq [w]$. In Sect. 4 we characterize all words $w^2 \in \mathcal{L}(\alpha)$ satisfying the square root condition. The result is that w^2 satisfies the square root condition if and only if w is a reversed standard or semistandard word or a reversed standard word with the two first letters exchanged.

The square root condition turns out to have another characterization in terms of specific solutions of the word equation $X_1^2 X_2^2 \cdots X_m^2 = (X_1 X_2 \cdots X_m)^2$. A word $w \in \mathcal{L}(\alpha)$ is a solution to this equation if $w^2 \in \mathcal{L}(\alpha)$ and it can be written that $w = X_1 X_2 \cdots X_m$ where the words X_i are roots of minimal squares satisfying the word equation. In Sect. 5 we prove that the word w is a primitive solution to the word equation if and only if w^2 satisfies the square root condition.

In the final section we show how to locate prefixes of $\sqrt{s}$ in s. As an important step in proving this, we provide necessary and sufficient conditions for a Sturmian word to be a product of reversed standard and semistandard words.

2   Notation and Preliminary Results

Due to space constraints we refer the reader to [5] for basic notation and results in words and for basic concepts such as *prefix, suffix, factor, primitive word, conjugate, ultimately periodic word* and *aperiodic word*.

A *period* of the word $w = a_1 a_2 \cdots a_n$ is an integer $p \geq 1$ such that $a_i = a_{i+p}$ for $1 \leq i \leq n - p$. If w has period p and $|w|/p \geq \alpha$ for some real $\alpha \geq 1$, then w is called an α-repetition. An α-repetition is *minimal* if it does not have an α-repetition as a proper prefix. If $w = u^2$, then w is a *square* with *square root* u. A square is *minimal* if it does not have a square as a proper prefix.

The *reversal* of a word $w = a_1 a_2 \cdots a_n$ is the word $\widetilde{w} = a_n \cdots a_2 a_1$. The shift operator on infinite words is denoted by T. In this paper we take binary words to be over the alphabet $\{0, 1\}$.

2.1   Optimal Squareful Words

In [8] Kalle Saari considers α-repetitive words. An infinite word is α-*repetitive* if every position in the word starts an α-repetition and the number of distinct minimal α-repetitions occurring in the word is finite. If $\alpha = 2$, then α-repetitive words are called *squareful words*. This means that every position of a squareful word begins with a minimal square. Saari proves that if the number of distinct minimal squares occurring in a squareful word is at most 5, then the word must be ultimately periodic. On the other hand, if a squareful word contains at least 6 distinct minimal squares, then aperiodicity is possible. Saari calls the aperiodic squareful words containing exactly 6 minimal squares *optimal squareful words*. Further, he shows that the six minimal squares take a very specific form:

Proposition 1. *Let s be an optimal squareful word. If 10^i1 occurs in s for some $i > 1$, then the roots of the six minimal squares in s are*

$$S_1 = 0, \qquad S_2 = 010^{a-1}, \qquad S_3 = 010^a, \qquad (1)$$
$$S_4 = 10^a, \qquad S_5 = 10^{a+1}(10^a)^b, \qquad S_6 = 10^{a+1}(10^a)^{b+1},$$

for some $a \geq 1$ and $b \geq 0$.

The optimal squareful words containing the minimal square roots of (1) are called *optimal squareful words with parameters a and b.* For the rest of this paper we reserve this meaning for the symbols a and b. Furthermore, we agree that the symbols S_i always refer to the minimal square roots (1).

Saari completely characterizes optimal squareful words:

Proposition 2. *An aperiodic infinite word s is optimal squareful if and only if (up to renaming of letters) there exists integers $a \geq 1$ and $b \geq 0$ such that s is an element of the language*

$$0^*(10^a)^*(10^{a+1}(10^a)^b + 10^{a+1}(10^a)^{b+1})^\omega = S_1^* S_4^* (S_5 + S_6)^\omega.$$

2.2 Continued Fractions

In this section we briefly review results on continued fractions needed in this paper. Good references on the subject are [1,3].

Every irrational real number α has a unique infinite continued fraction expansion

$$\alpha = [a_0; a_1, a_2, a_3, \ldots] = a_0 + \cfrac{1}{a_1 + \cfrac{1}{a_2 + \cfrac{1}{a_3 + \cdots}}} \qquad (2)$$

with $a_0 \in \mathbb{Z}$ and $a_k \in \mathbb{N}$ for all $k \geq 1$.

The *convergents* $c_k = \frac{p_k}{q_k}$ of α are defined by the recurrences

$$p_0 = a_0, \qquad p_1 = a_1 a_0 + 1, \qquad p_k = a_k p_{k-1} + p_{k-2}, \qquad k \geq 2,$$
$$q_0 = 1, \qquad q_1 = a_1, \qquad q_k = a_k q_{k-1} + q_{k-2}, \qquad k \geq 2.$$

The sequence $(c_k)_{k\geq0}$ converges to α. Moreover, the even convergents are less than α and form an increasing sequence. On the other hand, the odd convergents are greater than α and form a decreasing sequence.

If $k \geq 2$ and $a_k > 1$, then between the convergents c_{k-2} and c_k there are *semiconvergents* (called intermediate fractions in [1]) which are of the form

$$\frac{p_{k,l}}{q_{k,l}} = \frac{l p_{k-1} + p_{k-2}}{l q_{k-1} + q_{k-2}}$$

with $1 \leq l < a_k$. When the semiconvergents (if any) between c_{k-2} and c_k are ordered by the size of their denominators, the sequence obtained is increasing if k is even and decreasing if k is odd.

Note that we make a clear distinction between convergents and semiconvergents, i.e., convergents are not a specific subtype of semiconvergents.

2.3   Sturmian Words

Sturmian words are a well-known class of infinite, aperiodic binary words over $\{0, 1\}$ with minimal factor complexity. They are defined as the (right-)infinite words having $n + 1$ factors of length n for every $n \geq 0$. For our purposes it is more convenient to view Sturmian words equivalently as the infinite words obtained as codings of orbits of points in an irrational circle rotation with two intervals [5,7].

We identify the unit interval $[0, 1)$ with the unit circle $\mathbb{T}$. Let $\alpha \in (0, 1)$ be irrational. The map $R : [0, 1) \to [0, 1)$, $x \mapsto \{x + \alpha\}$, where $\{x\}$ stands for the fractional part of the number x, defines a rotation on $\mathbb{T}$. Divide the circle $\mathbb{T}$ into two intervals I_0 and I_1 defined by the points 0 and $1 - \alpha$. Then define the coding function ν by setting $\nu(x) = 0$ if $x \in I_0$ and $\nu(x) = 1$ if $x \in I_1$. The coding of the orbit of a point x is the infinite word $s_{x,\alpha}$ obtained by setting its $n^{\text{th}}, n \geq 0$, letter to equal $\nu(R^n(x))$. This word $s_{x,\alpha}$ is defined to be the Sturmian word of slope α and intercept x. To make the definition proper, we need to define how ν behaves in the endpoints 0 and $1 - \alpha$. We have two options: either take $I_0 = [0, 1 - \alpha)$ and $I_1 = [1 - \alpha, 1)$ or $I_0 = (0, 1 - \alpha]$ and $I_1 = (1 - \alpha, 1]$. The difference is seen in the codings of the orbits of the special points $\{-n\alpha\}$, and both options are needed for equivalence between this definition of Sturmian words and the usual definition. In this paper we are not concerned about this choice, and we adopt the notation $I_0 = I(0, 1 - \alpha), I_1 = I(1 - \alpha, 1)$ standing for either choice. Since the sequence $(\{n\alpha\})_{n \geq 0}$ is dense in $[0, 1)$—as is well-known—every Sturmian word of slope α has the same language (that is, the set of factors); this language is denoted by $\mathcal{L}(\alpha)$. For every factor $w = a_0 a_1 \ldots a_{n-1}$ of length n there exists a unique subinterval $[w]$ of $\mathbb{T}$ such that $s_{x,\alpha}$ begins with w if and only if $x \in [w]$. Clearly $[w] = I_{a_0} \cap R^{-1}(I_{a_1}) \cap \ldots \cap R^{-(n-1)}(I_{a_{n-1}})$. The points $0, \{-\alpha\}, \{-2\alpha\}, \ldots, \{-n\alpha\}$ partition the circle into $n + 1$ intervals which are in one-to-one correspondence with the words of $\mathcal{L}(\alpha)$ of length n. Among these intervals the interval containing the point $\{-(n+1)\alpha\}$ corresponds to the right special factor of length n. A factor w is *right special* if both $w0, w1 \in \mathcal{L}(\alpha)$. Similarly a factor is *left special* if both $0w, 1w \in \mathcal{L}(\alpha)$. In a Sturmian word there exists a unique right special and a unique left special factor of length n for all $n \geq 0$. The language $\mathcal{L}(\alpha)$ is mirror-invariant, that is, for every $w \in \mathcal{L}(\alpha)$ also $\widetilde{w} \in \mathcal{L}(\alpha)$. It follows that the right special factor of length n is the reversal of the left special factor of length n.

Given the continued fraction expansion of $\alpha \in (0, 1)$ as in (2) having convergents q_k and semiconvergents $q_{k,l}$, we define the corresponding *standard sequence* $(s_k)_{k \geq 0}$ of words by

$$s_{-1} = 1, \qquad s_0 = 0, \qquad s_1 = s_0^{a_1 - 1} s_{-1}, \qquad s_k = s_{k-1}^{a_k} s_{k-2}, \qquad k \geq 2.$$

As s_k is a prefix of s_{k+1} for $k \geq 1$, the sequence (s_k) converges to a unique infinite word c_α called the infinite standard Sturmian word of slope α, and it equals the word $s_{\alpha,\alpha}$. Inspired by the notion of semiconvergents, we define *semistandard* words for $k \geq 2$ by

$$s_{k,l} = s_{k-1}^l s_{k-2}$$

with $1 \leq l < a_k$. Clearly $|s_k| = q_k$ and $|s_{k,l}| = q_{k,l}$. Instead of writing "standard or semistandard", we often simply write "(semi)standard". The set of standard words of slope α is denoted by $Stand(\alpha)$, and the set of standard *and* semistandard words of slope α is denoted by $Stand^+(\alpha)$. (Semi)standard words are left special as prefixes of the word c_α. Every (semi)standard word is primitive [5, Proposition 2.2.3]. An important property of standard words is that the words s_k and s_{k-1} almost commute; namely $s_k s_{k-1} = wxy$ and $s_{k-1} s_k = wyx$ for some word w and distinct letters x and y. For more information about standard words see [2,5]. The only difference between the words c_α and $c_{\overline{\alpha}}$ where $\alpha = [0; 1, a_2, a_3, \ldots]$ and $\overline{\alpha} = [0; a_2 + 1, a_3, \ldots]$ is that the roles of the letters 0 and 1 are reversed. We may thus assume without loss of generality that $a_1 \geq 2$.

For the rest of this paper we make the convention that α stands for an irrational number in $(0, 1)$ having the continued fraction expansion as in (2) with $a_1 \geq 2$. The numbers q_k and $q_{k,l}$ refer to the denominators of the convergents of α, and the words s_k and $s_{k,l}$ refer to the standard or semistandard words of slope α.

2.4   Powers in Sturmian Words

In this section we review some known results on powers in Sturmian words, and prove helpful results for the next section.

If a square w^2 occurs in a Sturmian word of slope α, then the length of the word w must be a really specific number, namely a denominator of a convergent or a semiconvergent of α. The proof can be found in [4, Theorem 1] or [6, Proposition 4.1].

Proposition 3. *If $w^2 \in \mathcal{L}(\alpha)$ with w nonempty and primitive, then $|w| = q_0$, $|w| = q_1$ or $|w| = q_{k,l}$ for some $k \geq 2$ with $0 < l \leq a_k$.*

Next we need to know when conjugates of (semi)standard words occur as squares in a Sturmian word.

Proposition 4. *The following holds:*

(i) The square of every conjugate of s_k is in $\mathcal{L}(\alpha)$ for all $k \geq 0$.
(ii) Let w be a conjugate of $s_{k,l}$ with $k \geq 2$ and $0 < l < a_k$. Then $w^2 \in \mathcal{L}(\alpha)$ if and only if the intervals $[w]$ and $[s_{k,l}]$ have the same length.
(iii) Let $n = q_0$, $n = q_1$ or $n = q_{k,l}$ with $k \geq 2$ and $0 < l \leq a_k$, and let s be the (semi)standard word of length n. A factor w of length n is conjugate to s if and only if w and s have equally many occurrences of the letter 0.

3   The Square Root Map

In [8] Saari observed that every Sturmian word with slope α is an optimal squareful word with parameters $a = a_1 - 1$ and $b = a_2 - 1$. In particular, Saari's result means that every Sturmian word can be (uniquely) written as a product of the six *minimal squares of slope* α (1). Thus the square root map defined next is well-defined.

Definition 5. *Let s be a Sturmian word with slope α and write it as a product of minimal squares $s = X_1^2 X_2^2 X_3^2 \cdots$. The square root of s is then defined to be the word $\sqrt{s} = X_1 X_2 X_3 \cdots$.*

For some flavor, see the example at the beginning of Sect. 6. Note that this square root map can be defined for any optimal squareful word. However, here we only focus on Sturmian words.

We aim to prove the surprising fact that given a Sturmian word s the word $\sqrt{s}$ is also a Sturmian word having the same slope as s. Moreover, knowing the intercept of s, we can compute the intercept of $\sqrt{s}$.

In the proof we need a special function $\psi : \mathbb{T} \to \mathbb{T}, \psi(x) = \frac{1}{2}(x + 1 - \alpha)$. The mapping ψ moves a point x on the circle $\mathbb{T}$ towards the point $1 - \alpha$ by halving the distance between the points x and $1 - \alpha$. The distance to $1 - \alpha$ is measured in the interval I_0 or I_1 depending on which of these intervals the point x belongs to. The mapping ψ has the important property that $\psi(x + 2n\alpha) = \psi(x) + n\alpha$ for all $n \in \mathbb{Z}$.

We can now state the result.

Theorem 6. *Let $s_{x,\alpha}$ be a Sturmian word with slope α and intercept x. Then $\sqrt{s_{x,\alpha}} = s_{\psi(x),\alpha}$. In particular, $\sqrt{s_{x,\alpha}}$ is a Sturmian word with slope α.*

For a combinatorial version of the above theorem see Theorem 25 in Sect. 6.

The main idea of the proof is to demonstrate that the square root map is actually the symbolic counterpart of the function ψ. We begin with a definition.

Definition 7. *A square $w^2 \in \mathcal{L}(\alpha)$ satisfies the square root condition if $\psi([w^2]) \subseteq [w]$.*

Note that if the interval $[w]$ in the above definition has $1 - \alpha$ as an endpoint, then w automatically satisfies the square root condition. This is because ψ moves points towards the point $1 - \alpha$ but does not map them over this point. Actually, if w satisfies the square root condition, then necessarily the interval $[w]$ has $1 - \alpha$ as an endpoint (see Corollary 12).

We will only sketch the next proof of the following lemma.

Lemma 8. *The minimal squares of slope α satisfy the square root condition.*

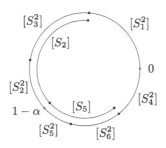

Fig. 1. The positions of the intervals on the circle in the proof sketch of Lemma 8.

Proof (Sketch). Since we already know the exact form of the minimal squares, we just show that $\psi(S_i^2) \subseteq [S_i]$ for $1 \leq i \leq 6$ in a straightforward manner. By a direct calculation $\psi(S_1^2) \subseteq [0] = [S_1]$. It is readily verified that $[0] = [S_1^2] \cup [S_2] = [S_1^2] \cup [S_3^2] \cup [S_2^2]$ (see Fig. 1). Since $[S_2]$ has $1 - \alpha$ as an endpoint, by the above remark $\psi([S_2^2]) \subseteq [S_2]$. As $[S_2] = [S_3]$, also $\psi([S_3^2]) \subseteq [S_3]$. Similarly $[1] = [S_4] = [S_4^2] \cup [S_5] = [S_4^2] \cup [S_6^2] \cup [S_5^2]$. Clearly $\psi([S_4^2]) \subseteq [S_4]$. Since $1 - \alpha$ is an endpoint of $[S_5]$, we have that $\psi([S_5^2]) \subseteq [S_5]$. Finally, $\psi([S_6^2]) \subseteq [S_6]$ as $[S_5] = [S_6]$. □

Proof (of Theorem 6). Write $s_{x,\alpha} = X_1^2 X_2^2 X_3^2 \cdots$ as a product of minimal squares. As $x \in [X_1^2]$, by Lemma 8 $\psi(x) \in [X_1]$. Hence both $\sqrt{s_{x,\alpha}}$ and $s_{\psi(x),\alpha}$ begin with X_1. Since $\psi(x + 2n\alpha) = \psi(x) + n\alpha$ for all $n \in \mathbb{Z}$, by shifting $s_{x,\alpha}$ the amount $2|X_1|$ and by applying Lemma 8, we conclude that $s_{\psi(x),\alpha}$ shifted by the mount $|X_1|$ begins with X_2. Therefore the words $\sqrt{s_{x,\alpha}}$ and $s_{\psi(x),\alpha}$ agree on their first $|X_1| + |X_2|$ letters. By repeating this procedure, we conclude that $\sqrt{s_{x,\alpha}} = s_{\psi(x),\alpha}$. □

Theorem 6 allows us effortlessly to characterize the Sturmian words which are fixed points of the square root map:

Corollary 9. *The only Sturmian words of slope α which are fixed by the square root map are the two words $01c_\alpha$ and $10c_\alpha$, both having intercept $1 - \alpha$.*

Proof. The only fixed point of the map ψ is the point $1 - \alpha$. Having this point as an intercept, we obtain two Sturmian words: either $01c_\alpha$ or $10c_\alpha$, depending on which of the intervals I_0 and I_1 the point $1 - \alpha$ belongs to. □

The set $\{01c_\alpha, 10c_\alpha\}$ is not only the set of fixed points but also the unique attractor of the square root map in the set of Sturmian words of slope α. When iterating the square root map on a fixed Sturmian word $s_{x,\alpha}$, the obtained word has longer and longer prefixes in common with either of the words $01c_\alpha$ and $10c_\alpha$ because $\psi^n(x)$ tends to $1 - \alpha$ as n increases.

4 One Characterization of Words Satisfying the Square Root Condition

In the previous section we saw that minimal squares satisfying the square root condition were crucial in proving that the square root of a Sturmian word is again Sturmian with the same slope. The minimal squares of slope α are not the only squares in $\mathcal{L}(\alpha)$ satisfying the square root condition; in this section we will characterize combinatorially such squares. To be able to state the characterization, we need to define $RStand(\alpha) = \{\widetilde{w} \colon w \in Stand(\alpha)\}$; the set of reversed standard words of slope α. Similarly we set $RStand^+(\alpha) = \{\widetilde{w} \colon w \in Stand^+(\alpha)\}$. We also need the operation L which exchanges the first two letters of a word (we do not apply this operation to too short words).

The main result of this section is the following:

Theorem 10. *A word $w^2 \in \mathcal{L}(\alpha)$ satisfies the square root condition if and only if $w \in RStand^+(\alpha) \cup L(RStand(\alpha))$.*

Theorem 10 can be proved using Corollary 12 and Proposition 13 below.

As was remarked in Sect. 3, a square $w^2 \in \mathcal{L}(\alpha)$ trivially satisfies the square root condition if the interval $[w]$ has $1 - \alpha$ as an endpoint. Our aim is to prove that the converse is also true. We begin with a technical lemma.

Lemma 11. *Let $n = q_1$ or $n = q_{k,l}$ for some $k \geq 2$ with $0 < l \leq a_k$, and let i be an integer such that $1 < i \leq n$.*

(i) *If $\{-i\alpha\} \in I_0$ and $\{-(i+n)\alpha\} < \{-i\alpha\}$, then $\psi(-(i+n)\alpha) > \{-i\alpha\}$.*
(ii) *If $\{-i\alpha\} \in I_1$ and $\{-(i+n)\alpha\} > \{-i\alpha\}$, then $\psi(-(i+n)\alpha) < \{-i\alpha\}$.*

With Lemma 11 we can prove the following:

Corollary 12. *If $w^2 \in \mathcal{L}(\alpha)$ with w primitive satisfies the square root condition, then $[w]$ has $1 - \alpha$ as an endpoint.*

Next we study in more detail the properties of squares $w^2 \in \mathcal{L}(\alpha)$ whose interval has $1 - \alpha$ as an endpoint.

Proposition 13. *Consider the intervals of factors in $\mathcal{L}(\alpha)$ of length $n = q_1$ or $n = q_{k,l}$ with $k \geq 2$ and $0 < l \leq a_k$. Let u and v be the two distinct words having intervals with endpoint $1 - \alpha$. Then the following holds:*

(i) *There exists a word w such that $u = xyw$ and $v = yxw = L(u)$ for distinct letters x and y.*
(ii) *Either u or v is right special.*
(iii) *If μ is the right special word among the words u and v, then $\mu^2 \in \mathcal{L}(\alpha)$.*
(iv) *If λ is the word among the words u and v which is not right special, then $\lambda^2 \in \mathcal{L}(\alpha)$ if and only if $n = q_1$ or $l = a_k$.*

5   Characterization by a Word Equation

It turns out that the squares of slope α satisfying the square root condition have also a different characterization in terms of specific solutions of the word equation

$$X_1^2 X_2^2 \cdots X_m^2 = (X_1 X_2 \cdots X_m)^2 \tag{3}$$

in the language $\mathcal{L}(\alpha)$. We are interested only in the solutions of (3) where all words X_i are *minimal square roots* (1), i.e., primitive roots of minimal squares. We thus define:

Definition 14. *A nonempty word w is a solution to (3) if w can be written as a product of minimal square roots $w = X_1 X_2 \cdots X_m$ which satisfy the equation (3). The solution is trivial if $X_1 = X_2 = \ldots = X_m$ and primitive if w is primitive. The word w is a solution to (3) in $\mathcal{L}(\alpha)$ if w is a solution to (3) and $w^2 \in \mathcal{L}(\alpha)$.*

All minimal square roots of slope α are trivial solutions to (3). One example of a nontrivial solution is $w = S_2 S_1 S_4$ in $\mathcal{L}(\alpha)$, where $\alpha = [0; 2, 1, 1, \ldots]$, since $w^2 = (01010)^2 = (01)^2 \cdot 0^2 \cdot (10)^2 = S_2^2 S_1^2 S_4^2$. It is clear that the presentation of a word as product of minimal squares is unique.

Our aim is to complete the characterization of Theorem 10 to:

Theorem 15. *Let $w \in \mathcal{L}(\alpha)$. The following are equivalent:*

(i) w *is a primitive solution to* (3) *in* $\mathcal{L}(\alpha)$,
(ii) w^2 *satisfies the square root condition,*
(iii) $w \in RStand^+(\alpha) \cup L(RStand(\alpha))$.

Note that a word w in the set $L(RStand^+(\alpha)) \setminus L(RStand(\alpha))$ is a solution to (3) but not in the language $\mathcal{L}(\alpha)$. Rather, w is a solution to (3) in $\mathcal{L}(\beta)$ where β is a suitable irrational such that $L(w)$ is a reversed standard word of slope β.

We begin with a definition:

Definition 16. *The language $\mathcal{L}(a, b)$ consists of all factors of the infinite words in the language*

$$(10^{a+1}(10^a)^b + 10^{a+1}(10^a)^{b+1})^\omega = (S_5 + S_6)^\omega.$$

Observe that by Proposition 2 every factor in $\mathcal{L}(a, b)$ is a factor of some optimal squareful word with parameters a and b. Moreover if $\alpha = [0; a + 1, b + 1, \ldots]$, then $\mathcal{L}(\alpha) \subseteq \mathcal{L}(a, b)$.

To prove Proposition 18 below, we need the following technical lemma:

Lemma 17. *Let w be a primitive solution to* (3) *having the word $s = \tilde{s}_{3,1} = 10^{a+1}(10^a)^{b+1}$ as a suffix such that $w^2, L(w) \in \mathcal{L}(a, b)$. Then $wL(w) = X_1^2 X_2^2 \cdots X_n^2$ and $w = X_1 X_2 \cdots X_n$ where the words X_i^2 are minimal squares.*

Proposition 18. *Let $w \in RStand^+(\alpha) \cup L(RStand(\alpha))$. Then the word w is a primitive solution to* (3) *in* $\mathcal{L}(\alpha)$.

From Proposition 18 we conclude the following interesting fact:

Corollary 19. *There exist arbitrarily long primitive solutions of* (3) *in* $\mathcal{L}(\alpha)$.

We can now prove Theorem 15 using Theorem 10 and Proposition 18.

6 A More Detailed Combinatorial Description of the Square Root Map

We saw in Sect. 3 that the square root of a Sturmian word s has the same factors as s has. The proofs were dynamical; we used the special mapping ψ on the circle. In this section we describe combinatorially why the language is preserved. We give a location for any prefix of $\sqrt{s}$ in s. As a side product, we are able to describe when a Sturmian word is uniquely expressible as a product of squares of reversed (semi)standard words.

Let us consider as an example the famous Fibonacci word f (see for instance [5]), having parameters $a = 1$ and $b = 0$, and its square root:

$$f = (010)^2(100)^2(10)^2(01)^2 0^2 (10010)^2(01)^2 \cdots ,$$
$$\sqrt{f} = 010 \cdot 100 \cdot 10 \cdot 01 \cdot 0 \cdot 10010 \cdot 01 \cdots .$$

Obviously the square root $X_1 = 010$ of $(010)^2$ occurs as a prefix of f. Equally clearly the word $010 \cdot 100 = \sqrt{(010)^2(100)^2}$ occurs, not as a prefix, but after the prefix X_1 of f. Thus the position of the first occurrence of $010 \cdot 100$ shifted $|X_1| = 3$ positions from the position of the first occurrence of X_1. However, when comparing the position of the first occurrence of $\sqrt{(010)^2(100)^2(10)^2}$ with the first occurrence of $010 \cdot 100$, we see that there is no further shift. By further inspection, the word $\sqrt{(010)^2(100)^2(10)^2(01)^2 0^2 (10010)^2}$ occurs for the first time at position $|X_1|$ of f (when indexing from zero). This is no longer true for the first seven minimal squares; the first occurrence of $X_1 X_2 = 010 \cdot 100 \cdot 10 \cdot 01 \cdot 0 \cdot 10010 \cdot 01$ is at position $|X_1 X_2| = 16$ of f. The amount of shift from the previous position $|X_1| = 3$ is $|X_2| = 13$; observe that both of these numbers are Fibonacci numbers. Thus the amount of shift was exactly the length of the square roots added after observing the previous shift. As an observant reader might have noticed, both of the words X_1 and X_2 are reversed standard words, or equivalently, primitive solutions to (3). Repeating similar inspections on other Sturmian words suggests that there is a certain pattern to these shifts and that knowing the pattern would make it possible to locate prefixes of $\sqrt{s}$ in the Sturmian word s. Thus it makes very much sense to "accelerate" the square root map by considering squares of solutions to (3) instead of just minimal squares. Next we make these somewhat vague observations more precise.

Every Sturmian word has a solution of (3) as a square prefix. Next we aim to characterize Sturmian words having infinitely many solutions of (3) as square prefixes. The next two lemmas are key results towards such a characterization.

Lemma 20. *Consider the reversed (semi)standard words of slope α. The set $[\tilde{s}_{k,l}] \setminus \{1 - \alpha\}$ equals the disjoint union*

$$\left(\bigcup_{i=0}^{\infty} \bigcup_{j=1}^{a_{k+2i}} [\tilde{s}_{k+2i,j}^2] \right) \setminus \bigcup_{i=1}^{l-1} [\tilde{s}_{k,i}^2]$$

for all $k \geq 2$ with $0 < l \leq a_k$. Analogous representations exist for the sets $[\tilde{s}_0] \setminus \{1 - \alpha\}$ and $[\tilde{s}_1] \setminus \{1 - \alpha\}$.

To put it more simply: for each $x \neq 1 - \alpha$ there exists a unique reversed (semi)standard word w such that $x \in [w^2]$.

Lemma 21. *Let $u \in RStand^+(\alpha)$ and $v \in RStand^+(\alpha) \cup L(RStand^+(\alpha))$. Then u^2 is never a proper prefix of v^2.*

Let s be a fixed Sturmian word of slope α. Since the largest power of a factor of a Sturmian word is bounded, Lemma 21 and Theorem 15 imply that if s has infinitely many solutions of (3) as square prefixes then no word in $RStand^+(\alpha)$ is a square prefix of s. We have now the proper tools to prove the following:

Proposition 22. *Let $s_{x,\alpha}$ be a Sturmian word of slope α and intercept x. Then $s_{x,\alpha}$ begins with a square of a word in $RStand^+(\alpha)$ if and only if $x \neq 1 - \alpha$.*

It follows that if s has infinitely many solutions of (3) as square prefixes, then $s \in \{01c_\alpha, 10c_\alpha\}$.

Next we take one extra step and characterize when s can be written as a product of squares of words in $RStand^+(\alpha)$.

Theorem 23. *A Sturmian word s of slope α can be written as a product of squares of words in $RStand^+(\alpha)$ if and only if s is not of the form $X_1^2 X_2^2 \cdots X_n^2 c$ where $X_i \in RStand^+(\alpha)$ and $c \in \{01c_\alpha, 10c_\alpha\}$. If such an expression exists, then it is unique.*

Suppose that $s \notin \{01c_\alpha, 10c_\alpha\}$. Then s has only finitely many solutions of (3) as square prefixes. We call the longest solution *maximal*. Note that the maximal solution is not necessarily primitive since any power of a solution to (3) is also a solution. Sturmian words of slope α can be classified into two types:

Type A. Sturmian words s of slope α which can be written as products of maximal solutions to (3). In other words, it can be written that $s = X_1^2 X_2^2 \cdots$ where X_i is the maximal solution occurring as a square prefix of the word $T^{h_i}(s)$ where $h_i = |X_1^2 X_2^2 \cdots X_{i-1}^2|$.

Type B. Sturmian words s of slope α which are of the form $s = X_1^2 X_2^2 \cdots X_n^2 c$ where $c \in \{01c_\alpha, 10c_\alpha\}$, and the words X_i are maximal solutions as above.

Proposition 22 and Lemma 21 imply that the words X_i in the above definitions are uniquely determined and that the primitive root of a maximal solution is in $RStand^+(\alpha)$. Consequently, a maximal solution is always right special. When finding the expression of a Sturmian word as a product of squares of maximal solutions, it is sufficient to detect at each position the shortest square of a word in $RStand^+(\alpha)$ and take its largest even power occurring in that position.

Keeping the Sturmian word s of slope α fixed, we define two sequences (μ_k) and (λ_k). We set $\mu_0 = \lambda_0 = \varepsilon$. Following the notation above, we define depending on the type of s as follows:

(A) If s is of type A, then we set for all $k \geq 1$ that

$$\mu_k = X_1^2 X_2^2 \cdots X_k^2 \text{ and}$$
$$\lambda_k = X_1 X_2 \cdots X_k.$$

(B) If s is of type B, then for $1 \leq k \leq n$ we define λ_k and μ_k as in the previous case. We let

$$\mu_{n+1} = X_1^2 X_2^2 \cdots X_n^2 c \text{ and}$$
$$\lambda_{n+1} = X_1 X_2 \cdots X_n c.$$

Compare these definitions with Theorem 25 and the example of the Fibonacci word (of type A); the words X_1 and X_2 are maximal solutions.

Proposition 24. *Suppose that s is a Sturmian word of type A. Then the word λ_k is right special and a suffix of the word μ_k for all $k \geq 0$.*

Note that even though λ_k is right special and always a suffix of μ_k, it is not necessary for μ_k to be right special.

We are finally in a position to formulate precisely the observations made at the beginning of this section and state the main result of this section.

Theorem 25. *Let s be a Sturmian word with slope α.*
(A) If s is of type A, then

$$\sqrt{s} = \lim_{k \to \infty} T^{|\lambda_k|}(s).$$

Moreover, the first occurrence of the prefix λ_{k+1} of $\sqrt{s}$ is at position $|\lambda_k|$ of s for all $k \geq 0$.
(B) If s is of type B, then

$$\sqrt{s} = T^{|\lambda_n|}(s).$$

Moreover, the first occurrence of the prefix λ_{k+1} with $0 \leq k \leq n-1$ is at position $|\lambda_k|$ of s, and the first occurrence of any prefix of $\sqrt{s}$ having length greater than $|\lambda_n|$ is at position $|\lambda_n|$ of s.
In particular, $\sqrt{s}$ is a Sturmian word with slope α.

The theorem only states where the prefixes λ_k of $\sqrt{s}$ occur for the first time. For the first occurrence of other prefixes we do not have a guaranteed location.

Acknowledgments. We thank our supervisors Juhani Karhumäki and Luca Zamboni for suggesting that the square root map might preserve the language of a Sturmian word.

References

1. Khinchin, A.Ya.: Continued Fractions. Dover Publications, New York (1997)
2. Berstel, J.: On the index of Sturmian words. In: Karhumäki, J., Maurer, H., Păun, G., Rozenberg, G. (eds.) Jewels Are Forever, pp. 287–294. Springer, Heidelberg (1999)
3. Cassels, J.W.S.: An Introduction to Diophantine Approximation. Cambridge Tracts in Mathematics and Mathematical Physics, vol. 45. Cambridge University Press, Cambridge (1957)
4. Damanik, D., Lenz, D.: Powers in Sturmian sequences. Eur. J. Comb. **24**, 377–390 (2003). doi:10.1016/S0304-3975(99)90294-3
5. Lothaire, M.: Algebraic Combinatorics on Words. Encyclopedia of Mathematics and Its Applications, vol. 90. Cambridge University Press, Cambridge (2002)

6. Peltomäki, J.: Characterization of repetitions in Sturmian words: a new proof. Inform. Process. Lett. **115**(11), 886–891 (2015). doi:10.1016/j.ipl.2015.05.011
7. Pytheas Fogg, N.: Substitutions in Dynamics, Arithmetics and Combinatorics. Lecture Notes in Mathematics, vol. 1794. Springer, Heidelberg (2002)
8. Saari, K.: Everywhere α-repetitive sequences and Sturmian words. Eur. J. Comb. **31**, 177–192 (2010). doi:10.1016/j.ejc.2009.01.004

Specular Sets

Valérie Berthé[1], Clelia De Felice[2], Vincent Delecroix[3], Francesco Dolce[4],
Julien Leroy[5], Dominique Perrin[4(✉)],
Christophe Reutenauer[6], and Giuseppina Rindone[4]

[1] CNRS, Université Paris 7, Paris, France
[2] Università Degli Studi di Salerno, Salerno, Italy
[3] CNRS, Université de Bordeaux, Bordeaux, France
[4] Université Paris Est, Champs-sur-marne, France
dominique.perrin@esiee.fr
[5] Université de Liège, Liège, Belgium
[6] Université du Québec à Montréal, Montreal, Canada

Abstract. We introduce specular sets. These are subsets of groups which
form a natural generalization of free groups. These sets are an abstract
generalization of the natural codings of interval exchanges and of linear
involutions. We prove several results concerning the subgroups generated
by return words and by maximal bifix codes in these sets.

1 Introduction

We have studied in a series of papers initiated in [1] the links between minimal
sets, subgroups of free groups and bifix codes. In this paper, we continue this
investigation in a situation which involves groups which are not free anymore.
These groups, named here specular, are free products of a free group and of a
finite number of cyclic groups of order two. These groups are close to free groups
and, in particular, the notion of a basis in such groups is clearly defined. It
follows from Kurosh's theorem that any subgroup of a specular group is specular.
A specular set is a subset of such a group which generalizes the natural codings
of linear involutions studied in [2].

The main results of this paper are Theorem 5.1, referred to as the Return
Theorem and Theorem 5.2, referred to as the Finite Index Basis Theorem. The
first one asserts that the set of return words to a given word in a uniformly
recurrent specular set is a basis of a subgroup of index 2 called the even subgroup.
The second one characterizes the monoidal bases of subgroups of finite index of
specular groups contained in a specular set S as the finite S-maximal symmetric
bifix codes contained in S. This generalizes the analogous results proved initially
in [1] for Sturmian sets and extended in [3] to the class of tree sets (this class
contains both Sturmian sets and interval exchange sets).

There are two interesting features of the subject of this paper.

In the first place, some of the statements concerning the natural codings of
interval exchanges and of linear involutions can be proved using geometric methods, as shown in a separate paper [2]. This provides an interesting interpretation

F. Manea and D. Nowotka (Eds.): WORDS 2015, LNCS 9304, pp. 210–222, 2015.
DOI: 10.1007/978-3-319-23660-5_18

of the groups playing a role in these natural codings (these groups are generated either by return words or by maximal bifix codes) as fundamental groups of some surfaces. The methods used here are purely combinatorial.

In the second place, the abstract notion of a specular set gives rise to groups called here specular. These groups are natural generalizations of free groups, and are free products of $\mathbb{Z}$ and $\mathbb{Z}/2\mathbb{Z}$. They are called *free-like* in [4] and appear at several places in [5].

The idea of considering recurrent sets of reduced words invariant by taking inverses is connected, as we shall see, with the notion of G-rich words of [6].

The paper is organized as follows. In Sect. 2, we recall some notions concerning words, extension graphs and bifix codes. In Sect. 3, we introduce specular groups, which form a family with properties very close to free groups. We prove properties of these groups extending those of free groups, like a Schreier's Formula (Formula (3.1)). In Sect. 4, we introduce specular sets. This family contains the natural codings of linear involutions without connection studied in [7]. We prove a result connecting specular sets with the family of tree sets introduced in [8] (Theorem 4.6). In Sect. 5, we prove several results concerning subgroups generated by subsets of specular groups. We first prove that the set of return words to a given word forms a basis of the even subgroup (Theorem 5.1 referred to as the Return Theorem). This is a subgroup defined in terms of particular letters, called even letters, that play a special role with respect to the extension graph of the empty word. We next prove the Finite Index Basis Theorem (Theorem 5.2).

2   Preliminaries

A set of words on the alphabet A and containing A is said to be *factorial* if it contains the factors of its elements. An *internal factor* of a word x is a word v such that $x = uvw$ with u, w nonempty.

Let S be a set of words on the alphabet A. For $w \in S$, we denote $L_S(w) = \{a \in A \mid aw \in S\}$, $R_S(w) = \{a \in A \mid wa \in S\}$ and $E_S(w) = \{(a,b) \in A \times A \mid awb \in S\}$. Further $\ell_S(w) = \mathrm{Card}(L_S(w))$, $r_S(w) = \mathrm{Card}(R_S(w))$, $e_S(w) = \mathrm{Card}(E_S(w))$. We omit the subscript S when it is clear from the context. A word w is *right-extendable* if $r(w) > 0$, *left-extendable* if $\ell(w) > 0$ and *biextendable* if $e(w) > 0$. A factorial set S is called *right-extendable* (resp. *left-extendable*, resp. *biextendable*) if every word in S is right-extendable (resp. left-extendable, resp. biextendable).

A word w is called *right-special* if $r(w) \geq 2$. It is called *left-special* if $\ell(w) \geq 2$. It is called *bispecial* if it is both left-special and right-special.

For $w \in S$, we denote

$$m_S(w) = e_S(w) - \ell_S(w) - r_S(w) + 1.$$

The word w is called *weak* if $m_S(w) < 0$, *neutral* if $m_S(w) = 0$ and *strong* if $m_S(w) > 0$.

We say that a factorial set S is *neutral* if every nonempty word in S is neutral. The *characteristic* of S is the integer $1 - m_S(\varepsilon)$. Thus a neutral set of characteristic 1 is such that all words (including the empty word) are neutral. This is what is called a neutral set in [8].

A set of words $S \neq \{\varepsilon\}$ is *recurrent* if it is factorial and if for any $u, w \in S$, there is a $v \in S$ such that $uvw \in S$. An infinite factorial set is said to be *uniformly recurrent* if for any word $u \in S$ there is an integer $n \geq 1$ such that u is a factor of any word of S of length n. A uniformly recurrent set is recurrent.

The *factor complexity* of a factorial set S of words on an alphabet A is the sequence $p_n = \operatorname{Card}(S \cap A^n)$. Let $s_n = p_{n+1} - p_n$ and $b_n = s_{n+1} - s_n$ be respectively the first and second order differences sequences of the sequence p_n.

The following result is from [9] (see also [10], Theorem 4.5.4).

Proposition 2.1 *Let S be a factorial set on the alphabet A. One has $b_n = \sum_{w \in S \cap A^n} m(w)$ and $s_n = \sum_{w \in S \cap A^n} (r(w) - 1)$ for all $n \geq 0$.*

Let S be a biextendable set of words. For $w \in S$, we consider the set $E(w)$ as an undirected graph on the set of vertices which is the disjoint union of $L(w)$ and $R(w)$ with edges the pairs $(a, b) \in E(w)$. This graph is called the *extension graph* of w. We sometimes denote $1 \otimes L(w)$ and $R(w) \otimes 1$ the copies of $L(w)$ and $R(w)$ used to define the set of vertices of $E(w)$.

If the extension graph $E(w)$ is acyclic, then $m(w) = 1 - c$, where c is the number of connected components of the graph $E(w)$. Thus w is weak or neutral.

A biextendable set S is called *acyclic* if for every $w \in S$, the graph $E(w)$ is acyclic. A biextendable set S is called a *tree set* of characteristic c if for any nonempty $w \in S$, the graph $E(w)$ is a tree and if $E(\varepsilon)$ is a union of c trees (the definition of tree set in [8] corresponds to a tree set of characteristic 1). Note that a tree set of characteristic c is a neutral set of characteristic c.

As an example, a Sturmian set is a tree set of characteristic 1 (by a Sturmian set, we mean the set of factors of a strict episturmian word, see [1]).

Let S be a factorial set of words and $x \in S$. A *return word* to x in S is a nonempty word u such that the word xu is in S and ends with x, but has no internal factor equal to x. We denote by $\mathcal{R}_S(x)$ the set of return words to x in S. The set of *complete return words* to $x \in S$ is the set $x\mathcal{R}_S(x)$.

Bifix Codes. A *prefix code* is a set of nonempty words which does not contain any proper prefix of its elements. A *suffix code* is defined symmetrically. A *bifix code* is a set which is both a prefix code and a suffix code (see [11] for a more detailed introduction).

Let S be a recurrent set. A prefix (resp. bifix) code $X \subset S$ is S-maximal if it is not properly contained in a prefix (resp. bifix) code $Y \subset S$. Since S is recurrent, a finite S-maximal bifix code is also an S-maximal prefix code (see [1], Theorem 4.2.2). For example, for any $n \geq 1$, the set $X = S \cap A^n$ is an S-maximal bifix code.

Let X be a bifix code. Let Q be the set of words without any suffix in X and let P be the set of words without any prefix in X. A *parse* of a word w with respect to a bifix code X is a triple $(q, x, p) \in Q \times X^* \times P$ such that $w = qxp$. We denote by $d_X(w)$ the number of parses of a word w with respect to X. The S-degree of X, denoted $d_X(S)$, is the maximal number of parses with respect to X of a word of S. For example, the set $X = S \cap A^n$ has S-degree n.

Let S be a recurrent set and let X be a finite bifix code. By Theorem 4.2.8 in [1], X is S-maximal if and only if its S-degree is finite. Moreover, in this case, a word $w \in S$ is such that $d_X(w) < d_X(S)$ if and only if it is an internal factor of a word of X.

3 Specular Groups

We consider an alphabet A with an involution $\theta : A \to A$, possibly with some fixed points. We also consider the group G_θ generated by A with the relations $a\theta(a) = 1$ for every $a \in A$. Thus $\theta(a) = a^{-1}$ for $a \in A$. The set A is called a *natural* set of generators of G_θ.

When θ has no fixed point, we can set $A = B \cup B^{-1}$ by choosing a set of representatives of the orbits of θ for the set B. The group G_θ is then the free group on B. In general, the group G_θ is a free product of a free group and a finite number of copies of $\mathbb{Z}/2\mathbb{Z}$, that is, $G_\theta = \mathbb{Z}^{*i} * (\mathbb{Z}/2\mathbb{Z})^{*j}$ where i is the number of orbits of θ with two elements and j the number of its fixed points. Such a group will be called a *specular group* of type (i, j). These groups are very close to free groups, as we will see. The integer $\mathrm{Card}(A) = 2i + j$ is called the *symmetric rank* of the specular group $\mathbb{Z}^{*i} * (\mathbb{Z}/2\mathbb{Z})^{*j}$. Two specular groups are isomorphic if and only if they have the same type. Indeed, the commutative image of a group of type (i, j) is $\mathbb{Z}^i \times (\mathbb{Z}/2\mathbb{Z})^j$ and the uniqueness of i, j follows from the fundamental theorem of finitely generated Abelian groups.

Example 3.1. Let $A = \{a, b, c, d\}$ and let θ be the involution which exchanges b, d and fixes a, c. Then $G_\theta = \mathbb{Z} * (\mathbb{Z}/2\mathbb{Z})^2$ is a specular group of symmetric rank 4.

By Kurosh's Theorem, any subgroup of a free product $G_1 * G_2 * \cdots * G_n$ is itself a free product of a free group and of groups conjugate to subgroups of the G_i (see [12]). Thus, we have, replacing the Nielsen-Schreier Theorem of free groups, the following result.

Theorem 3.1 *Any subgroup of a specular group is specular.*

It also follows from Kurosh's theorem that the elements of order 2 in a specular group G_θ are the conjugates of the j fixed points of θ and this number is thus the number of conjugacy classes of elements of order 2.

A word on the alphabet A is *θ-reduced* (or simply reduced) if it has no factor of the form $a\theta(a)$ for $a \in A$. It is clear that any element of a specular group is represented by a unique reduced word.

A subset of a group G is called *symmetric* if it is closed under taking inverses. A set X in a specular group G is called a *monoidal basis* of G if it is symmetric, if the monoid that it generates is G and if any product $x_1 x_2 \cdots x_m$ of elements of X such that $x_k x_{k+1} \neq 1$ for $1 \leq k \leq m-1$ is distinct of 1. The alphabet A is a monoidal basis of G_θ and the symmetric rank of a specular group is the cardinality of any monoidal basis (two monoidal bases have the same cardinality since the type is invariant by isomorphism).

If H is a subgroup of index n of a specular group G of symmetric rank r, the symmetric rank s of H is

$$s = n(r-2) + 2. \tag{3.1}$$

This formula replaces Schreier's Formula (which corresponds to the case $j = 0$). It can be proved as follows. Let Q be a Schreier transversal for H, that is, a set of reduced words which is a prefix-closed set of representatives of the right cosets Hg of H. Let X be the corresponding Schreier basis, formed of the paq^{-1} for $a \in A$, $p, q \in Q$ with $pa \notin Q$ and $pa \in Hq$. The number of elements of X is $nr - 2(n-1)$. Indeed, this is the number of pairs $(p, a) \in Q \times A$ minus the $2(n-1)$ pairs (p, a) such that $pa \in Q$ with pa reduced or $pa \in Q$ with pa not reduced. This gives Formula (3.1).

Any specular group $G = G_\theta$ has a free subgroup of index 2. Indeed, let H be the subgroup formed of the reduced words of even length. It has clearly index 2. It is free because it does not contain any element of order 2 (such an element is conjugate to a fixed point of θ and thus is of odd length).

A group G is called *residually finite* if for every element $g \neq 1$ of G, there is a morphism φ from G onto a finite group such that $\varphi(g) \neq 1$.

It follows easily by considering a free subgroup of index 2 of a specular group that any specular group is residually finite. A group G is said to be *Hopfian* if any surjective morphism from G onto G is also injective. By a result of Malcev, any finitely generated residually finite group is Hopfian (see [13], p. 197). Thus any specular group is Hopfian.

As a consequence, one has the following result, which can be obtained by considering the commutative image of a specular group.

Proposition 3.2 *Let G be a specular group of type (i, j) and let $X \subset G$ be a symmetric set with $2i + j$ elements. If X generates G, it is a monoidal basis of G.*

4   Specular Sets

We assume given an involution θ on the alphabet A generating the specular group G_θ. A *specular set* on A is a biextendable symmetric set of θ-reduced words on A which is a tree set of characteristic 2. Thus, in a specular set, the extension graph of every nonempty word is a tree and the extension graph of the empty word is a union of two disjoint trees.

The following is a very simple example of a specular set.

Example 4.1. Let $A = \{a, b\}$ and let θ be the identity on A. Then the set of factors of $(ab)^\omega$ is a specular set (we denote by x^ω the word x infinitely repeated).

The following result shows in particular that in a specular set the two trees forming $E(\varepsilon)$ are isomorphic since they are exchanged by the bijection $(a, b) \mapsto (b^{-1}, a^{-1})$.

Proposition 4.1 *Let S be a specular set. Let T_0, T_1 be the two trees such that $E(\varepsilon) = T_0 \cup T_1$. For any $a, b \in A$ and $i = 0, 1$, one has $(1 \otimes a, b \otimes 1) \in T_i$ if and only if $(1 \otimes b^{-1}, a^{-1} \otimes 1) \in T_{1-i}$*

Proof. Assume that $(1 \otimes a, b \otimes 1)$ and $(1 \otimes b^{-1}, a^{-1} \otimes 1)$ are both in T_0. Since T_0 is a tree, there is a path from $1 \otimes a$ to $a^{-1} \otimes 1$. We may assume that this path is reduced, that is, does not use consecutively twice the same edge. Since this path is of odd length, it has the form $(u_0, v_0, u_1, \ldots, u_p, v_p)$ with $u_0 = 1 \otimes a$ and $v_p = a^{-1} \otimes 1$. Since S is symmetric, we also have a reduced path $(v_p^{-1}, u_p^{-1}, \cdots, u_1^{-1}, v_0^{-1}, u_0^{-1})$ which is in T_0 (for $u_i = 1 \otimes a_i$, we denote $u_i^{-1} = a_i^{-1} \otimes 1$ and similarly for v_i^{-1}). Since $v_p^{-1} = u_0$, these two paths have the same origin and end. But if a path of odd length is its own inverse, its central edge has the form (x, y) with $x = y^{-1}$ a contradiction with the fact that the words of S are reduced. Thus the two paths are distinct. This implies that $E(\varepsilon)$ has a cycle, a contradiction.

The next result follows easily from Proposition 2.1.

Proposition 4.2 *The factor complexity of a specular set on the alphabet A is $p_n = n(k - 2) + 2$ for $n \geq 1$ with $k = \mathrm{Card}(A)$.*

Doubling Maps. We now introduce a construction which allows one to build specular sets.

A *transducer* is a graph on a set Q of vertices with edges labeled in $\Sigma \times A$. The set Q is called the set of states, the set Σ is called the *input* alphabet and A is called the *output* alphabet. The graph obtained by erasing the output letters is called the *input automaton* (with an unspecified initial state). Similarly, the *output automaton* is obtained by erasing the input letters.

Let $\mathcal{A}$ be a transducer with set of states $Q = \{0, 1\}$ on the input alphabet Σ and the output alphabet A. We assume that

1. The input automaton is a group automaton, that is, every letter of Σ acts on Q as a permutation,
2. The output labels of the edges are all distinct.

We define two maps $\delta_0, \delta_1 : \Sigma^* \to A^*$ corresponding to initial states 0 and 1 respectively. Thus $\delta_0(u) = v$ (resp. $\delta_1(u) = v$) if the path starting at state 0 (resp. 1) with input label u has output label v. The pair $\delta = (\delta_0, \delta_1)$ is called a *doubling map* on $\Sigma \times A$ and the transducer $\mathcal{A}$ a *doubling transducer*. The *image* of a set T on the alphabet Σ by the doubling map δ is the set $S = \delta_0(T) \cup \delta_1(T)$.

If $\mathcal{A}$ is a doubling transducer, we define an involution $\theta_\mathcal{A}$ as follows. For any $a \in A$, let (i, α, a, j) be the edge with input label α and output label a. We define

$\theta_A(a)$ as the output label of the edge starting at $1-j$ with input label α. Thus, $\theta_A(a) = \delta_i(\alpha) = a$ if $i+j = 1$ and $\theta_A(a) = \delta_{1-i}(\alpha) \neq a$ if $i = j$.

The *reversal* of a word $w = a_1a_2 \cdots a_n$ is the word $\tilde{w} = a_n \cdots a_2a_1$. A set S of words is closed under reversal if $w \in S$ implies $\tilde{w} \in S$ for every $w \in S$. As is well known, any Sturmian set is closed under reversal (see [1]). The proof of the following result can found in [7].

Proposition 4.3 *For any tree set T of characteristic 1 on the alphabet Σ, closed under reversal and for any doubling map δ, the image of T by δ is a specular set relative to the involution θ_A.*

We now give an example of a specular set obtained by a doubling map.

Example 4.2. Let $\Sigma = \{\alpha, \beta\}$ and let T be the Fibonacci set, which is the Sturmian set formed of the factors of the fixed point of the morphism $\alpha \mapsto \alpha\beta, \beta \mapsto \alpha$. Let δ be the doubling map given by the transducer $\mathcal{A}$ of Fig. 1 on the left.

Then θ_A is the involution θ of Example 3.1 and the image of T by δ is a specular set S on the alphabet $A = \{a, b, c, d\}$. The graph $E_S(\varepsilon)$ is represented in Fig. 1 on the right.

Note that S is the set of factors of the fixed point $g^\omega(a)$ of the morphism $g : a \mapsto abcab, b \mapsto cda, c \mapsto cdacd, d \mapsto abc$. The morphism g is obtained by applying the doubling map to the cube f^3 of the Fibonacci morphism f in such a way that $g^\omega(a) = \delta_0(f^\omega(\alpha))$.

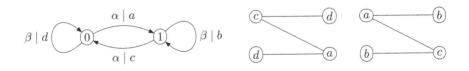

Fig. 1. A doubling transducer and the extension graph $E_S(\varepsilon)$.

Odd and Even Words. We introduce a notion which plays, as we shall see, an important role in the study of specular sets. Let S be a specular set. Since a specular set is biextendable, any letter $a \in A$ occurs exactly twice as a vertex of $E(\varepsilon)$, one as an element of $L(\varepsilon)$ and one as an element of $R(\varepsilon)$. A letter $a \in A$ is said to be *even* if its two occurrences appear in the same tree. Otherwise, it is said to be *odd*. Observe that if S is recurrent, there is at least one odd letter.

Example 4.3. Let S be the specular set of Example 4.2. The letters a, c are odd and b, d are even.

A word $w \in S$ is said to be *even* if it has an even number of odd letters. Otherwise it is said to be *odd*. The set of even words has the form $X^* \cap S$ where $X \subset S$ is a bifix code, called the *even code*. The set X is the set of even words without a nonempty even prefix (or suffix).

Proposition 4.4 *Let S be a recurrent specular set. The even code is an S-maximal bifix code of S-degree 2.*

Proof. Let us verify that any $w \in S$ is comparable for the prefix order with an element of the even code X. If w is even, it is in X^*. Otherwise, since S is recurrent, there is a word u such that $wuw \in S$. If u is even, then wuw is even and thus $wuw \in X^*$. Otherwise wu is even and thus $wu \in X^*$. This shows that X is S-maximal. The fact that it has S-degree 2 follows from the fact that any product of two odd letters is a word of X which is not an internal factor of X and has two parses. ∎

Example 4.4. Let S be the specular set of Example 4.2. The even code is $X = \{abc, ac, b, ca, cda, d\}$.

Denote by $\mathcal{T}_0, \mathcal{T}_1$ the two trees such that $E(\varepsilon) = \mathcal{T}_0 \cup \mathcal{T}_1$. We consider the directed graph $\mathcal{G}$ with vertices $0, 1$ and edges all the triples (i, a, j) for $0 \le i, j \le 1$ and $a \in A$ such that $(1 \otimes b, a \otimes 1) \in \mathcal{T}_i$ and $(1 \otimes a, c \otimes 1) \in \mathcal{T}_j$ for some $b, c \in A$. The graph $\mathcal{G}$ is called the *parity graph* of S. Observe that for every letter $a \in A$ there is exactly one edge labeled a because a appears exactly once as a left (resp. right) vertex in $E(\varepsilon)$.

Note that, when S is a specular set obtained by a doubling map using a transducer $\mathcal{A}$, the parity graph of S is the output automaton of $\mathcal{A}$.

Example 4.5. The parity graph of the specular set of Example 4.2 is the output automaton of the doubling transducer of Fig. 1.

The proof of the following result can be found in [7].

Proposition 4.5 *Let S be a specular set and let $\mathcal{G}$ be its parity graph. Let $S_{i,j}$ be the set of words in S which are the label of a path from i to j in the graph $\mathcal{G}$.*

(1) The family $(S_{i,j} \setminus \{\varepsilon\})_{0 \le i,j \le 1}$ is a partition of $S \setminus \{\varepsilon\}$.
(2) For $u \in S_{i,j} \setminus \{\varepsilon\}$ and $v \in S_{k,\ell} \setminus \{\varepsilon\}$, if $uv \in S$, then $j = k$.
(3) $S_{0,0} \cup S_{1,1}$ is the set of even words.
(4) $S_{i,j}^{-1} = S_{1-j,1-i}$.

A *coding morphism* for a prefix code X on the alphabet A is a morphism $f : B^* \to A^*$ which maps bijectively B onto X. Let S be a recurrent set and let f be a coding morphism for an S-maximal bifix code. The set $f^{-1}(S)$ is called a *maximal bifix decoding* of S.

The following result is the counterpart for uniformly recurrent specular sets of the main result of [14, Theorem 6.1] asserting that the family of uniformly recurrent tree sets of characteristic 1 is closed under maximal bifix decoding. The proof can be found in [7].

Theorem 4.6. *The decoding of a uniformly recurrent specular set by the even code is a union of two uniformly recurrent tree sets of characteristic 1.*

Palindromes. The notion of palindromic complexity originates in [15] where it is proved that a word of length n has at most $n + 1$ palindrome factors. A word of length n is full if it has $n + 1$ palindrome factors and a factorial set is *full* (or *rich*) if all its elements are full. By a result of [16], a recurrent set S closed under reversal is full if and only if every complete return word to a palindrome in S is a palindrome. It is known that all Sturmian sets are full [15] and also all natural codings of interval exchange defined by a symmetric permutation [17]. In [6], this notion was extended to that of H-fullness, where H is a finite group of morphisms and antimorphisms of A^* (an antimorphism is the composition of a morphism and reversal) containing at least one antimorphism. As one of the equivalent definitions of H-full, a set S closed under H is H-full if for every $x \in S$, every complete return word to the H-orbit of x is fixed by a nontrivial element of H (a complete return word to a set X is a word of S which has exactly two factors in X, one as a proper prefix and one as a proper suffix).

The following result connects these notions with ours. If δ is a doubling map, we denote by H the group generated by the antimorphism $u \mapsto u^{-1}$ for $u \in G_\theta$ and the morphism obtained by replacing each letter $a \in A$ by $\tau(a)$ if there are edges (i, b, a, j) and $(1 - i, b, \tau(a), 1 - j)$ in the doubling transducer. Actually, we have $H = \mathbb{Z}/2\mathbb{Z} \times \mathbb{Z}/2\mathbb{Z}$. The proof of the following result can be found in [7]. The fact that T is full generalizes the results of [15,17].

Proposition 4.7 *Let T be a recurrent tree set of characteristic 1 on the alphabet Σ, closed under reversal and let S be the image of T under a doubling map. Then T is full and S is H-full.*

Example 4.6. Let S be the specular set of Example 4.2. Since it is a doubling of the Fibonacci set (which is Sturmian and thus full), it is H-full with respect to the group H generated by the map σ taking the inverse and the morphism τ which exchanges a, c and b, d respectively. The H-orbit of $x = a$ is the set $X = \{a, c\}$ and $\mathcal{CR}_S(X) = \{ac, abc, ca, cda\}$.

All four words are fixed by $\sigma\tau$. As another example, consider $x = ab$. Then $X = \{ab, bc, cd, da\}$ and $\mathcal{CR}_S(X) = \{abc, bcab, bcd, cda, dab, dacd\}$. Each of them is fixed by some nontrivial element of H.

5   Subgroup Theorems

In this section, we prove several results concerning the subgroups generated by subsets of a specular set.

The Return Theorem. By [8, Theorem 4.5], the set of return words to a given word in a uniformly recurrent tree set of characteristic 1 containing the alphabet A is a basis of the free group on A. We will see a counterpart of this result for uniformly recurrent specular sets.

Let S be a specular set. The *even subgroup* is the group generated by the even code. It is a subgroup of index 2 of G_θ with symmetric rank $2(\mathrm{Card}(A) - 1)$ by (3.1). Since no even word is its own inverse (see Proposition 4.5), it is a free group. Thus its rank is $\mathrm{Card}(A) - 1$. The proof can be found in [7].

Theorem 5.1. *Let S be a uniformly recurrent specular set on the alphabet A. For any $w \in S$, the set of return words to w is a basis of the even subgroup.*

Note that this implies that $\mathrm{Card}(\mathcal{R}_S(x)) = \mathrm{Card}(A) - 1$.

Example 5.1. Let S be the specular set of Example 4.2. The set of return words to a is $\mathcal{R}_S(a) = \{bca, bcda, cda\}$. It is a basis of the even subgroup.

Finite Index Basis Theorem. The following result is the counterpart for specular sets of the result holding for uniformly recurrent tree sets of characteristic 1 (see [3, Theorem 4.4]). The proof can be found in [7].

Theorem 5.2. *Let S be a uniformly recurrent specular set and let $X \subset S$ be a finite symmetric bifix code. Then X is an S-maximal bifix code of S-degree d if and only if it is a monoidal basis of a subgroup of index d.*

Note that when X is not symmetric, the index of the subgroup generated by X may be different of $d_X(S)$.

Note also that Theorem 5.2 implies that for any uniformly recurrent specular set and for any finite symmetric S-maximal bifix code X, one has $\mathrm{Card}(X) = d_X(S)(\mathrm{Card}(A) - 2) + 2$. This follows actually also (under more general hypotheses) from Theorem 2 in [18].

The proof of the Finite Index Basis Theorem needs preliminary results which involve concepts like that of incidence graph which are interesting in themselves.

Saturation Theorem. The *incidence graph* of a set X, is the undirected graph $\mathcal{G}_X$ defined as follows. Let P be the set of proper prefixes of X and let Q be the set of its proper suffixes. Set $P' = P \setminus \{1\}$ and $Q' = Q \setminus \{1\}$. The set of vertices of $\mathcal{G}_X$ is the disjoint union of P' and Q'. The edges of $\mathcal{G}_X$ are the pairs (p, q) for $p \in P'$ and $q \in Q'$ such that $pq \in X$. As for the extension graph, we sometimes denote $1 \otimes P', Q' \otimes 1$ the copies of P', Q' used to define the set of vertices of $\mathcal{G}_X$.

Example 5.2. Let S be a factorial set and let $X = S \cap A^2$ be the bifix code formed of the words of S of length 2. The incidence graph of X is identical with the extension graph $E(\varepsilon)$.

Let X be a symmetric set. We use the incidence graph to define an equivalence relation γ_X on the set P of proper prefixes of X, called the *coset equivalence* of X, as follows. It is the relation defined by $p \equiv q \bmod \gamma_X$ if there is a path (of even length) from $1 \otimes p$ to $1 \otimes q$ or a path (of odd length) from $1 \otimes p$ to $q^{-1} \otimes 1$ in the incidence graph $\mathcal{G}_X$. It is easy to verify that, since X is symmetric, γ_X is indeed an equivalence. The class of the empty word ε is reduced to ε.

The following statement is the generalization to symmetric bifix codes of Proposition 6.3.5 in [1]. We denote by $\langle X \rangle$ the subgroup generated by X.

Proposition 5.3 *Let X be a symmetric bifix code and let P be the set of its proper prefixes. Let γ_X be the coset equivalence of X and let $H = \langle X \rangle$. For any $p, q \in P$, if $p \equiv q \bmod \gamma_X$, then $Hp = Hq$.*

We now use the coset equivalence γ_X to define the *coset automaton* $\mathcal{C}_X$ of a symmetric bifix code X as follows. The vertices of $\mathcal{C}_X$ are the equivalence classes of γ_X. We denote by $\hat{p}$ the class of p. There is an edge labeled $a \in A$ from s to t if for some $p \in s$ and $q \in t$ (that is $s = \hat{p}$ and $t = \hat{q}$), one of the following cases occurs (see Fig. 2):

(i) $pa \in P$ and $pa \equiv q \mod \gamma_X$
(ii) or $pa \in X$ and $q = \varepsilon$.

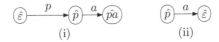

(i)   (ii)

Fig. 2. The edges of the coset automaton.

The proof of the following statement can be found in [7].

Proposition 5.4 *Let X be a symmetric bifix code, let P be its set of proper prefixes and let $H = \langle X \rangle$. If for $p, q \in P$ and a word $w \in A^*$ there is a path labeled w from the class $\hat{p}$ to the class $\hat{q}$, then $Hpw = Hq$.*

Let A be an alphabet with an involution θ. A directed graph with edges labeled in A is called *symmetric* if there is an edge from p to q labeled a if and only if there is an edge from q to p labeled a^{-1}. If $\mathcal{G}$ is a symmetric graph and v is a vertex of $\mathcal{G}$, the set of reductions of the labels of paths from v to v is a subgroup of G_θ called the subgroup *described* by $\mathcal{G}$ with respect to v.

A symmetric graph is called *reversible* if for every pair of edges of the form $(v, a, w), (v, a, w')$, one has $w = w'$ (and the symmetric implication since the graph is symmetric).

Proposition 5.5 *Let S be a specular set and let $X \subset S$ be a finite symmetric bifix code. The coset automaton $\mathcal{C}_X$ is reversible. Moreover the subgroup described by $\mathcal{C}_X$ with respect to the class of the empty word is the group generated by X.*

Prime Words with Respect to a Subgroup. Let H be a subgroup of the specular group G_θ and let S be a specular set on A relative to θ. The set of *prime* words in S with respect to H is the set of nonempty words in $H \cap S$ without a proper nonempty prefix in $H \cap S$. Note that the set of prime words with respect to H is a symmetric bifix code. One may verify that it is actually the unique bifix code X such that $X \subset S \cap H \subset X^*$.

The following statement is a generalization of Theorem 5.2 in [8] (Saturation Theorem). The proof can be found in [7].

Theorem 5.6. *Let S be a specular set. Any finite symmetric bifix code $X \subset S$ is the set of prime words in S with respect to the subgroup $\langle X \rangle$. Moreover $\langle X \rangle \cap S = X^* \cap S$.*

A Converse of the Finite Index Basis Theorem. The following is a converse of Theorem 5.2. For the proof, see [7].

Theorem 5.7. *Let S be a recurrent and symmetric set of reduced words of factor complexity $p_n = n(\mathrm{Card}(A) - 2) + 2$. If $S \cap A^n$ is a monoidal basis of the subgroup $\langle A^n \rangle$ for all $n \geq 1$, then S is a specular set.*

Acknowledgments. The authors thank Laurent Bartholdi and Pierre de la Harpe for useful indications. This work was supported by grants from Région Île-de-France and ANR project Eqinocs.

References

1. Berstel, J., De Felice, C., Perrin, D., Reutenauer, C., Rindone, G.: Bifix codes and sturmian words. J. Algebra **369**, 146–202 (2012)
2. Berthé, V., Delecroix, V., Dolce, F., Perrin, D., Reutenauer, C., Rindone, G.: Return words of linear involutions and fundamental groups. Ergodic Th. Dyn. Syst. (2015, To appear). http://arxiv.org/abs/1405.3529
3. Berthé, V., De Felice, C., Dolce, F., Leroy, J., Perrin, D., Reutenauer, C., Rindone, G.: The finite index basis property. J. Pure Appl. Algebra **219**, 2521–2537 (2015)
4. Bartholdi, L.: Growth of groups and wreath products (2014, preprint)
5. de la Harpe, P.: Topics in Geometric Group Theory. Chicago Lectures in Mathematics. University of Chicago Press, Chicago, IL (2000)
6. Pelantová, E., Starosta, Š.: Palindromic richness for languages invariant under more symmetries. Theoret. Comput. Sci. **518**, 42–63 (2014)
7. Berthé, V., De Felice, C., Delecroix, V., Dolce, F., Perrin, D., Reutenauer, C., Rindone, G.: Specular sets. In: Manea, F., Nowotka, D. (eds.) WORDS 2015. Lecture Notes in Computer Science, vol. 9304, pp. xx–yy. Springer, Heidelberg (2015). https://arXiv.org/abs/1505.00707
8. Berthé, V., De Felice, C., Dolce, F., Leroy, J., Perrin, D., Reutenauer, C., Rindone, G.: Acyclic, connected and tree sets. Monats. Math. **176**, 521–550 (2015)
9. Cassaigne, J.: Complexité et facteurs spéciaux. Bull. Belg. Math. Soc. Simon Stevin **4**(1), 67–88 (1997). Journées Montoises (Mons, 1994)
10. Berthé, V., Rigo, M.: Combinatorics, Automata and Number Theory. Encyclopedia of Mathematics and its Applications, vol. 135. Cambridge University Press, Cambridge (2010)
11. Berstel, J., Perrin, D., Reutenauer, C.: Codes and Automata. Cambridge University Press, Cambridge (2009)
12. Magnus, W., Karrass, A., Solitar, D.: Combinatorial Group Theory, 2nd edn. Dover Publications Inc, Mineola, NY (2004)
13. Lyndon, R.C., Schupp, P.E.: Combinatorial Group Theory. Classics in Mathematics. Springer, Heidelberg (2001). Reprint of the 1977 edition
14. Berthé, V., De Felice, C., Dolce, F., Leroy, J., Perrin, D., Reutenauer, C., Rindone, G.: Maximal bifix decoding. Discrete Math. **338**, 725–742 (2015)
15. Droubay, X., Justin, J., Pirillo, G.: Episturmian words and some constructions of de Luca and Rauzy. Theoret. Comput. Sci. **255**(1–2), 539–553 (2001)
16. Glen, A., Justin, J., Widmer, S., Zamboni, L.Q.: Palindromic richness. Eur. J. Combin. **30**(2), 510–531 (2009)

17. Baláži, P., Masáková, Z., Pelantová, E.: Factor versus palindromic complexity of uniformly recurrent infinite words. Theoret. Comput. Sci. **380**(3), 266–275 (2007)
18. Dolce, F., Perrin, D.: Enumeration formulæ in neutral sets (2015, submitted). http://arxiv.org/abs/1503.06081

On the Tree of Ternary Square-Free Words

Elena A. Petrova[✉] and Arseny M. Shur

Ural Federal University, Ekaterinburg, Russia
{elena.petrova,arseny.shur}@urfu.ru

Abstract. We contribute to the study of the set of ternary square-free words. Under the prefix order, this set forms a tree, and we prove two results on its structure. First, we show that non-branching paths in this tree are short: such a path from a node of nth level has length $O(\log n)$. Second, we prove that any infinite path in the tree has a lot of branching points: the lower density of the set of such points is at least $2/9$.

1 Introduction

Power-free words and languages are studied in lots of papers starting with the seminal work by Thue [14], but the number of challenging open problems is still quite big. One group of problems concerns the internal structure of power-free languages. Such a language can be viewed as a poset with respect to prefix, suffix, or factor order; by "internal structure" we mean the structure of these posets. In the case of prefix or suffix order, the diagram of the poset is a tree; since power-free languages are closed under reversal, these two trees are isomorphic. Each node of a prefix tree generates a subtree and is a common prefix of its descendants. In this paper, we study the prefix tree of the language of ternary square-free words.

There are several papers discussing the structure of (prefix) trees of k-power-free languages. For all these languages, the subtree generated by any word has at least one leaf [1]. Further, Currie and Shelton [2,4] showed that it is always decidable whether a given word generates finite or infinite subtree and, moreover, every infinite subtree branches infinitely often. All other results concern particular languages. Among these languages, the binary overlap-free language has the simplest structure due to its slow (polynomial) growth and immense connection to the Thue-Morse word. The finiteness problem for subtrees of the tree of binary overlap-free words was solved by Restivo and Salemi [8] in a constructive way (the general solution by Currie and Shelton is non-constructive); they also proved the existence of arbitrarily large finite subtrees in this tree. Furthermore, for this language it is decidable (in linear time!) whether the subtrees generated by two given words are isomorphic [12]. *En passant*, the deciding procedure checks finiteness of a given subtree in a way different from that of [8] and computes its depth if it is finite. The existence of finite subtrees of arbitrary

E.A. Petrova and A.M. Shur—Supported by the grant 13-01-00852 of the Russian Foundation of Basic Research.

F. Manea and D. Nowotka (Eds.): WORDS 2015, LNCS 9304, pp. 223–236, 2015.
DOI: 10.1007/978-3-319-23660-5_19

depth in the prefix trees of binary cube-free words and of ternary square-free words was proved, in a constructive way also, in [6,7] respectively.

Shelton and Sony [9,10] proved that in any infinite part of the tree T of ternary square-free words, infinite branches are not "too sparse": if all infinite paths from a node u contain the same node uv, then the word v has length at most $Kn^{2/3}$, where n is the length of u and K is an absolute constant. We continue this study, focusing on all, not only infinite, branches of T. Our main results are the following two theorems. By a *fixed context* of a square-free word u we mean a word v such that the subtree generated by u contains the node uv and has no branches above it.

Theorem 1. *A fixed context of a ternary square-free word w has length $O(\log |w|)$.*

Theorem 2. *Let w be an infinite ternary square-free word and let $b(n)$ be the number of branching points in the $(\lambda, w[1..n])$-path in the tree T. Then one has $\liminf_{n\to\infty} \frac{b(n)}{n} \geq \frac{2}{9}$.*

Theorem 1 clearly calls for an improvement over the result by Shelton and Sony, but such an improvement can be quite tricky. See Sect. 5 for the discussion of a more feasible extension of Theorem 1.

Theorem 2 demonstrates that T is reasonably "uniform" in terms of branching. Since the ternary square-free language grows exponentially at rate $\alpha \approx 1.30176$ (see [13]), a node in T has, on average, α children. The lower bound in Theorem 2 is not too far from this average.

After short preliminaries, we prove a series of lemmas clarifying the interaction of periodic factors inside square-free words. Then we prove Theorems 1 and 2 and finish the paper with a short discussion.

2   Definitions and Notation

We study words over the ternary alphabet $\{a, b, c\}$ and auxiliary finite alphabets. The empty word is denoted by λ. Finite [infinite] words over an alphabet Σ are treated as functions $w : \{1, \ldots, n\} \to \Sigma$ (resp., $w : \mathbb{N} \to \Sigma$). We often refer to the argument of this function as *time*, using passages like "u occurs in w earlier than v". We write $[i..j]$ for the range $i, i+1, \ldots, j$ of positive integers and $w[i..j]$ for the factor of the word w occupying this range; $w[i..i] = w[i]$ is just the ith letter of w. A word w has *period* $p < |w|$ if $w[1..|w|-p] = w[p+1..w]$.

Standard notions of factor, prefix, and suffix are used. A *square* is a nonempty word of the form uu. A word is *square-free* if it has no squares as factors; a square is *minimal* if it contains no other squares as factors. The only period of a minimal square uu is $|u|$. A *p-square* is a minimal square of period p. Words u and v are *conjugates* if $u = ws$, $v = sw$ for some words w, s; conjugacy is an equivalence relation. Linking up the ends of a finite word, we obtain a *circular word*. A circular word represents a conjugacy class in an obvious way. The factors of a circular word are just words, so one can speak about square-free circular words.

The set of ternary square-free words can be represented by the *prefix tree* T, in which each square-free word is a node, λ is the root, and any two nodes w and wx are connected by a directed edge labeled by the letter x. By a subtree in T we mean a tree consisting of a node and all its descendants in T. The *index* of a subtree S of T is the length of its root.

In a square-free word w, we say that a letter $w[i]$ is *fixed by a p-square* if $w[i-2p+1..i-1]y$ is a p-square, where y is the letter distinct from both $w[i]$ and $w[i-1]$. Note that $w[1..i]$ in this case is the only child of the word $w[1..i-1]$ in the tree T. A *fixed context* of w is any word v such that in the word wv each letter of v is fixed by some square.

3   Letters Fixed by Short Squares

We call squares with periods ≤ 17 *short* (the choice of the constant is caused by the properties discussed below). Our goal in this section is to give an upper bound on the number of letters in a square-free word that are fixed by short squares. Namely, we will prove the following

Lemma 1. *Suppose that $t, l \geq 1$ are integers and w is a square-free word such that either $|w| \geq t+l$ or w is infinite. Then the range $[t+1..t+l]$ contains at most $\frac{2l+5}{3}$ positions in which the letters of w are fixed by the squares with periods ≤ 17.*

To study the structure of square-free words, we use a special binary encoding which makes the reasoning about squares and periods more convenient. Any ternary word w containing no squares of letters can be encoded by a binary *Pansiot codeword* $\mathsf{cwd}(w)$ of length $|w| - 2$:

$$\mathsf{cwd}(w)[i] = \begin{cases} 0 & \text{if } w[i] = w[i+2], \\ 1 & \text{otherwise;} \end{cases} \quad \text{for example,} \quad \begin{aligned} w &= a\,b\,c\,b\,a\,c\,b\,c\ldots, \\ \mathsf{cwd}(w) &= 1\ 0\ 1\ 1\ 1\ 0 \quad \ldots. \end{aligned}$$

It is easy to see that a word can be reconstructed from its first two letters and its Pansiot codeword; thus, the codeword represents the encoded word up to a permutation of the alphabet. This type of encoding was proposed in [5] for bigger alphabets and studied in [11] for the ternary alphabet. Let us recall necessary facts from [11]. First, Pansiot's encoding can be naturally extended for circular words of length ≥ 3. The Pansiot codeword for a circular word (w) is the binary circular word $(\mathsf{cwd}(w))$ of length $|w|$ obtained as in following example:

$$\begin{matrix} & & c & & \\ b & & & b & \\ a & & & & a \\ & c & & & c \\ & & b & & \end{matrix} \quad \longrightarrow \quad \begin{matrix} & 0 & & 1 & \\ & & & & 1 \\ 1 & & & & 1 \\ 1 & & & & 0 \\ & & 1 & & \end{matrix}$$

The codewords of square-free (ordinary or circular) words are also called *square-free*. Square-free circular words and minimal squares are connected by

Proposition 1 ([11]). *The word w^2 is a minimal square if and only if (w) is square-free.*

There are no minimal squares with roots of lengths 5, 7, 9, 10, 14, and 17 over the ternary alphabet. For any other length ≤ 17, there is a unique square-free circular word; the Pansiot codewords of these words are as follows:

$$
\begin{array}{llllll}
(111) & 3 & (01110111) & 8 & (0101110110111) & 13 \\
(0101) & 4 & (01011010111) & 11 & (010110101101011) & 15 \\
(011011) & 6 & (010110111011) & 12 & (0101101101011011) & 16
\end{array}
\tag{1}
$$

To obtain the codeword of a minimal square from a square-free circular codeword (u) one has to double u or any of its conjugates and delete the two last letters. For example, construct such a codeword u' from the circular square-free codeword $(u) = (01110111)$. Taking the conjugate 11011101 of u, doubling it and deleting the two last letters we get $u' = 11011101\,110111$. The word u' has length 14 and period 8, so the decoded word will have length 16 and the same period 8. In addition, it is easy to see that any 2-square has the codeword 00.

Let w be a square-free word. If some letter $w[i]$ is fixed by a p-square, the factor $\mathsf{cwd}(w)[i-2p+1..i-2]$ differs from the codeword of a minimal square in the last letter only. We say that this letter $\mathsf{cwd}(w)[i-2]$ is also *fixed by a p-square* and call these square-free codewords *broken p-squares*. Below, the broken 8-square is obtained from the word u':

$$
\begin{aligned}
w &= \cdots a\,b\,c\,a\,c\,b\,a\,c\,a\,b\,c\,a\,c\,b\,a\,b\cdots \\
\mathsf{cwd}(w) &= \cdots 1\,1\,0\,1\,1\,1\,0\,1\,1\,1\,0\,1\,1\,0 \cdots
\end{aligned}
$$

A full list of short broken p-squares (see Table 1) can be obtained from (1): one builds the codewords of squares and changes the last letter; if the resulting word contains no codeword of a square as a suffix, it is a broken square.

For each codeword u we define its *fixing sequence* $\mathsf{fix}(u)$ to be a word of length $|u|$ over $\{2,3,4,6,8,11,12,13,15,16,?\}$ such that $\mathsf{fix}(u)[i] = p$ if $u[i]$ is fixed by a p-square with $p \leq 17$, and $\mathsf{fix}(u)[i] =?$ otherwise (i.e., if $u[i]$ is either non-fixed or fixed by a long square). For example,

$$
\begin{aligned}
u &= 1\,1\,0\,1\,1\,1\,0\,1\,1\,1\,0\,1\,1\,0 \\
\mathsf{fix}(u) &= ?\,?\,?\,2\,?\,?\,3\,2\,?\,?\,3\,2\,?\,8
\end{aligned}
\tag{2}
$$

Lemma 1 says exactly that

($\sharp$) for any codeword u and integers t, l, the factor $\mathsf{fix}(u)[t+1..t+l]$ contains at most $\frac{2l+5}{3}$ numbers.

Note that $\mathsf{fix}(u)[1] =?$ for any u. A sequence v of numbers is *maximal* if $?v?$ is a factor of some fixing sequence. The number of numbers in a factor v of a fixing sequence is denoted by $N(v)$. To prove ($\sharp$), we first find all maximal sequences.

Let $w[i]$ be fixed by a p-square, $u = \mathsf{cwd}(w)$, $i > 2p$. Then $u[i-2p+1..i-2]$ is a broken square and, in addition, $u[i-2p] \neq u[i-p]$. Otherwise, $u[i-2p..i-3]$ would encode a p-square. We call the broken square $u[i-2p+1..i-2]$ *regular*. Note that not all broken squares are regular: appending a letter from the left can produce a codeword of some square. The list of short regular broken squares is given in the first column of Table 1.

Table 1. All regular (left column) and non-regular (right column) short broken squares. In the left column, the letter written in boldface extends a broken square to the left. The codewords from the list (1) are underlined.

1 01	2	0110110111	6_2
0 1110	3	11101110111010	8_2
1 101011	4	011010111010111010110	11_1
0 1011011010	6_1	0101110101101011011011	11_2
0 11011101110110	8_1	111010110101011101010111	11_3
1 110111011010110111011	12_1	101011010111010111011	11_4
0 101101011011101011010111	12_2	101011011101101011011	12_4
1 110101101110110101011010	12_3	011011101101011011011010	12_5
1 110111011011010111011010	13_1	011101101011011011011	12_6
0 101101110101110110101111	13_2	101011101101110101110111	13_5
0 101110101101101110110110	13_3	011101101110101110110110	13_6
0 110101110110111010111010	13_4	010110101101011010101011011	15_2
1 110101101011010110101101011	15_1	011011010110110101101101011010	16_2
1 11011010110110101101011010110111	16_1	010110110101101101011011010111	16_3

All maximal sequences can be found as follows. If $v = \mathsf{fix}(u)[i..j]$ is a maximal sequence, then u contains a broken square from the list in Table 1 ending in each position of the range $[i..j]$. We search through this list in the length-increasing order. Processing a broken p-square u, we search for its latest letter which is not fixed by a q-square with $q \leq p$ in any occurrence of u in a square-free codeword. After this, we extend u to the right while the added letters are fixed by squares of period $\leq p$. If both processes are finite for each broken p-square, the search gives us the full list of maximal sequences. It is the case, and the list is

$$
\begin{array}{llllllll}
2 & 6_2\,3\,2 & 3\,2\,11_2 & 2\,4\,12_2\,3\,2 & 2\,12_6 & 3\,2\,13_4 & 2\,15_2 \\
3\,2 & 8_1\,2 & 2\,4\,11_3\,3\,2 & 2\,12_3\,2\,4 & 2\,13_1\,2\,4 & 13_5\,3\,2 & 16_1\,3\,2 \\
2\,4 & 3\,2\,8_2\,2\,4 & 2\,11_4 & 2\,12_4 & 3\,2\,13_2\,2\,4 & 13_6 & 2\,16_2\,2\,4 \\
2\,6_1\,2\,4 & 2\,4\,11_1\,2 & 12_1\,3\,2 & 3\,2\,12_5\,2\,4 & 2\,4\,13_3\,2 & 2\,4\,15_1\,3\,2 & 2\,4\,16_3\,3\,2
\end{array} \quad (3)
$$

The argument is similar for all words, so we consider just one of them, 12_2:

$$u = 101101011011101101\underline{010}\overline{111} \leftarrow 01$$

We see two broken squares (underlined and overlined) ending in the two penultimate positions of u, but no such square matches the previous position. Further, u has a fixed context 01, giving us two more numbers in the maximal sequence, but the next position after these two cannot be fixed by a short square. So we finally obtain the maximal sequence $2\,4\,12_2\,3\,2$.

Now we finalize the proof of ($\sharp$) and then of Lemma 1. Let us take a square-free codeword u and a factor $v = \mathsf{fix}(u)[t+1..t+l]$. Since we want to bound $N(v)$ from above, we assume w.l.o.g. that v begins and ends with maximal sequences, i.e., $v = v_1?^{j_1}v_2?^{j_2}\cdots v_k$, where each v_i is a maximal sequence and each $j_i \geq 1$. The proof is by induction on k. The base case is $k = 1$. Here v is a maximal sequence itself, $l \leq 5$ by (3) and so $N(v) = l = \frac{2l+l}{3} \leq \frac{2l+5}{3}$, as required.

For the inductive step, let us consider v_k. If $|v_k| \leq 2$, then the word $v' = v_1?^{j_1}v_2?^{j_2}\cdots v_{k-1}$ satisfies $N(v') = N(v) - 2$ and $|v'| \leq l - 3$. By the inductive assumption, $N(v) = 2 + N(v') \leq 2 + \frac{2(l-3)+5}{3} = \frac{2l+5}{3}$. Let $|v_k| > 2$. Nineteen candidates for v_k satisfy this condition, see (3). Each of them encodes some suffix of the codeword $u[1..l]$ (the whole word $u[1..l]$ in case if a non-regular broken square presents). From this factor of u one can restore several previous elements of the word v, using two rules: (i) if u contains a broken p-square from Table 1, assign p to the last position of this factor; (ii) if no word from Table 1 can end in a given position of u, assign the questionmark to it. The aim is to reconstruct a factor of the fixing sequence which ends with v_k and has at most $2/3$ of positions occupied by numbers. The results are gathered in Table 2. The aim is achieved for all words v_k except for two; in these two special cases, exclamation marks indicate the positions where we need questionmarks to reach the desired $2/3$ bound, but cannot guarantee them.

If $v_k = 2\,6_1\,2\,4$ [resp., $v_k = 2\,4\,15_1\,3\,2$], then there are two broken squares, 11_4 and 15_2 [resp., one broken square 11_1], that can end at the position of "!". All these squares are not regular, so we know the whole codeword $u[1..l]$ in each case and prove ($\sharp$) by a direct check of the corresponding rows of Table 2.

After excluding these special cases, we can replace exclamation marks by questionmarks. Let $\hat{v}$ be the reconstructed part of $\mathsf{fix}(u)$. If $\hat{v}$ contains v, then the inequality $N(v) \leq \frac{2l+5}{3}$ is checked directly. If not, then $\hat{v} =?^{j_i}v_{i+1}j^{j_{i+1}}\cdots v_k$. Let $v = v'\hat{v}$. We have $N(\hat{v}) \leq 2|\hat{v}|/3$ from Table 2 and $N(v') \leq \frac{2|v'|+5}{3}$ by inductive assumption. Hence, $N(v) \leq \frac{2l+5}{3}$. The inductive step is proved, and so are ($\sharp$) and Lemma 1.

4   Long Squares

Lemma 2. *Suppose that $t, l \geq 1$, $p, q \geq 2$ are integers, w is a word of length $t+l$ such that $w[1..t+l-1]$ is square-free, the letter $w[t]$ is fixed by a p-square, and w ends with a q-square. Then one of the following three conditions is satisfied:*

(1) $2q \geq 2p + l$ and $q \geq 2p$;
(2) $p + l < 2q < 2p + l$ and $q \leq l$;
(3) $2q \leq p + l$.

Proof. Let $UxUy$ be the suffix of $w[1..t]$ of length $2p$, where $y = w[t] \neq w[t-p] = x$, and let XX be the suffix of w of length $2q$. Condition 1 means that XX begins

Table 2. Reconstructed fragments of fixing sequences. Maximal sequence (1st column) defines a factor of the codeword (2nd column, top) which, in turn, determines a factor of the fixing sequence (2nd column, bottom); the ratio between the number of numbers in the latter factor and its length (3rd column) is less or equal to 2/3. In two special cases, '!' can be replaced either by the period of a non-regular broken square or by '?'.

Maximal sequence	Codeword and a suffix of its fixing sequence	Share of short periods
$2\,6_1\,24$	$! \in \{11_4, 15_2, ?\}$ 0 1 0 1 1 0 1 1 0 1 0 1 1 ! ? 2 6 2 4	**2/3**
$6_2\,32$	0 1 1 0 1 1 0 1 1 1 0 1 ? 2 ? ? 2 ? ? ? 2 ? 6 3 2	1/2
$3\,2\,8_2\,24$	1 1 1 0 1 1 1 0 1 1 1 0 1 0 1 1 ? ? ? 3 2 ? ? 3 2 ? ? 3 2 8 2 4	9/16
$24\,11_1\,2$	0 1 1 0 1 0 1 1 1 0 1 0 1 1 0 1 0 1 1 1 0 1 ? 2 ? ? 2 ? 2 4 ? 3 2 ? 2 4 ? 2 ? 2 4 11 1 2	13/21
$3\,2\,11_2$	0 1 0 1 1 1 0 1 0 1 1 0 1 0 1 1 1 0 1 1 ? 2 ? 2 4 ? 3 2 ? 2 4 ? 2 ? 2 4 ? 3 2 11	13/22
$24\,11_3\,32$	1 1 1 0 1 0 1 1 0 1 0 1 1 1 0 1 0 1 1 1 0 1 ? ? ? 3 2 ? 2 4 ? 2 ? 2 4 ? 3 2 ? 2 4 11 3 2	7/12
$12_1\,32$	1 1 1 0 1 1 1 0 1 1 0 1 0 1 1 0 1 1 1 0 1 1 1 0 1 ? ? 3 2 ? 12 3 2	5/8
$24\,12_2\,32$	0 1 0 1 1 0 1 0 1 1 0 1 1 1 0 1 1 0 1 0 1 1 1 0 1 ? ? 2 ? 2 4 12 3 2	2/3
$2\,12_3\,24$	1 1 1 0 1 0 1 1 0 1 1 1 0 1 1 0 1 0 1 1 0 1 0 1 1 ? ? 2 ? 2 4 ? 2 12 2 4	7/11
$3\,2\,12_5\,24$	0 1 1 0 1 1 1 0 1 1 0 1 0 1 1 0 1 1 1 0 1 0 1 1 ? 2 ? ? 2 ? ? 3 2 ? ? 2 ? 2 4 ? 2 ? ? 3 2 12 2 4	13/24
$2\,13_1\,24$	1 1 0 1 1 1 0 1 1 0 1 1 1 0 1 0 1 1 1 0 1 1 0 1 0 1 1 ? ? 2 13 2 4	2/3
$3\,2\,13_2$	0 1 0 1 1 0 1 1 1 0 1 0 1 1 1 0 1 1 0 1 1 1 0 1 1 ? ? 3 2 13	3/5
$24\,13_3\,2$	0 1 0 1 1 1 0 1 0 1 1 1 0 1 1 0 1 1 1 0 1 0 1 1 0 1 ? ? 3 2 ? 2 4 13 2	2/3
$3\,2\,13_4\,24$	0 1 1 0 1 0 1 1 1 0 1 1 0 1 1 1 0 1 0 1 1 1 0 1 0 1 1 ? 2 ? ? 3 2 ? 2 4 ? 3 2 13 2 4	5/8
$13_5\,32$	1 0 1 0 1 1 1 0 1 1 0 1 1 1 0 1 0 1 1 1 0 1 1 1 0 1 ? ? 2 ? 2 4 ? 3 2 ? ? 2 ? ? 3 2 ? 2 4 ? 3 2 ? 13 3 2	15/26
$24\,15_1\,32$	1 1 1 0 1 0 1 1 0 1 0 1 1 0 1 0 1 1 0 1 0 1 1 0 1 0 1 1 1 0 1 $! \in \{11_1, ?\}$! 2 ? 2 4 ? 2 ? 2 4 ? 2 ? 2 4 15 3 2	**2/3**
$16_1\,32$	1 1 0 1 1 0 1 0 1 1 0 1 1 0 1 0 1 1 0 1 1 0 1 0 1 1 0 1 1 1 0 1 ? 2 ? 16 3 2	2/3
$2\,16_2\,24$	0 1 1 0 1 1 0 1 0 1 1 0 1 1 0 1 0 1 1 0 1 1 0 1 0 1 1 0 1 0 1 1 ? 2 ? ? 2 ? ? 2 ? 2 4 ? 2 ? ? 2 ? 2 4 ? 2 ? ? 2 ? 2 16 2 4	17/32
$24\,16_3\,32$	0 1 0 1 1 0 1 1 0 1 0 1 1 0 1 1 0 1 0 1 1 0 1 1 0 1 0 1 1 1 0 1 ? 2 ? 2 4 ? 2 ? ? 2 ? 2 4 ? 2 ? ? 2 ? 2 4 ? 2 ? ? 2 ? 2 4 16 3 2	9/16
$2\,11_4$	1 0 1 0 1 1 0 1 0 1 1 1 0 1 0 1 1 0 1 1 ? ? 2 ? 2 4 ? 2 ? 2 4 ? 3 2 ? 2 4 ? 2 11	3/5
$2\,15_2$	0 1 0 1 1 0 1 0 1 1 0 1 0 1 1 0 1 0 1 1 0 1 0 1 1 0 1 1 ? 2 ? 2 4 ? 2 ? 2 4 ? 2 ? 2 4 ? 2 ? 2 4 ? 2 ? 2 4 ? 2 15	17/28

in w at the same time as $UxUy$, or earlier. Two cases are possible depending on whether the second X begins earlier or later than Uy (see Fig. 1 a,b; if the second X begins outside $UxUy$, then the required inequality $q \geq 2p$ holds trivially). The case where these two words begin at the same time is attached to case b (so in each case the factor u can be empty).

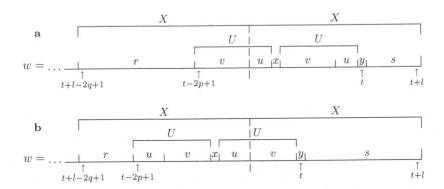

Fig. 1. Mutual location of factors under the condition $2q \geq 2p + l$ (Lemma 2).

In case **a** we have $U = vu$, $X = rv = uxvuys$. If $|r| \leq |uxvu|$, then the factor U of the first X touches or overlaps the occurrence of U in the middle of XX; this contradicts square-freeness of $w[1..t+l-1]$. Hence, $|r| \geq |uxvuy|$. So $q = |X| = |r| + |v| \geq |uxvuy| + |v| = |UxUy| = 2p$, as required.

Case **b** is similar: $U = uv$, $X = ruvxu = vys$, and if $|r| \leq |v|$, then the factor U of the second X touches or overlaps the occurrence of U in the middle of XX. Hence $|r| \geq |vy|$, and so $q = |X| = |r| + |uvxu| \geq |vy| + |uvxu| = |UxUy| = 2p$.

Now turn to condition 2. We must prove that if XX begins inside Ux, then the first X ends not earlier than $UxUy$. To do this, we rule out the only alternative: the first X ends inside Uy. Let us study this alternative. If $p = q$, then $x = y$, which is impossible. So we consider the cases $q < p$ and $q > p$ (see Fig. 2a,b). In case **a** we have $U = ruv$, $X = vxr = uvys$. Then $UxUy = ruvxr\,uvy = ru\,uvys\,uvy$, contradicting square-freeness. In case **b** X begins and ends with u, implying that XX contains uu and so is not a minimal square. This contradiction finishes the proof.   □

Lemma 3. *Suppose that w is a square-free word, $t > 0$, and the letter $w[t]$ is fixed by a p-square. If for some $t' > t$ the letter $w[t']$ is fixed by a q-square, then the point (t', q) of the Euclidean plane does not lie inside the polygon bounded by the lines*

$$y = 2p, \quad y = x - t, \quad y = \frac{x + p - t}{2}, \qquad (4)$$

where x is the time axis and y is the period axis.

The statement of the lemma is illustrated by Fig. 3; the impossible positions for the pair (t', q) are marked by crosses. In what follows, we refer to the polygon formed by the lines (4) as the (t, p)-*polygon*, and to the three lines (4) as its *upper*, *right*, and *lower boundaries*, respectively.

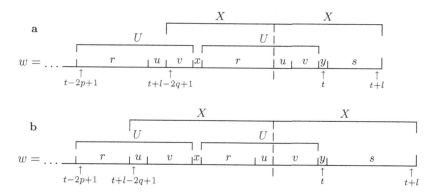

Fig. 2. Mutual location of factors under the condition $p + l < 2q < 2p + l$ (Lemma 2).

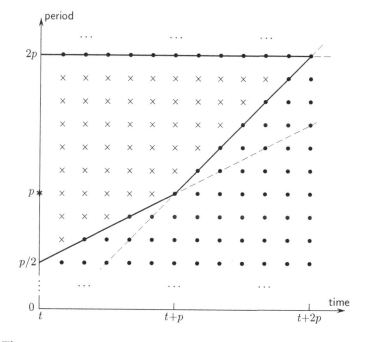

Fig. 3. The restrictions on fixing the letters after the letter $w[t]$ was fixed by a p-square. The picture is drawn for $p = 6$. Possible [resp., impossible] pairs (time, period) are marked with dots [resp., crosses]. The borders of the polygon of impossible squares are given by (4).

Proof. The three boundaries of a (t, p)-polygon correspond to the three restrictions on the location of squares given in Lemma 2 for $l = t' - t$. Consider the q-square XX fixing the position $w[t']$. Then $XX = (w[t'-2q+1..t'-q])^2$, while the position t is fixed by the square $(w[t-2p+1..t-p])^2$. If $t'-2q+1 \leq t-2p+1$, then $q \geq 2p$ by Lemma 2(1), so the point (t', q) is not lower then the upper boundary. If $t-2p+1 < t'-2q+1 < t-p+1$, then $t'-q \geq t$ by Lemma 2(2), so (t', q) cannot be on the left from the right boundary. Finally, if $t'-2q+1 \geq t-p+1$, (case 3 of Lemma 2), then $q \leq \frac{t'+p-t}{2}$, i.e., (t', q) is not higher than the lower bound. Thus, the point (t', q) is not in the interior of the (t, p)-polygon in all three cases. □

Let $p \geq 2$ be an integer. A sequence $(t_0, q_0), (t_1, q_1), \ldots, (t_s, q_s)$ of pairs of positive integers is a *p-schedule* if (i) $t_0 < t_1 < \cdots < t_s$, $p \leq q_i < 2p$ for all $i = 0, \ldots, s$ and (ii) for any $j < i$, the pair (t_i, q_i) is not inside the (t_j, q_j)-polygon. For convenience, we still refer to t's and q's as "times" and "periods" respectively. Note that any subsequence of a p-schedule is also a p-schedule.

Lemma 4. *Let* $(t_0, q_0), \ldots, (t_s, q_s)$ *be a p-schedule. Then* $t_s - t_0 > (s-1)p$.

Proof. We call a p-schedule $(t_0, q_0), \ldots, (y_s, q_s)$ *tight* if for any $i > 0$, the point (t_i-1, q_i) is inside the (t_j, q_j)-polygon for some $j < i$. Thus, in a tight schedule each period q_i is used immediately when it becomes available. If the schedule $(t_0, q_0), \ldots, (t_s, q_s)$ is not tight, then for some i the point (t_i-1, q_i) is not inside the polygons of the previous points. Hence

$$(t_0, q_0), \ldots, (t_{i-1}, q_{i-1}), (t_i-1, q_i), (t_{i+1}, q_{i+1}), \ldots, (t_s, q_s)$$

is a schedule. Indeed, for each $j > i$ the point (t_j, q_j) is not inside the polygons of the previous points because one of these polygons moved one unit left while all others remained in place. Using this "tightening" procedure repeatedly, we get the following claim:

(*) for any p-schedule $(t_0, q_0), (t_1, q_1), \ldots, (t_s, q_s)$, there is a tight p-schedule $(t_0, q_0), (t'_1, q_1), \ldots, (t'_s, q_s)$ with the same sequence of periods and such that $t'_s \leq t_s$.

By (*), it is enough to verify the statement of the lemma for tight p-schedules; so the p-schedules below are assumed tight. Note that the upper boundary of any (t_i, q_i)-polygon is outside the range $[p..2p-1]$, so we look at the other two boundaries only. A closer examination of (4) and Fig. 3 reveals an important feature: the boundaries of a (t_{i+1}, q_{i+1})-polygon never intersect the boundaries of the (t_i, q_i)-polygon. Indeed, the right boundary moves to the right with time, independently of the period; and the lower boundary also moves to the right: for the (t_i, q_i)-polygon, the line forming the lower boundary takes the value $y = p$ for some $x \leq t_{i+1}$, and for the (t_{i+1}, q_{i+1})-polygon such a line takes this value for some $x > t_{i+1}$. The observed feature implies another useful property: for any $i > 0$, the point (t_{i+1}, q_{i+1}) of a tight p-schedule lies *on the boundary of* *the* (t_i, q_i)-*polygon.* If $q_{i+1} \geq q_i$ [resp., $q_{i+1} \leq q_i$], this is the right [resp., lower] boundary. So we have by (4)

$$t_{i+1} = \begin{cases} t_i + q_{i+1}, & \text{if } q_{i+1} \geq q_i, \\ t_i + 2q_{i+1} - q_i, & \text{if } q_{i+1} \leq q_i. \end{cases} \tag{5}$$

Now let us consider three successive points, $(t_i, q_i), (t_{i+1}, q_{i+1}), (t_{i+2}, q_{i+2})$, in a p-schedule. We will show that (t_{i+1}, q_{i+1}) can be replaced by the point (t', p), where $t' = t_i + 2p - q_i$, and the new sequence will still be a p-schedule (not necessarily tight, but we can then tighten it). By (5), the point (t', p) lies on the boundary of the (t_i, q_i)-polygon. Due to the non-intersection of the boundaries of polygons for different points in a p-schedule, it remains to show that the point (t'', q_{i+2}) on the boundary of the (t', p)-polygon satisfies $t'' \leq t_{i+2}$. By (5), $t'' = t' + q_{i+2} = t_i + q_{i+2} - q_i + 2p$. For t_{i+2}, four cases are to be considered:

$q_i \leq q_{i+1} \leq q_{i+2}: \quad t_{i+2} = t_i + q_{i+2} + q_{i+1} \geq t'', \qquad$ since $q_i + q_{i+1} \geq 2p$;

$q_i \leq q_{i+1} \geq q_{i+2}: \quad t_{i+2} = t_i + 2q_{i+2} \geq t'', \qquad$ since $q_i + q_{i+2} \geq 2p$;

$q_i \geq q_{i+1} \leq q_{i+2}: \quad t_{i+2} = t_i + q_{i+2} + 2q_{i+1} - q_i \geq t'', \qquad$ since $q_{i+1} \geq p$;

$q_i \geq q_{i+1} \geq q_{i+2}: \quad t_{i+2} = t_i + 2q_{i+2} + q_{i+1} - q_i \geq t'', \qquad$ since $q_{i+1} + q_{i+2} \geq 2p$.

Thus we have proved that the proposed replacement leads to a valid p-schedule. Using such replacements repeatedly and tightening the p-schedule after each replacement, we replace all periods $q_1, \ldots, q_{s-1}$ with p. After this, we also replace q_s with p (this moves a point on the boundary of a polygon down, i.e., outside the polygon) and tighten the resulting p-schedule. Finally, we have a tight p-schedule of the form

$$(t_0, q_0), (\hat{t}_1, p), \ldots, (\hat{t}_s, p),$$

where $\hat{t}_s \leq t_s$. By (5), $t_{i+1} = t_i + p$ for each $i \geq 1$. Hence, $t_s - t_0 > \hat{t}_s - \hat{t}_1 = (s-1)p$.

Lemmas 3 and 4 readily imply the following result on the number of positions fixed by squares from a given range.

Lemma 5. *Suppose that* $t, l \geq 1$, $p \geq 2$ *are integers, and* w *is a square-free word such that either* $|w| \geq t+l$ *or* w *is infinite. Then the range* $[t+1..t+l]$ *contains at most* $1 + \lceil (l-1)/p \rceil$ *positions in which the letters of* w *are fixed by squares with periods in the range* $[p..2p-1]$.

Proof (of Theorem 1). Let v be the fixed context of w, $|w| = n$, $|v| = l$. Thus, the square-free word wv ends with l fixed letters. By Lemma 1, at most $\frac{2l+5}{3}$ of these positions are fixed by short squares (with periods ≤ 17). Let us partition the range $[18..\lfloor \frac{n+l}{2} \rfloor]$ of all possible longer periods into ranges of the form $[p..2p-1]$ starting from the left (the last range can be incomplete). The number of ranges is the minimal number k such that $2^k \cdot 18 - 1 \geq \lfloor \frac{n+l}{2} \rfloor$, i.e., $k = \lceil \log \frac{1}{18} \lfloor \frac{n+l+2}{2} \rfloor \rceil$. According to Lemma 5, the periods from the ith smallest range fix less than $2 + \frac{l-2}{18 \cdot 2^{i-1}}$ positions. Since the total number of fixed positions is l, we have

$$l \leq \frac{2l}{3} + \frac{5}{3} + 2k + \frac{l-2}{18}\left(\frac{1 - 1/2^k}{1 - 1/2}\right) = \frac{7l}{9} + \frac{13}{9} + 2k - \frac{l-2}{18 \cdot 2^{k-1}}. \tag{6}$$

Observing that $k \leq \lceil \log \frac{n+l+2}{36} \rceil$, $2^{k-1} < \frac{n+l}{36}$, we get from (6)

$$\frac{2l}{9} < 2 \left\lceil \log \frac{n+l+2}{36} \right\rceil + \frac{13}{9} - \frac{2(l-2)}{n+l} \text{ and thus } l < 9 \left\lceil \log \frac{n+l+2}{36} \right\rceil + \frac{13}{2} - \frac{9(l-2)}{n+l}.$$
$$\tag{7}$$

If $l \geq n$, then (7) implies $l < 9 \lceil \log \frac{l+1}{18} \rceil + 2 + \frac{9}{l}$, but this inequality clearly has no integer solutions. Therefore $l < n$ for any n, and we have, from (7),

$$l < 9 \left\lceil \log \frac{n}{18} \right\rceil + \frac{13}{2} < 9 \log \frac{n}{9} + \frac{13}{2} < 9 \log n - \frac{57}{2} + \frac{13}{2} = 9 \log n - 22,$$

whence the result[1].

Proof (of Theorem 2). Assume to the contrary that $\liminf_{n \to \infty} \frac{b(n)}{n} < \frac{2}{9}$. Then there exist $\alpha < 2/9$, $\{l_i\}_1^\infty$ such that $b(l_i) \leq \alpha l_i$. Consider the $(\lambda, w[1..l_i])$-path. The letters in it are partitioned in three sets: fixed by short squares, fixed by long squares, and non-fixed. Lemmas 1 and Lemma 5 and our assumption give upper bounds for the numbers of letters in each part. Similar to (6), (7), we get the following formula (recall that now we have $n = 0$):

$$l_i < \frac{2l_i + 5}{3} + 2 \left\lceil \log \frac{l_i + 2}{36} \right\rceil + \frac{l_i - 2}{9} + \alpha l_i.$$

Then $(2/9 - \alpha)l_i < 2 \log l_i - O(1)$. This inequality should hold for any l_i, but it has only finitely many integer solutions. This contradiction proves the theorem.

5   Discussion

A weaker version of Theorem 1, with $O(\log n)$ replaced by $O(n^{2/3})$, is a particular case of the result by Shelton and Sony [9,10], mentioned in the introduction[2]: *if the tree T contains a (w,wv)-path with $|w| = n$ and $|v| \geq Kn^{2/3}$, where K is a constant, then an infinite branch is attached to some internal point of this path.* An important corollary of this result concerns finite subtrees of T: the depth of any finite subtree of index n is $O(n^{2/3})$. We suggest that a much stronger result holds for subtrees.

Conjecture 1. In T, the size of any finite subtree of index n is $O(\log n)$.

Note that Conjecture 1 cares about the *size* (the number of nodes), not the *depth*, of a subtree: the result by Shelton and Sony gives only a superpolynomial upper bound on the size of these subtrees. The statement of the conjecture holds for finite paths as a corollary of Theorem 1. Proposition 2 below gives more support to Conjecture 1, showing that indices of even very small finite subtrees are big enough. We call a subtree S with the root w *regular* if T contains infinitely many isomorphic copies of S with roots of the form vw. The subtrees that are

---

[1] For big n, the bound can be lowered to $8.5 \log n - 28$ by some additional argument.
[2] The original proof contained a flaw, later fixed by Currie (see [2,3]).

not regular do not count in asymptotic studies: they appear only close to the root of T, cannot be parts of bigger finite subtrees, and so on. For example, the word *abacaba* is a root of a one-node non-regular finite subtree, which appears in T only once; on the other hand, a one-node tree with the root *abcacbabcac* is regular, because this root can be preceded by infinitely many words ending with *bac*, each of them generating the same subtree.

Proposition 2. *The lower bounds for the indices of regular finite subtrees with at most three nodes are as indicated in Fig. 4.*

The proof is by an exhaustive search.

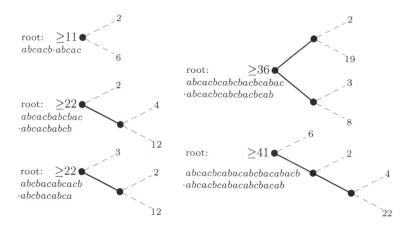

Fig. 4. Minimal indices of small regular subtrees. Bounds on the length of roots and the shortest possible roots are indicated. For reader's convenience, long roots are written in two lines. The numbers at the ends of dotted lines are periods of squares appearing when appending a letter to a word in the tree.

Acknowledgement. We thank the anonymous referee for useful comments.

References

1. Bean, D.A., Ehrenfeucht, A., McNulty, G.: Avoidable patterns in strings of symbols. Pac. J. Math. **85**, 261–294 (1979)
2. Currie, J.D.: On the structure and extendibility of k-power free words. Eur. J. Comb. **16**, 111–124 (1995)
3. Currie, J.D., Pierce, C.W.: The fixing block method in combinatorics on words. Combinatorica **23**, 571–584 (2003)
4. Currie, J.D., Shelton, R.O.: The set of k-power free words over σ is empty or perfect. Eur. J. Comb. **24**, 573–580 (2003)
5. Pansiot, J.J.: A propos d'une conjecture de F. Dejean sur les répétitions dans les mots. Discrete Appl. Math. **7**, 297–311 (1984)

6. Petrova, E.A., Shur, A.M.: Constructing premaximal binary cube-free words of any level. Int. J. Found. Comp. Sci. **23**(8), 1595–1609 (2012)
7. Petrova, E.A., Shur, A.M.: Constructing premaximal ternary square-free words of any level. In: Rovan, B., Sassone, V., Widmayer, P. (eds.) MFCS 2012. LNCS, vol. 7464, pp. 752–763. Springer, Heidelberg (2012)
8. Restivo, A., Salemi, S.: Some decision results on non-repetitive words. In: Apostolico, A., Galil, Z. (eds.) Combinatorial Algorithms on Words. NATO ASI Series, vol. 12, pp. 289–295. Springer, Heidelberg (1985)
9. Shelton, R.: Aperiodic words on three symbols. II. J. Reine Angew. Math. **327**, 1–11 (1981)
10. Shelton, R.O., Soni, R.P.: Aperiodic words on three symbols. III. J. Reine Angew. Math. **330**, 44–52 (1982)
11. Shur, A.M.: On ternary square-free circular words. Electron. J. Comb. **17**, R140 (2010)
12. Shur, A.M.: Deciding context equivalence of binary overlap-free words in linear time. Semigroup Forum **84**, 447–471 (2012)
13. Shur, A.M.: Growth properties of power-free languages. Comput. Sci. Rev. **6**, 187–208 (2012)
14. Thue, A.: Über unendliche Zeichenreihen. Norske vid. Selsk. Skr. Mat. Nat. Kl. **7**, 1–22 (1906)

Author Index

Avgustinovich, Sergey V.   59

Barton, Carl   73
Berthé, Valérie   210
Blažek, José Eduardo   85

Day, Joel D.   97
De Felice, Clelia   210
Delecroix, Vincent   210
Dolce, Francesco   210
Dumitran, Marius   147

Endrullis, Jörg   1, 109

Fici, Gabriele   122
Frid, Anna E.   59

Grabmayer, Clemens   109

Hendriks, Dimitri   109

I, Tomohiro   135

Klop, Jan Willem   1
Köppl, Dominik   135

Leroy, Julien   210
Lohrey, Markus   14

Manea, Florin   147, 160
Mignosi, Filippo   122
Mousavi, Hamoon   170

Néraud, Jean   27

Parshina, Olga G.   191
Peltomäki, Jarkko   197
Perrin, Dominique   35, 210
Petrova, Elena A.   223
Pissis, Solon P.   73
Puzynina, Svetlana   59

Reidenbach, Daniel   97
Reutenauer, Christophe   210
Rindone, Giuseppina   210

Saarela, Aleksi   1
Seki, Shinnosuke   160
Shallit, Jeffrey   170
Shur, Arseny M.   223
Stoll, Thomas   47

Whiteland, Markus   1, 197

Zantema, Hans   109

Printed in the United States
By Bookmasters